# ELSEWHERE

# ELSEWHERE

## HOW THE US FOOD SYSTEM CULTIVATES, CONSUMES, AND CONCEALS ITS VIOLENCE

STIAN RICE

WEST VIRGINIA UNIVERSITY PRESS | MORGANTOWN

First edition published 2026 by West Virginia University Press
Printed in the United States of America
ISBN 978-1-959000-91-4 (paperback) / ISBN 978-1-959000-92-1 (epub) / ISBN 978-1-959000-93-8 (PDF)

Library of Congress Cataloging-in-Publication Data

Names: Rice, Stian author
Title: Elsewhere : how the US food system cultivates, conceals, and consumes its violence / Stian Rice.
Description: Morgantown : West Virginia University Press, 2026. | Includes bibliographical references and index.
Identifiers: LCCN 2026026779 | ISBN 9781959000914 paperback | ISBN 9781959000921 epub | ISBN 9781959000938 pdf
Subjects: LCSH: Food supply—United States—History | Food industry and trade—Moral and ethical aspects—United States | Japanese Americans—Forced removal and internment, 1942-1945 | Hunger—United States—History | Food relief—International cooperation—History | Agriculture—Economic aspects—United States | Environmental degradation—United States
Classification: LCC HD9000.9.U5 R53 2026
LC record available at https://lccn.loc.gov/2026026779

Cover design by Kimberly Glyder
Book design by Kelley Galbreath
Cover photo credit: Dorothea Lange, Power farming displaces tenants from the land in the western dry cotton area, Childress County, Texas Panhandle. 1938. Courtesy Library of Congress. LCCN: 2017770664



For EU safety/GPSR concerns, please direct inquiries to WVUPress@mail.wvu.edu or our physical mailing address at West Virginia University Press / PO Box 6295 / West Virginia University / Morgantown

# CONTENTS

*List of Figures | vii*

*Preface | ix*

| | | |
|---|---|---|
| | **Introduction:** "The Food System Is Broken" | 1 |
| 1 | **Divide and Cultivate:** Removals, Reservations, and the Making of Elsewhere | 16 |
| 2 | **Scatological Imperialism:** Squandering Soil and Grabbing Guano for Ravenous Monocultures | 51 |
| 3 | **Grand Theft Agriculture:** Japanese American Incarceration and California's Produce Empire | 76 |
| 4 | **Grain, Grain Everywhere:** Technology, Policy, and the Making of Excess | 105 |
| 5 | **Invade, Aid, Trade:** Making the World Safe for Profit | 136 |
| 6 | **Of Meat and Migrants:** Domesticating Humans and Animals Through Economies of Scale | 175 |
| 7 | **Hunger Is Not a Polygon:** Making Place for Malnutrition on the National Map | 215 |
| | **Conclusion:** If It Ain't Broke, Break It | 245 |

*Acknowledgments | 265*

*Notes | 269*

*Bibliography | 303*

*Index | 335*

# LIST OF FIGURES

**Figure 1:** Schoolgirls attend sewing class at Albuquerque Indian School, 1910. 37

**Figure 2:** An oil painting of Navassa Island, circa 1870. 69

**Figure 3:** Johnny Fukazawa weeds a cornfield at the Manzanar internment camp, 1942. 95

**Figure 4:** A Mississippi Delta plantation owner near Clarksdale, Mississippi in 1936. 116

**Figure 5:** Map of United Fruit Company steamship routes, 1924. 146

**Figure 6:** Union Stockyards, Chicago, circa 1947. 195

**Figure 7:** A girl in the London Asylum suffering from pellagra, 1925. 222

# PREFACE

"Eating is an act of violence." This was the thought that preoccupied me on a snowy night in March 2012 as I walked home from a meeting of food activists in Kent, Ohio. I was at the point in my graduate studies where I had learned just enough to be annoying but not enough to be useful, enough to want change but not enough to know how to achieve it. I was searching for clues about my role in the world, and like many college towns, Kent offered a mix of good leads and dead ends.

The meeting that night was about the advantages of "localism," a movement that encourages consumers to buy and eat food that is produced nearby. As an alternative to the corporate-controlled supply chains that dominate industrial agriculture, localism promotes farmers markets, community-supported agriculture, gardens, and regional food labeling. A popular expression of localism is the "locavore" diet by which consumers limit their consumption to food produced within a hundred miles.[1]

At the meeting, a panel of activists described the horrors of the current system: how a handful of large agribusinesses exploit producers and consumers through vertical integration, how globe-spanning supply chains deliver cheap food onto our dinner plates at extraordinary social and environmental cost. One of the panelists, an investment banker turned organic farmer, explained how localism would benefit small farms, make the country less dependent on foreign producers, and inevitably lead to a more sustainable and less violent food system. In the audience, heads nodded and pens scribbled.

I raised my hand. I had been juggling an ethical dilemma ever since he mentioned foreign farms, and in my haste to resolve it, I threw it to the room: "If my food purchases currently support distant producers, and I stop making those purchases, won't I be harming those producers? What if they depend more on my purchases for their survival than local farmers? What is the humane thing for me to do?"

Heads and pens stopped moving. Panelists gazed down at the table. As the ensuing silence grew awkward, I realized with embarrassment that I had violated an unwritten rule of local food discourse. I had introduced something nonlocal, a distant and *different* local—not the familiar evil of a corporate-owned, globe-spanning supply chain but the often-overlooked human at the start of it. A moment earlier, we had all been picturing backyard gardens filled with heirloom kale. Now, thanks to my naïve interjection, we were all picturing starving *campesinos* with malnourished children. The sacred boundaries of the group's spatial ethos had been defiled by an insouciant reminder of geography.

I was about to withdraw my question when the banker-farmer-panelist broke the silence: "Yes, whatever we choose to do will harm someone. But given the choice between people in your own community and those on the other side of the world, wouldn't you rather support people more like yourself?" His question was rhetorical and the panel moved on. I had been put in my *place*, and that place was *here*, not *there*.

As the discussion returned to kale, I thought about his response. There was something tantalizing about the promise of localism—that by eschewing grocery stores and committing myself to the Saturday farmers market I might give back to the community some of the benefits I had received. It felt a bit like "closing the loop"; my food dollars would enrich the lives of those whose food had enriched me. I need only fence in my empathy at the radius of a hundred miles and this whole violent food system thing would disappear.

But the image of the campesino didn't disappear. What makes nearby people more "like me" than distant ones? What makes the people in my own community more worthy of my food choices than those I've never met? Why do I have to choose in the first place? More to the point, *what is it that is making me choose?* This book is about the social, economic, and political structures that present us with this Hobson's choice.

The locavores are right, of course: The food system is built on long supply chains that distance us from those who grow our food and obscure the diverse harms baked into our meals. The coffee you drink was grown by slaves in Brazil. Pesticides on your spinach burned the pickers' hands. Your chicken was deliberately deprived of sleep to help it bulk up. The underaged workers who slaughtered it were prohibited from taking sick leave. And the grapes in your wine were picked by convicts.[2] Sometimes the violence occurs at a distance—on Indonesian palm oil plantations or Ghanaian cacao farms. Sometimes it is right there on the plate—in pesticide residue and high-fructose corn syrup. Making dinner these days has become an act of violence, against farmers, farmworkers, animals, the environment, and our own bodies. And yet we all must eat.

This book examines how the food system perpetuates violence even when it appears not to, how it harms even when it claims to nurture. Most of all, it reveals a geographic sleight of hand that has redirected systemic violence into out-of-the-way places: reservations, island colonies, prison camps, inner cities, and war-torn countries. Some of these places were built for this purpose. For the past 250 years, the US food system has been perfecting this trick, amassing extraordinary wealth and abundance in one place as it devalues, colonizes, and exploits others. This book is about those elsewheres and the people who occupy them.

Recent commentaries have put the consumer at the fulcrum of change in the food system: If only we make different choices, eat what our great-grandmothers ate, shop in other aisles, or abandon the grocery store altogether, the food system will obediently follow our dollars into greener, less violent pastures. If only we encourage new technologies or pass stricter regulations, we can train our agribusinesses to be more sustainable, equitable, and humane.[3]

The cases in this book suggest otherwise. Gains in one part of the system have always come at a cost to others. This is how the food system has managed its crises, crushed resistance, concentrated its benefits, and disposed of its harms. Violence is redirected from privileged farms and farmers onto others, from privileged eaters onto others. Even the chance to "fix" the system has been offered to some consumers and not others. Health and wealth continue to follow the contours of race, class, gender, and geography.

This book focuses on the US food system, but the historical processes it describes are now global. Many were perfected in the United States before being exported elsewhere, just as the strategies used by European settlers in the New World were perfected in the Old World's agrarian battlegrounds. My hope in revisiting the past is that it might be put to work in the present, as food movements (like localism) work to define what people and places they care about. Each case maps a path already taken in the hope that different, future paths might be found. In the words of Eduardo Galeano, "History is a prophet who looks back: because of what was, and against what was, it announces what will be."[4] So, too, with geography, which invites us to look elsewhere to understand what is happening here.

# ELSEWHERE

# INTRODUCTION

*"The Food System Is Broken"*

Today, this refrain echoes in farmers markets, food coalition meetings, legislative assemblies, and the press rooms of transnational institutions. It adorns tweets from angry consumers and UN agency directors alike. The narrative of a broken food system underpins the work of activists and analysts, lobbyists and legislators, scholars and celebrity chefs. And it animates the throughlines for countless documentaries, books, and exposés.

Those who embrace this narrative have reason to be concerned. Around the world, a growing number of families find it impossible to put healthy food on the table. Malnourishment and starvation continue, despite the world producing more food per person than at any time in history.[1] For the poor, the bodily consequences of malnutrition are the daily grind, as stunting, anemia, and blindness steal vitality. National populations are experiencing unprecedented rates of obesity, diabetes, and heart disease as processed foods and refined sugars replace traditional diets. In fields and packing plants, laborers endure physical and mental suffering through exposure to hazardous chemicals, harsh working conditions, and low wages. Small farms navigate diminishing returns, predatory lending, and bankruptcy. And further afield, forests burn to make way for burgers, hormones turn chickens into flightless monsters, and fish die from breathing pig feces.

The refrain of a broken food system emerges from this glaring contradiction: The complex network of growers, processors, transporters, manufacturers, and retailers upon which our lives depend often behaves in ways that are *hostile* to life.

## THIS OLD SNAKE

In the United States, this hostility has taken many forms over the past 250 years. Beneath amber waves of grain is stolen land—its Indigenous people, villages, fields, orchards, and livestock systematically destroyed in a coordinated plan we now call genocide.[2] Plantation agriculture and the slave labor system sanctioned extraordinary bloodshed on both sides of the Atlantic. After Emancipation, slave labor in the South was replaced by exploitative sharecropping and a convict leasing system that killed up to 16 percent of its inmates each year in the 1880s.[3] Since the 1920s, falling farmgate prices and rising input costs have transformed the countryside from a mosaic of small farms to a uniform landscape of large agribusinesses.[4] Decades of racist lending practices have forced unwanted groups off the land and into tenancy, and from tenancy into unemployment.

Over time, other systemic harms have come to light. Upton Sinclair's 1906 exposé *The Jungle* brought the horrific conditions of Chicago's stockyards to the attention of an increasingly meat-obsessed public.[5] Carey McWilliams's 1939 *Factories in the Field* revealed the shifting currents of racism and labor exploitation in the growth of California's agricultural empire.[6] John Steinbeck's *Grapes of Wrath* (1939) exposed the food system's senseless response to natural resource crisis, while his lesser-known *In Dubious Battle* (1936) introduced readers to the deadly conflict between pickers, growers, and the state.[7] These and other works conveyed, with visceral effect, the absurdity of a system that demands working bodies one day and disposes of them the next. Little has changed: Today's domestic and migrant food workers endure dangerous working conditions, wage suppression, threats of bodily harm, and public opinions that vacillate between apathy and enmity.[8]

Hunger and malnutrition have been constant companions. In the nineteenth century, high retail prices starved the urban poor as low farmgate prices starved rural families.[9] Today, over nine million people in the United States can barely afford to eat, and the food they *can* afford may not be good for them. Public health experts have linked emerging diseases with food that is increasingly produced through industrial means.[10] Mechanical and chemical revolutions altered not only how food is made but the substance of food itself, removing nutrient-rich components (like the bran and germ of wheat) and adding ingredients with long-term negative health impacts (like sodium nitrite and processed sugar). In another paradox of the food system, what lengthens the shelf life of food may shorten the lives of those who eat it.

These revolutions also accelerated the collapse of ecosystems and the suffering of animals. Throughout the United States, communities struggle to manage the impacts of soil fertility loss, toxic runoff, biodiversity loss, increasing temperatures, and groundwater depletion.[11] Meanwhile, growing concern for animal welfare has highlighted the suffering of living creatures raised and slaughtered for food.[12]

Since the colonial period, the food system has enlisted a diverse cast of perpetrators and victims of violence. Evolving geographic and historical contexts have collaborated to orchestrate new forms of harm and reprise old ones. Well into the twenty-first century, this dissonant symphony continues.

Frederick Douglass once remarked that "slavery has been fruitful in giving itself names" and warned us to "wait and see what new form this old monster will assume, in what new skin this old snake will come forth next."[13] The system of slavery to which Douglass was referring is exemplary of the kinds of violence the food system can produce.[14] But it is by no means the only form. With each economic crisis, labor shortage, geopolitical pivot, and technological innovation, old monsters have emerged in new skins. As the following chapters attest, our food system has been busy giving its violence new names, too.

## A BAD RELATIONSHIP

It is tempting to equate violence with physical or mental suffering. With respect to food, this would include starvation, malnourishment, ailments related to farmwork, and chronic diseases exacerbated by diet, among others. But this list omits many of the food system's most pervasive harms, ones that manifest slowly, create conditions for future suffering, or limit us without our knowledge. Left out are harms with spiritual, cultural, economic, and social impacts, harms that affect loosely defined populations and extend across time and space.[15] Violence is present whenever a child is forced to eat foods that increase the chance of adult diseases, whenever a fieldworker must choose between eating and paying rent, whenever a person gives up his dream of farm ownership for fear his racist neighbors will turn on him. By this measure, the unraveling of social safety nets, wage suppression, and white supremacy create violent conditions as surely as unhealthy food and nasty chemicals.

It is also tempting to assume that violence is measurable. Some harms can be observed, categorized, and counted: For example, the number of malnourished children in a county can be quantified, graphed to show change from one year to the next, and mapped to reveal spatial patterns. Malnourishment can be made into a discrete phenomenon upon which analytical tools (like statistics) and institutions (like social services) may be brought to bear.[16] But not all violence can be cut to size in this way. Counting the victims of racial terror lynching will never truly reflect the fear that permeated the South. Counting burned villages will never reflect the spiritual deprivation experienced by Indigenous people forced off their land.

Finally, it is tempting to think that violence always has a clear perpetrator—a person who knowingly and intentionally commits a harmful act. But this, too, is overly narrow. Left out are harms where the perpetrators are difficult to identify, did not anticipate the consequences, or committed no formal crime. Many of the food system's slow-emerging harms fall into this category: "unintended consequences," "coincidences," or "necessary evils" resulting from multiple, nebulous, or distant institutions going about their business as usual. For forensic investigators of systemic violence, working backward from corpses to culprits often leads into an ethical *terra nullius* where the trail of responsibility goes cold.

But this is precisely where we must go. To examine violence within the US food system, we must account for harms that emerge slowly, occur without our knowledge, and lead to future suffering: *conditions that materially limit what is possible.*[17] We must also account for violences that emanate from the structure of the system, from complex networks of participants and processes operating over vast distances and long time spans. Such violences emerge *from the social, economic, and political relationships that exist between actors*, not just the nefarious or negligent decisions of a few bad apples.[18] In the end, a meatpacker who dies of a communicable disease is killed not only by the supervisor who denied her sick leave, but by the architects of company policy, the competitors who normalized such mistreatment in the industry, the officials who failed to enforce regulations, the legislators who voted against worker protections, and the consumers who responded to her illness with apathy. All these actors are related to the victim by way of the food system, and all these conspirators have names. Only after we untangle this intricate network can we understand why the food system produces violence in the first place.

This book examines the relationships that have historically produced and sustained violence in the food system. But what, exactly, is the food system?

## A WEALTH-GENERATING MACHINE THAT ALSO MAKES FOOD

It has been described as a jigsaw puzzle, a web, a battlefield, a rubber band, a vast machine, a complex adaptive system, and a runaway train.[19] It demands a good metaphor, yet it evades one. It is sufficiently vague, yet immensely powerful. It occupies space, yet its edges are hard to find. It is seemingly everywhere, yet its operation is easily overlooked. It keeps us alive, yet it sometimes kills us.

The food system is the *network of actors, processes, and relationships that connect consumers to farmers, ranchers, fishers, and the edible products of their labor.*[20] In its entirety, the system encompasses many interlocking subsystems that manage workers, provide inputs for farming, transport materials, mill, slaughter, purify, enrich, package, distribute, and sell food. Still other subsystems define and enforce the rules by which the system operates, create incentives and deterrents to shape the behavior of actors, provide financial credit to producers and consumers, and offer wealthy speculators the chance to gamble on the system's behavior. These social arrangements are entangled with ecological and biological systems that grow the plants and flesh we call food.[21]

For the most part, the US food system is capitalist—organized around a set of relationships that include private ownership of land and the means of production, a wage labor system, and free markets. Nearly all food is produced as a commodity—that is, not for direct consumption but for sale.[22] These arrangements impose certain obligations: Workers must sell their labor to survive, and businesses must pursue greater profits to remain competitive. If we judge the system based on what it does (and not what it claims to do), this complex arrangement of agribusinesses, banks, markets, and laws is less of a *food-producing* system than a *wealth-generating* system that also makes food. It makes a lot of food along the way, but its top priority—and the priority of its most powerful actors—is profit. As the following chapters demonstrate, the system has distributed wealth to the powerful far more consistently than it has distributed food to the hungry.

## THE VIOLENCE OF ABUNDANCE

One quirk of a capitalist food system is that it tends to *over*produce. We are used to thinking that abundance is a good thing, that harm comes from having too little. After all, food scarcity is a persistent concern in the United States and around the world. But under capitalism, abundance can be just as deadly.

A common way to increase profits is to increase production. In agriculture, this may involve expanding the cultivated area or increasing the yield.[23] Settler agriculture in the United States used both strategies. Before 1900, growth was achieved by taking Indigenous land and putting it under the plow. After 1900, new tractors, fertilizers, and pesticides helped farms squeeze more out of each acre. The resulting gains led to chronic *overproduction*, a condition in which more commodities were produced than were needed to meet demand.[24]

Overproduction can be deadly. As farmers produce more (relative to demand), food prices begin to fall, and as they fall, profits shrink. From the farmer's perspective, the logical response is to produce even *more* food—plant more, harvest more, and sell more to make up for smaller margins. If enough farms do this, the ensuing cycle of feedback can lead to bankruptcies. From 1870 to 1900, total agricultural output in the United States grew 135 percent and farmgate prices for wheat and corn collapsed. Rural destitution and starvation ensued, a deadly combination that would return to torment small farms in the 1930s.[25] Overproduction is so common and so dangerous it has been called the "curse" of American agriculture.[26]

Aware of these dangers, the government takes protective measures. To prop up farmgate prices, the US Department of Agriculture (USDA) buys and stockpiles farm products then looks for ways to dispose of them. It may place quotas on how much can be sold or how many acres can be cultivated. Some policies are downright bizarre: During the Great Depression, the USDA paid farmers to destroy their crops and kill their livestock, all while the hungry looked on. The history of capitalism is rife with inhumanity, but few moments can top the willful destruction of food as people starve. Steinbeck captured the absurdity of this moment in *Grapes of Wrath*:

> There is a crime here that goes beyond denunciation. . . . The fertile Earth, the straight tree rows, the sturdy trunks, and the ripe fruit. And the children dying of hunger must die because a profit cannot be taken from an orange. And coroners must file in the certificates—died of malnutrition—because the food must rot. . . . The people come with nets to fish for potatoes in the river, and the guards hold them back; they come in rattling cars to get the dumped oranges, but the kerosene is sprayed. And they stand still and watch the potatoes float by, listen to the screaming pigs being killed in a ditch and covered with quicklime, watch the mountains of oranges slop down to a putrefying ooze; and in the eyes of the people there is failure; and in the eyes of the hungry there is a growing wrath.[27]

Selling the food would have lowered the price. Giving it away would have allowed people to satisfy their hunger without contributing to the economy. Such violence is born not of scarcity but of *abundance*—a blessing for a system designed to feed the hungry, but a curse for one built to accumulate wealth.

## BUT IS IT BROKEN?

There is clearly something wrong with a system that starves, murders, displaces, enslaves, bankrupts, malnourishes, pollutes, and tortures in the interest of nurturing life. But is the food system, as the oft-tweeted refrain insists, *broken*?

"Broken" suggests it is not performing its primary function. Like a roof that leaks or a car that won't start, a broken system is one that does not do what it was designed to do. Broken also suggests that at some point in the past, the system was working. Like a roof that once shed rain or a car that was once drivable, something happened and the system no longer "does its job." It has been damaged, degraded, or inappropriately modified. If the food system's purpose is to support life—but it behaves in ways that are hostile to life—perhaps it makes sense to call it broken. But what if its purpose is to support something else?[28]

This book makes three arguments. First, the food system is not broken, but functioning as designed. This is old news for those who have lived with its inequities and whose stories appear in this book: the descendants of Indigenous farmers, Black sharecroppers, Hispanic meatpackers, and Guatemalan smallholders.[29] In 2020, agriculture, food, and related industries contributed $1.06 trillion to the US economy, not including industrial sectors that make agricultural inputs.[30] The lion's share of this haul benefited those who were already wealthy. From capital's perspective, the food system is operating more or less as designed, amassing extraordinary wealth with exceptional efficiency—far less broken than its critics would suggest.

Second, a consequence of this design is that it produces violence—a lot of it. This is an essential feature of a capitalist food system, not an accidental side effect. This violence causes trouble: The meteoric growth of US agriculture over the past 250 years coincided with growing populism, labor movements, and food riots as mobilizing populations pushed for change. Those who work for and depend on the food system have never taken its inherent violence lying down. The "growing wrath" in the eyes of Steinbeck's hungry onlookers is still here today.

Finally, the system protects its powerful actors and inflates their influence by moving this violence around, away from privileged places and into out-of-the-way "elsewheres." As the following chapters illustrate, this redirection of harm has allowed these actors to achieve other objectives: to maintain white supremacy, gentrify urban neighborhoods, build an overseas empire, fight communism, contain domestic enemies, and support foreign dictators.

## A GLOBAL SHELL GAME

The history of violence in the food system has shaped an uneven geography. The violence of overproduction has been diverted into Indigenous boarding schools, Southern mill towns, and commodity food programs. Good health for some is bought by diverting disease burden onto others. Today, landscapes of scarcity and abundance are often contiguous: Abandoned farms surround flourishing cities and hungry neighborhoods abut gated communities where organic meal kits are delivered. In other cases, suffering has been projected across borders and oceans: Free trade agreements that open markets for domestic growers crush smallholders half a world away; the extraction of fertilizer for domestic gardens enslaves workers on distant islands.

Elsewhere is not just a place to hide the violent by-products of accumulation, but where new rounds of profit making begin: a near-continuous rearrangement of scarcity and abundance, accumulation and exploitation, peace and violence. This book tells the 250-year story of this shell game, the processes that drive it, and the geographic ideas that continue to justify it.

Capital has long sought geographic solutions to its problems. Known collectively as *spatial fixes*, these processes move capital and labor around the world in search of resources and profitable markets. Factories close in one place only to reopen where labor costs are lower. Farmworkers deprived of land hit the road in search of fields where labor is needed. Governments—wary that surpluses will suppress prices—concoct new ways to broadcast unwanted grain around the globe. Within the food system, spatial fixes are literally everywhere.[31]

The United States was born from a spatial fix. The settler colonies of North America emerged at a time when European land reforms were displacing the peasant class. Confronted with the loss of the commons, limited opportunities in cities, and "poor laws" that criminalized unemployment, many former farmers left Europe for New World colonies and the possibility of title to land.

The violent repercussions of that spatial fix are still being felt today among Indigenous peoples, many of whom are fighting to rebuild healthy food systems after centuries of settler colonialism. Spatial fixes for capital and labor are also fixes for violence: Europe mitigated the risk of social unrest by exporting its discontented lower classes, maintaining domestic order at the expense of New World societies.

As spatial fixes altered the landscape, a pattern began to emerge. Parts of the food system amassed enormous wealth as other parts languished, a polarization that reinforced narratives of who belonged and who didn't. Some kinds of farms and farmers were welcomed while others were excluded. Some laborers were protected while others were ignored. Some consumers were satisfied while others were left to fight over the scraps. These inequities came from conscious decisions over where (and onto whom) systemic violence would be directed. Capitalism made this redirection necessary: Food production grew not *in spite of* failing farms, racial discrimination, and starving workers, but *because of* them.

The question, then, is not whether the food system is becoming more violent. The question is *where* its systemic violence is being directed and where its systemic benefits are being amassed. This book is about how the food system bought peace and prosperity for some by moving violence and scarcity elsewhere. It is also about what happens next.

## ORGANIZATION OF THIS BOOK

Throughout the book, I use "Indigenous" and "Native" to refer to the people, societies, political entities, and food systems present in North America at the time of European colonization. I use "Indian" to refer to Indigenous people in the context of US laws, policies, government actions, and institutions; for example, Indian reservations, Indian treaties, and the Bureau of Indian Affairs.[32] I use "Euro-American" or "settler" to refer to the people of European ancestry who began colonizing North America in the sixteenth century, and I refer to the country that emerged at the end of the Revolutionary War as the "United States." I avoid using "America" or "American" in connection with the United States or its population as the term applies to the land and people of an entire hemisphere. One exception is in the names of some populations descended from immigrants—for example, Japanese Americans—where the historical *omission* of "American" has been used to signify exclusion from US society.[33] Finally, I use "Black" to refer to people of African ancestry and "Latino" or "Latina" to refer to people from Latin America (Central America, South America, and the Caribbean) irrespective of race.[34] Wherever possible, I use terms that would be used by each group to refer to itself.[35]

This is not an encyclopedia of violence, nor is it an exposé: All of these stories have been told elsewhere, and all are well-known to the communities that experienced them. Unlike traditional histories, this one is not chronological: There is no steady progression from one event to the next, nor any momentous turning points between "eras." Instead, this book examines violent processes that span the timeline, disappear and reappear, and shape the food system's uneven geography in unexpected ways.

In chapter 1, the US food system is born with European genes. It quickly learns to solve its class conflicts and natural resource crises through territorial expansion, a solution that pits the United States against Indigenous North Americans. Later, as chronic overproduction sets in, the system invents new strategies for managing Indigenous people: *carceral fixes* that divide up and cordon off populations in ways that put them to use.

As settlers march west, Eastern plantations shoot themselves in the foot. Decades of intensive monocropping steal vitality from the soil, the subject of chapter 2. The hasty search for fertilizer inspires the country's first foray into overseas empire. This chapter introduces the problem of *metabolic rift*—how it accompanies capitalist food production and how nineteenth-century imperialism offered the country a profitable (albeit bloody) solution.

In chapter 3, powerful companies, officials, and white supremacists manipulate public anxieties to capture California's most fertile farmland. As the curtain rises on World War II, Japanese Americans are dispossessed, locked up, and forced to grow food for their captors. This chapter explores carceral fixes as an expression of *racial capitalism.*

After the war, US agriculture flexes its technological muscle. Mechanical and chemical innovations generate staggering food surpluses and contribute to the collapse of sharecropping, the subject of chapter 4. As the country drowns in grain, Black children starve. This chapter discusses how *domestic rearrangements* move violence between existing spaces to manage overproduction and maintain social hierarchies.

In chapter 5, the food system pursues tropical commodities and concocts new strategies to dispose of domestic surpluses. In a dramatic pivot, the United States employs *extraterritorial fixes* to build elsewheres in foreign countries. This chapter explores a half-century of bloody intervention in Guatemala by the United Fruit Company, the CIA, and the Agency for International Development (USAID).

Military and economic interventions drive the migration of poor smallholders out of Central America and into the worst jobs in the US food system. Chapter 6 traces the evolution of one industry that has historically "welcomed" these migrants: meatpacking. This chapter weaves together the violent lives of food animals and the humans tasked with disassembling them, examining how strategies of *aggregation* and *segregation* redistribute old harms and generate new *violences of scale.*

Finally, chapter 7 traces the emergence, investigation, and treatment of two epidemics: a mysterious disease in the early 1900s that affected poor miners and factory workers, and today's obesity "epidemic" that has been linked to so-called food deserts. This is the story of how diseases come to be associated with particular people and places, and how these associations reinforce racial, gender, and class prejudices. It is the story of how geographic *metaphors* write off possible solutions long before a problem is understood.

At one level, the seven cases are unique, differing by population, geography, and historical context. But as expressions of a food system designed to accumulate wealth, these diverse harms are all symptoms of a common, chronic condition. Different skin, same old snake.

## PITFALLS AND PROSPECTS

In its various echoes, tweets, and throughlines, the refrain of a broken food system is often followed by suggestions for fixing it. Food movements propose to end corporate consolidation, end animal cruelty, and end food waste. Activists ask us to go organic, go vegan, and go local. We are told to honor the animal, honor the farmer, and honor the earth. Listen long enough to the language of reform and you might hear clashing messages. There is plenty of food and not enough food, we must increase trade and restrict trade, improve industrial agriculture and end industrial agriculture, consume more (presumably from an ethical part of the food system) and consume less altogether. Which prescription you choose depends on what aspect of the system you think is broken, and that depends on what you see the purpose of the system to be.

The conclusion returns to the locavore meeting in Kent, Ohio, to reflect on some contemporary proposals for change. What are the geographic implications for the suggested reforms? If successful, what places will benefit and what places will bear the cost? In the end, how do we build a system that fiercely protects food *as nourishment* and opposes its expropriation *as profit*? This last question is at the heart of this book and motivates the writing of these histories.

There are many questions we might ask of the system going forward: How can we get it to produce more food, produce healthier food, distribute its food and profits more equitably, and become more environmentally friendly? Each of these suggests certain solutions and discourages others. Each assumes a different purpose behind the complex network of relationships that connects us to what we eat. I offer that a critical question is how we can make the food system nonviolent. This question assumes the purpose of the system is to provide sufficient, nutritious, and appropriate food to every person—*and first do no harm*. What, then, does a nonviolent food system look like, and how do we build one? Understanding how the current one came to be violent is the first step.

# 1

# DIVIDE AND CULTIVATE

## *Removals, Reservations, and the Making of Elsewhere*

Placenames can be scars, monuments, and hopes. Albuquerque, New Mexico, where I spent my childhood, is filled with such names. In the 1980s, I attended Montezuma Elementary School, named after the last great Aztec king whose empire was destroyed by Spanish conquistadors in 1520. Albuquerque was named after an ostentatious Spanish viceroy in a transparent attempt by the town's residents to garner his favor. "New Mexico" is the anglicized form of Nuevo México, a province wrested from Spain in 1820 during Mexico's War of Independence. And every morning at school, we pledged allegiance to the flag of the United States, the country that invaded Nuevo México in 1846, crushed its homegrown independence movement, and rounded up its Indigenous inhabitants. Placenames like these are the scars of conquest.

Had my family lived in other parts of Albuquerque, I might have attended Apache, Comanche, Zuni, or Navajo Elementary, names that commemorate some of the twenty-two Indigenous nations and pueblos that resisted colonialism in New Mexico. By contrast, if my family had lived a few miles to the east, I would have enrolled at Andrew Jackson Middle School; a few miles to the west, Kit Carson Elementary. Jackson was the architect of the 1830 Indian Removal

Act that called for the expulsion of Indigenous nations east of the Mississippi River. Carson led a ruthless military campaign against the Navajo in 1863 in which he poisoned wells, burned fields, cut down orchards, and killed livestock. Starving Navajo were rounded up and force-marched 350 miles across the desert to a concentration camp. Oral histories of the "Long Walk" report how captives "suffered from cold, starvation, slave raids, rape and brutality from the soldiers." Pregnant women and the elderly who could not keep pace were shot.[1] The Navajo were incarcerated until a treaty in 1868 allowed their return. During this period, known as *Hwéeldi* ("a time and place of great suffering"), two thousand Navajo died, most from starvation. Some placenames are monuments to genocide.[2]

The meaning of a placename can change. For white settlers during the colonial period, "Indian country" was everything west of the frontier. Later, the term defined a bounded territory in the middle of the continent into which Eastern tribes were removed, a territory steadily eroded by legislation, court decisions, and fraud. In the settler imagination of the late 1800s, Indian country referred to the vast, untamed West, an expanse of prairie and sage where US cavalry protected homesteaders from rampaging enemies.[3] Still later, it referred to the reservations where those enemies were confined. Through it all, Indian country has been synonymous with "elsewhere," an amorphous periphery where unwanted things could be disposed of and valuable things extracted. A place that might be both exploited and forgotten. Most importantly, a place quite different from "here."

But placenames are also hopes. For many Indigenous people, Indian Country (capital C) refers not to an excluded periphery, but to an essential and sacred place. It is the literal and figurative soil that anchors Native identity. In the words of the National Congress of American Indians (NCAI), "Indian Country is wherever American Indian spirit, pride, and community are found. . . . It resides not only in law books, legislation, and historical treatises, but also on ancestral homelands, within our homes, and in the hearts of American

Indian and Alaska Native people everywhere."[4] Quite the opposite of elsewhere.

This is a story about how the settler food system helped build a particular version of Indian country, an elsewhere that benefited white settlement and mitigated chronic agricultural and political crises. At the heart of this story are questions of where "American agriculture" is located and who "American farmers" are. Of all the placenames we take for granted, "America" might be the most challenging one to unravel.[5]

• • •

At the time of Jamestown's settlement by Europeans in 1607, some six hundred Indigenous nations farmed, hunted, fished, cooked, and ate on the land that would eventually be claimed by the United States. Each had its own identity, language, traditions, and practices. Each shared different stories and received different names from the land. And each responded to European colonization in its own way. There was and is no singular "Native" identity, but a multitude of cultural histories and cosmologies. Neither was there a monolithic settler identity that represented all European arrivals. By 1776, wealthy elites, slave owners, middle-class farmers, and indentured servants pursued diverse opportunities on the continent, sometimes in violent opposition, sometimes with a semblance of unity.

Both Natives and settlers lived within economic, environmental, and political structures that shaped their options and influenced their particular experiences. That said, North America's Indigenous people were universally and negatively impacted through this encounter, enduring extraordinary loss of land, life, and culture. Settler and Native outcomes came to be oriented around *settler colonialism*, a political and economic structure that sought—and continues to seek—the dispossession and extinction of Indigenous people.[6]

The forms of colonialism that European empires deployed in Africa and South Asia forced Native farmers and pastoralists into wage labor or abducted them to work as slaves on distant plantations. By contrast, settler societies like the United States pursued Native *extermination* through diverse mechanisms, including direct killing, starvation, biological warfare, forced sterilization, relocation, assimilation, and cultural appropriation.

At the heart of settler colonialism are beliefs about racial, cultural, and religious supremacy.[7] There are also material goals. In the colonial period, dispossession and extinction fed the demands of settler economies, propped up unstable class structures in settler society, and addressed natural resource crises. Later, reservations provided an outlet for surplus commodities and a reserve of land and labor. Today, settler colonialism is expressed in the classroom through historical erasure, in the courts through the erosion of treaty rights, and in public policy through both active exploitation and calculated neglect. It continues to approach Indigenous knowledge the way it approached Indigenous land: by taking it, assimilating it into the public domain, and claiming it as its own.[8] Like the settler food system, settler colonialism evolves in response to changing demands, public sentiments, and forms of resistance. Underlying that evolution is geography.

## LAND HUNGER IS NOT HUNGER

Settler colonialism is also rooted in stories. Well-curated narratives remind each new generation of the country's revolutionary beginnings, celebrate its national character, and justify its shortcomings. A litany of myths purports to explain why Europeans came to dispossess Indigenous North Americans: For example, the land was empty; Natives did not know what they were giving up; and Natives gave way to settlers to achieve short-term gains. Over the past few decades, historians have challenged these misconceptions as new generations take greater account of Indigenous perspectives and settler-colonial legacies.

But one myth has been tough to shake. In popular narratives, the impetus for expansion is often expressed in a single metaphor:

Euro-Americans pushed west because they were "hungry for land."[9] Land hunger is presented as a characteristic of being a settler, a natural compulsion that colonists could never fully satisfy, and one that routinely aggravated government efforts to manage territory. The choice of "hunger" is telling: Unlike "desire," which can go unrelieved, hunger is a biological imperative. Hunger kills. Arguably, those who stand in the way of relieving hunger might rightly be called killers. Its use is particularly ironic (some might say offensive) in that *actual* hunger was commonplace among the Indigenous people who opposed settler expansion, and against whom starvation was used as a weapon.

But this is more than just sloppy word choice. Settler land hunger is an integral part of the US origin story. According to this narrative, homesteaders fought Indians and advanced the frontier to secure survival. Atrocities were committed, but these were necessary evils. After all, settlers were trying to satisfy an unrelenting need, gobbling up acres the same way a hungry person might gobble up food. The destruction of Indigenous villages and farms was an act of necessity and desperation, not racism or greed. Meanwhile, Indigenous defenders proved their malevolence by standing in the way of settler survival. In this way, the land hunger narrative inverts the historical record by turning *settlers* into the victims of colonialism and pushing the horrors of Indigenous starvation into the background. Over time, land hunger has accompanied other national myths that normalize genocide, like Manifest Destiny or conquering for freedom.[10]

The hunger myth also places responsibility for Indigenous-European conflict on poor settlers, not elites or the state: The new republic tried its best to hold back the ravenous masses, but ultimately settler appetites could not be contained.[11] Framed in this way, the historical record is once again inverted, with the state becoming the ill-fated *protector* of Native peoples, not an abettor of genocide. State-sponsored violence—like Kit Carson's campaign against the Navajo—recedes into the shadows as storytellers highlight the territorial appetite of homesteaders.

To be sure, poor farmers and ranchers wanted land, but not from any intrinsic "hunger." They sought fertile acres at the frontier in response to a national ideology that equated land with liberty, a territorial policy that limited its supply, and an economic system that profited from its sale. Poor farmers pushed west to stay ahead of two structural crises in the food system: class antagonism and dwindling natural resources.

## CHERISHED CULTIVATORS AND PLANTATION AGRICULTURE

Agrarianism is the belief that an agricultural life, more than any other occupation, can provide positive political and social goods: to farm is to become a better person and citizen, an independent thinker who understands the value of hard labor and embraces patriotic duties.[12] By the late 1800s, agrarianism had been elevated to civic religion, thanks in part to its greatest prophet, Thomas Jefferson.[13] "Cultivators of the earth are the most valuable citizens," wrote Jefferson to John Jay in 1785. "They are the most vigorous, the most independent, the most virtuous, and they are tied to their country, and wedded to its liberty and interests, by the most lasting bonds."[14]

At the time, 80 percent of Euro-American settlers, including all of the new republic's founders, were "cultivators of the earth." The vast majority were smallholders who owned one hundred acres or less, grew crops to feed themselves, and sold their surplus at the market. Jefferson dreamed of building an agrarian society around this "yeoman" class, promoting land ownership as the path to social advancement and political representation. In the new republic, land ownership guaranteed middle-class males the right to vote and run for public office. It was both a symbol of virtuous political principles and a license to exercise them.[15]

But if land was key to middle-class advancement, where could it be found? The yeomanry may have been central to Jefferson's agrarian vision, but the center of agricultural power belonged to an elite class of plantation owners (including Jefferson himself). Along the

East Coast, large slave-based agricultural businesses raised tobacco in Virginia and Maryland, indigo and rice in South Carolina and Georgia. Most of these crops were earmarked for sale in Europe and its colonies, and as foreign demand grew, so did the value of North American farmland. By the Revolution, the best acres on the East Coast were already under plantation control. Wealthy landowners invested in large swaths of undeveloped land, pushing prices out of reach for many would-be smallholders. Affordable land was on or beyond the frontier, far from navigable rivers and with limited access to markets. As increasing numbers of immigrants and formerly indentured whites sought access to the land market, political elites hoped to relieve yeoman-plantation tensions by opening up Western lands.

Expansion was also profitable. George Washington and other national patriarchs speculated on frontier territory in the hope of selling land to farmers escaping inflated prices, all while extoling the moral virtues of land ownership.[16] The government's tepid constraints on expansion served only to limit the supply, driving up speculators' profits. It was in elite interests to create a land frenzy, and for the next half century, a land frenzy is what they got.

Had plantation companies been content (or constrained) to monopolize the lowlands, the westward push of Euro-American settlement might have slowed. But corporate capital was facing its own crisis. By 1800, East Coast plantations had plowed the lowlands into disrepair: Decades of intensive tobacco cultivation had depleted once-fertile soils. Tobacco is a nutrient-intensive crop, and as European plantations in other parts of the world discovered, continuous tobacco cultivation requires long fallow periods to replace lost fertility.[17] But with cheap land in the wings, US plantation owners were loath to wait. It was more profitable to buy and clear new acres than to care for old ones.

Capital sought greener pastures to the west and south, switched from tobacco to cotton in response to changing markets, and left exhausted land in its wake. By 1803, Mid-Atlantic plantation capital had sunk roots in Kentucky (from *Kain-tuck-ee*, Shawnee for "head of the river") and Tennessee (from *Tanasse*, a local Cherokee settlement). By 1840, planters were cultivating the fertile soils of Alabama (from the name of a Creek tribe) and Mississippi (from *Mish sha Sup-pukni*, Choctaw for "the river beyond any age").[18] In the Southeast, the yeomanry sought lands farther west to stay ahead of rapidly expanding cotton- and speculation-inflated prices.[19] For a time, two lines of strife marched in lockstep across the continent: the Indian-yeoman frontier to the west, and the yeoman-plantation frontier to the east. With corporate capital not far behind, immigrants, former indentured servants, and poor white people became the literal bleeding edge of Euro-American settlement.

## A BUSINESS PLAN PREDICATED ON INVASION

Class conflict drove poor Europeans west across the Atlantic. Later, it drove poor Euro-Americans west toward the Mississippi. It is tradition in US historiography to draw a line under 1776 as the dawn of a new political and social era—and there is some basis to this: Unlike in Europe, access to land in North America no longer depended on the last name of your ancestors. But it still depended on having *European* ancestry: Land ownership was notably off-limits to descendants of Piscataway and Cherokee, Yoruba and Ibo. Moreover, land ownership depended increasingly on a capricious market dominated by elite interests. Insofar as it forced poor white households into tough geographic decisions, the Euro-American approach was more a continuation of Europe's agrarian revolution than a new beginning.

The United States addressed its early class conflicts and resource crises with geography, a template it would follow for the next two centuries. There were alternatives that might have slowed the flow of settlers and curbed growth in plantation slavery—alternatives vigorously debated before and after independence. But in the end, the founders of the new republic committed to a blueprint that pitted smallholders and plantation owners against each other. Public land sales would help the treasury, a mobilized (and armed) yeomanry would help secure territory, and land speculators would help themselves. It was a national business plan predicated on invasion.

Of course, invasion is a bad look. For a country founded on lofty humanist principles, dead bodies and burned villages can create some cognitive dissonance. Over time, it benefited the architects of state history to claim it was poor homesteaders—not plantation owners or speculators—who were the ones hungry for land.[20]

## A DISPOSSESSION IN FIVE ACTS (OF CONGRESS)

As a structure, settler colonialism seeks the dispossession and extermination of Native peoples. But the tactics and strategies it uses to achieve these goals often change. In the 1830s, the government was eager to create Indian reservations; by the 1950s, it was keen to eliminate them. In the 1830s, settlers refused to accept Indians into white society; in the 1950s, assimilation was promoted as the only way forward. Each shift in attitude and policy reflected changes in the country's political and economic landscape. Between the 1790s and 1980s, the Indigenous-settler relationship evolved through five periods, each reflecting a different technology of dispossession and a different definition of Indian country.[21] Critically, each shift emerged in response to the changing needs of the settler food system.

### ACT 1: REMOVAL

In settler histories of Native America, the hallowed placenames are battlefields: Tippecanoe, Little Big Horn, and Wounded Knee. Less familiar are Greenville, Fort Jackson, and New Echota, places where the spoils were divided. It was here that treaties were negotiated and signed, promises made, and territories ceded. The 1795 Treaty of Greenville ceded to the United States much of what is now Ohio (the Iroquois word for "beautiful river"). The 1814 Treaty of Fort Jackson ceded southern Georgia and most of Alabama. And the 1835 Treaty of New Echota ceded northern Georgia and parts of southern Tennessee.[22]

Even less is remembered about the earlier treaties these agreements nullified. Early Indian treaties followed a script: Common articles included assurances of perpetual peace and friendship, the establishment of boundaries, guarantees of protection against encroaching settlers, the return of prisoners, and the return of property (often horses) stolen by both sides. Early treaties with the Cherokees and the Creeks stipulated that US citizens must carry passports to enter Indian land and required both sides to extradite suspected criminals.[23] Early treaties read like agreements between sovereign nations because that is what they were.[24]

But in the late 1790s, the government changed its view of Indigenous nations and the intent behind its treaties. The Indigenous-settler relationship had become increasingly complex as plantation capital pushed southwest and the yeomanry sought cheap acres. As the United States renegotiated treaties to expand its territory, the language of perpetual peace and recognition was replaced by language granting the president discretion to "retrace, connect, and plainly remake" boundary lines.[25] Land was "relinquished to the United States forever."[26] Treaties read like terms of surrender, not agreements between political equals. And for good reason: Indigenous delegates often arrived at the negotiating table after their families had been killed and their villages and farms destroyed by settler militia.

One decorated militia leader was a Tennessee slave owner and land speculator named Andrew Jackson. Jackson's political career was launched in 1801 when he led attacks against the Muskogee Nation in Georgia and Alabama. In the 1814 Battle of Horseshoe Bend, Jackson led his cavalry against one thousand Muskogee warriors, killed eight hundred of them, and held women and children hostage to force a surrender. Afterward, his cavalry "fashioned reins for their horses' bridles from skin stripped from the Muskogee bodies" and gave "souvenirs from the corpses . . . to the ladies of

Tennessee."[27] Spectacles like this propelled Jackson to the presidency in 1828 on the votes of landless whites who hoped Indian clearances would open up cheap farmland.

In 1830, President Jackson rolled out his scorched earth policy nationwide with the infamous Indian Removal Act. Under the act, the government granted itself the right to round up Indigenous people living in Eastern states (on treaty-protected territory) and relocate them to Western lands. Over the next twenty years, one hundred thousand people—mostly Cherokee, Choctaw, Creek, Chickasaw, and Seminole—were forcibly removed to territories west of the Mississippi River. The infamous Trail of Tears, during which four thousand Cherokee died from starvation and exposure, was just one episode of forced removal authorized under the act.

Jackson's policies have come to epitomize the worst of settler colonialism, and understandably so. But tactically and strategically, Jackson was following a much older blueprint. In 1779, George Washington ordered attacks against the Iroquois in which soldiers set fire to forty towns and destroyed food stores. One Onondaga chief reported that US soldiers "put to death all the Women and Children, excepting some of the Young Women, whom they carried away for the use of their soldiers and were afterwards put to death in a more shameful manner."[28] Thomas Jefferson envisioned a firm Western boundary between Indian and settler territories, with Indians cleared from the East through treaty accessions. Before the ink had dried on the Louisiana Purchase in 1803, Jefferson proposed deporting the Cherokees of Georgia into the newly acquired territory through "voluntary or forced removal."[29]

Jefferson's go-to strategy was economic dependence. As he explained to his secretary of war, Henry Dearborn, "the Indians being once closed in between strong settled countries on the Missisipi [*sic*] & Atlantic, will, for want of game, be forced to agriculture . . . and will be continually parting with portions of [land], for money to buy stock, utensils & necessaries for their farms & families."[30] Jefferson knew he could use this growing dependency to leverage further concessions: "There is perhaps no method more irresistible of obtaining lands from them than by letting them get in debt, which when too heavy to be paid, they are always willing to lop off by a cession of lands."[31] Failing removal by destitution, Jefferson threatened force. "If ever we are constrained to lift the hatchet against any tribe," Jefferson warned, "we will never lay it down till that tribe is exterminated, or is driven beyond the Mississippi . . . in war, they will kill some of us; we shall destroy all of them."[32]

In the Southeast, the United States cut off Chickasaw and Choctaw access to Spanish merchants, forcing them to trade only with settlers under exploitative terms. The Choctaw and Chickasaw nations quickly fell into debt. The impact of financial coercion is apparent in treaties of the period. The 1805 Treaty with the Chickasaw explains that the Indians, having been "embarrassed by heavy debts [and] destitute of funds to effect important improvements in their country . . . hereby agree to cede" all lands north of the Tennessee River.[33] In the Fort Confederation Treaty of 1802, the Choctaw agreed to sell fifteen hundred square miles (an area the size of Rhode Island) to the government in exchange for *one dollar*.[34] That year, the starting price at auction for newly acquired land was $1.64 an acre—a potential return on investment of 157 million percent. Who says war is bad for business?

The land cession business was a buyer's market: Indigenous nations were prohibited from selling land to any party other than the US government.[35] If a dollar was what the United States offered, a dollar was what the Choctaw would get. It is taken for granted that the government wanted land for nutrient-intensive plantations and a growing white yeomanry, but there was another pressing concern: The new republic was broke. In 1790, government debt was 40 percent of GDP thanks in large part to the Revolutionary War. Land accessions and public auctions offered a way out. By the 1830s, public land sales were the government's second largest source of revenue. In 1836 alone, the government sold off twenty million acres (an area the size of South Carolina). That same year the country paid off its debt. It is ironic that the first and last time the United States was solvent was when it sold off stolen land—land it acquired, in part, by pushing the owners into debt.[36]

By the time Jackson signed the Indian Removal Act in 1830, previous administrations had perfected an array of strategies for removing Native people. In the East, Jackson's land clearances vastly expanded settler agriculture—less for the landless settlers who voted him into office than for the slave-owning class to which Jackson belonged.[37] In the West, Jackson envisioned a vast region encompassing nearly half of US territory that would be "governed by a confederacy of Indian polities and given a delegate in Congress." But his political opponents—like former President John Quincy Adams—objected, claiming that a nation divided between whites and Natives smacked of imperialism and despotism.[38] Little did they know the diverse forms of despotism that were to come.

## ACT 2: RESERVATIONS

By 1850, the United States was not content with an amorphous Indian country in the western half of its claimed territory, especially when that territory included some of the richest topsoil in the world. Bolstered by new efficiency-improving technologies like John Deere's polished steel plow and McCormick's grain reaper, homesteaders headed west, eager to "bust the sod" of the Great Plains and get on planting the country's amber waves of grain.

Meanwhile, congressional battles raged over the balance of slave versus free states, a crisis perpetuated by expanding plantation capital. Territories could achieve statehood only when settlers outnumbered Indigenous people, and political interests on both sides of the slavery debate recruited settlers to squat on Indigenous land in the hope of breaking the deadlock. In search of a territorial solution to the demands of its rapacious agricultural model, the United States pushed its program of dispossession and extermination west.

The strategies employed east of the Mississippi were replicated in the West, with one crucial difference. Indefinite Western displacement was no longer an option: The United States had designs in California and Oregon Territory, and white settlers to protect in the lands surrendered by Mexico in 1848.[39] The government needed a structure for *enclosing* Indians, not just displacing them, a structure that could be progressively cinched shut as settler farming and ranching expanded—something more flexible than early treaties and less of a sledgehammer than the Removal Act. In 1851, Congress passed the Indian Appropriations Act that provided a budgetary basis to confine Indians to reservations, establishing a system that remains in place today.

As before, direct violence characterized both military and militia actions in this period. US soldiers and volunteers did not distinguish combatants from noncombatants. Like the Colorado volunteers who killed 105 Cheyenne and Arapaho women and children at Sand Creek in 1864, then "decorated their weapons with body parts" before marching back to Denver to display their bloody bounty at the Apollo Theater.[40] Or the detachment of US Seventh Cavalry that killed 300 Lakota at Wounded Knee in 1890, then dumped their frozen bodies in a mass grave three days later. Popular histories characterize the United States as a country reluctant to wage a foreign war. And yet, for most of the nineteenth century, that is exactly what it did. In 1860, six of the US Army's seven regional commands (known as "departments") were located *west* of the Mississippi.

As in the East, military policy in the West targeted Indigenous food sources, none more so than the enormous rolling herds of American bison, a primary source of food for Plains Indians. Working together, military and civilian bison hunters used the expanding rail network to access the vast grasslands of the continental interior, where docile herds were mowed down with technologies perfected in the Civil War. In some cases, carcasses were harvested for pelts or meat. Others were left to rot in place—after all, the animal was worth more as territorial leverage. As Lieutenant John M. Schofield, a military commander from Missouri, explained, "With my cavalry and carbined artillery camped in front, I wanted no other occupation in life than to ward off the savage and kill off his food until there should no longer be an Indian frontier in our beautiful country."[41] Once numbering in the tens of millions, fewer than a thousand bison remained by 1890. Facing widespread hunger, the Sioux surrendered to punishing treaty concessions at Fort Laramie, despite possessing considerable military strength. In exchange for confinement to reservation lands that were often unsuitable for farming or ranching, Indigenous people across the West became dependent on government guarantees of food.

But "guarantee" vastly overstates the reality of government food aid. A healthy adult needs roughly 2,000 calories per day. In 1860, at Round Valley Reservation, Indians received from 480 to 910 calories of government rations daily. Two years later, it was a scant 160 to 390 calories. When the Dakota were removed to Crow Creek Reservation in 1863, a quarter died from starvation within the year.[42] The year before, the government failed to provide rations to the Santee Dakota in Minnesota. When starving Dakota attacked encroaching white settlers in the Minnesota River Valley, the Union Army crushed the uprising and sentenced 300 Santee captives to death. Concerned that executing so many prisoners of war could impact the treatment of Union soldiers in Confederate prisons, President Lincoln ordered the number reduced. On December 26, 1862, 28 Dakota captives were hanged in the largest mass execution in US history. Two months later, the United States annulled all treaties with the Dakota, releasing reservation lands for settlement, canceling government annuities, and ending promises of food aid.

Congress passed a flurry of acts in the 1860s and 1870s to speed the settlement of ceded lands. Under the Homestead Act of 1862 (the first of several "Homestead Acts"), US citizens over the age of twenty-one could request an allotment of public land and, after farming it for five years, obtain title for free. Under the Desert Land Act of 1877, citizens could buy 640 acres at a steep discount so long as they irrigated 80 of them. And through the Timber Culture Act, citizens could get 160 acres for free if they planted and maintained trees on 40.[43] These acts helped establish 57 percent of frontier farms, mostly in Kansas (named after the Kanza Sioux tribe), Nebraska (from the Sioux word for "broad water"), and Dakota Territory. Between 1870 and 1890, the settler population in this area expanded from 500,000 to 3 million.[44]

Other acts helped secure settlement through states and private companies. The Morrill Act of 1862 granted public land or scrips to states for "the endowment, support, and maintenance" of land-grant colleges, mandated to "teach such branches of learning as are related to agriculture and the mechanic arts."[45] The Pacific Railway Acts of 1862 and 1864 gave railroad companies 110 million acres of public land—ten miles on either side of the proposed track. Companies recruited immigrants to purchase land and settle along the railway lines to defray construction costs.[46]

By 1900, half a billion acres of public land had been disposed of and the total area under settler agriculture doubled. Grain monoculture, cattle ranching, and white settlers replaced prairie grass, bison, and Plains Indians one legislative decision at a time.

Like the 1830 Indian Relocation Act that aimed to ease class conflict between plantation owners and the yeomanry, pro-settlement acts aimed to address another growing conflict, this time in the country's urban areas. Over 350,000 immigrants arrived each year in the early 1850s, the majority bound for factory jobs in the Northeast and the food processing centers of Cincinnati, Buffalo, and Chicago (from *shikaakwa*, the Myami word for a swampy area where wild onions grow). Some politicians became concerned with overcrowding, poor working conditions, labor actions, and interethnic violence within an increasingly diverse workforce. As early as 1840, social reformers lobbied to distribute Western lands to poor urbanites free of charge.[47] But for twenty years, homesteading bills languished in Congress: Northern elites feared that relocation would deprive their factories of cheap labor, while Southern plantation owners feared that handouts of land would constrain plantation growth. In the end, the prospect of more free states in the West won over Northern abolitionists, while Southern opponents turned toward secession. As the Union Army called up its Western companies to fight the Confederacy, Congress enlisted an army of settlers to take their place.

The treaties that established reservations also created a new food system for Indian country. Through Army rations and government food aid, "beef and salt pork replaced bison, deer, and fish; wheat flour replaced cornmeal; refined sugar replaced honey, maple syrup, and dried fruits."[48] Navajo were given wheat flour and bacon while imprisoned in eastern New Mexico, not the traditional staples of beans, squash, and mutton. At the time, the government justified its food program as a way to keep Indians from starving as they switched to self-sufficient farming. But self-sufficiency was never a realistic possibility. Most Western reservations were on marginal lands, and Congress showed little sustained interest in financing Indian agriculture. In fact, of the $744,000 appropriated by Congress in 1851 to pay its Indian obligations, a mere $15,500 (2 percent) was allocated for agricultural purposes—to buy implements, grain, seed, and livestock, to support farmers, or to pay the salaries of "persons to aid them in agriculture."[49]

There were economic reasons for maintaining Native dependence. As settler agriculture converted Indigenous land into extensive ranches and monocultures, farms produced far more food than the market could bear. Overproduction led to falling prices: In 1866, a bushel of wheat sold for $2.06; by 1894, it fetched a mere $0.49. Over that same period, corn fell from $0.66 to $0.21 a bushel. Farmers produced these surpluses not by increasing yields but by adding acreage, a strategy made possible by the reservation system.[50] By undermining Native self-sufficiency, the government not only kept Indians from competing with settler farms but ensured Indian dependence on the commodities those farms hoped to dispose of.

Precisely how much surplus was dumped on Indigenous populations during this period is difficult to calculate. At different times and places, food was acquired and distributed by the War Department, the Department of the Interior, sectarian organizations, and sometimes all three. Tons of beans, barley, corn, hard bread, hominy, oatmeal, and salt (among other goods) were purchased from private suppliers and shipped to reservations, where agents stored, divided, and distributed the rations. Much of this accounting is now lost. But to take a single year as a snapshot, in 1891, with farm prices at historic lows, the US government purchased 461,000 pounds of corn, 8 million pounds of flour, and 35.7 million pounds of beef to feed the reservation bound.[51] Long before New Deal policies authorized the USDA to manage overproduction, the US government was using its treaty obligations to prop up falling prices. Once again, the historical narrative is reversed: commodity food programs did not save Indian bodies so much as Indian bodies saved settler farms.

By the 1870s, the frontier was closing. With fewer spaces to hide the country's unwanted people, "Indian country" changed from an amorphous territory to a precisely demarcated elsewhere. In 1871, Congress declared that "no Indian nation or tribe within the territory of the United States shall be acknowledged or recognized as an independent nation, tribe, or power with whom the United States may contract by treaty."[52] The reservation system, under the management of the Commissioner for Indian Affairs, replaced the treaty system as the means to organize Indian space. The geographic relationship between Indians and settlers would no longer be negotiated by both sides but determined unilaterally. The settler state would no longer conquer land with calvary but with cartography.

## ACT 3: ALLOTMENT

Before 1887, most Indigenous land was communally owned. Some treaties designated areas of private property, but the majority did not. The allocation and use of Indigenous lands fell to tribal authorities and followed traditional practices. For the most part, settlers did not care so long as cleared land was readily available. But with the closing of the frontier, settler society showed its capitalist and Calvinist roots. Reformists framed communal ownership as "communistic and backward," lacking the entrepreneurial impetus to make land and bodies "productive." Private property was a hallmark of a civilized society and would give Indians a "reason to stay in one place, cultivate land, disregard the cohesiveness of the tribe, and adopt the habits, practices and interests of the American settler population."[53] It would also facilitate the transfer of treaty-guaranteed lands to settlers, without the messy business of war.

Under the Dawes Act of 1887, reservation land was surveyed and divided into allotments of 40 to 160 acres each. Tribal households were assigned an allotment or allowed to select one by the agency superintendent. In theory, the Dawes Act was supposed to protect Indians by giving them clear title; in practice, it was a poorly disguised land grab. Under the act, if the amount of reservation land exceeded the area needed for allotments, this "surplus" was immediately ceded to the US government for sale to non-Indian homesteaders and companies. The most fertile reservation land was often the first to be declared surplus, with marginal lands parceled out for Indian ownership. Reservations with lower population densities saw enormous tracts of prime agricultural land lopped off with the stroke of the surveyor's pen.

Even where population densities were high, the Bureau of Indian Affairs (BIA) had a trick to shrink the number of qualifying Indians: a centuries-old settler calculus known as "blood quantum." Blood quantum refers to the fraction of "Indian blood" that a person has, determined by adding the blood quantum of the person's parents and dividing by two.[54] After centuries of contact with non-Indians, many Indigenous people fell short of blood quantum criteria for "Indianness." In fact, the criteria were designed to do just that. In the early 1800s, blood quantum was used to shrink the number of people receiving government benefits; under the Dawes Act, it was used to shrink *territory*. With fewer "statistical Indians," fewer allotments were prescribed and more land was carved out as surplus. Blood quantum and the surplus land stipulation resulted in the loss of sixty million acres of Indian-owned land (an area the size of Wyoming).

It got worse. Those Indians who qualified for an allotment received not simple titles like those given to homesteaders but titles held in trust by the government. Only after the twenty-five-year trust period expired did allotment titles convert to standard ones. On the upside, allotees did not have to pay tax on the land so long as it was in trust. On the downside, Indian landowners were prevented from selling or transferring their land without government approval. The "Dead Indian" Act of 1904 authorized the secretary of the interior to designate which heirs would inherit property if the allotee died before the trust period ended. Those heirs selected by the secretary were often the ones most likely to sell. The Burke Act of 1906 allowed the secretary to cancel a trust if he felt the allottee was "incompetent." Competence was often based on whether the allottee was putting the land to productive use—farming, ranching, or some other profitable activity. If not, the trust would be canceled and taxes assessed, *all without informing the owner.* Many Indians lost their land to non-Indians through foreclosure auctions because they owed taxes and did not realize it. In 1908, the secretary gained the right to sell the allotments of deceased Indians whose heirs he deemed incompetent.[55] By then, the secretary of the interior had become the most powerful landlord in US history.

Through the legislative chicanery of the Allotment Era, Natives lost an additional twenty-seven million acres (an area the size of Tennessee). By 1920, nearly all the best grazing land on Indian reservations was occupied by non-Indian ranchers, and by the 1930s, 75 percent of those living on Sioux land were non-Indian.[56] Historian Patrick Wolfe writes, "Ostensibly, this programme provided for a cultural transformation whereby the magic of private property ownership would propel Indians from the collective inertia of tribal membership into the progressive individualism of the American dream. . . . This cultural procedure turned out to yield a faster method of land transference than the US Cavalry had previously provided."[57]

The magic of private property was not the only strategy used to remedy Indian collectivism. The Allotment Era coincided with a general shift from removal and war toward what settler histories call "assimilation": a policy of integrating Indians into mainstream society. In standard narratives, assimilation is presented as a painful but inevitable transition, an improvement over the direct violence of earlier years. Within the reformist (and white-supremacist) mindset of the late nineteenth century, assimilation masqueraded as charity. Captain Richard Henry Pratt's famous call to "kill the Indian in him and save the man" energized a movement to "civilize" and "Americanize" Indigenous people through agricultural, dietary, educational, and religious interventions—all designed to permanently erase Indigenous culture, foodways, and language.[58] Despite its beneficent overtones, assimilation was a euphemism for cultural genocide. Killing the Indian, metaphorically or otherwise, was still the goal.[59]

One notorious assimilation technology was the boarding school system. Back in Albuquerque, my elementary school was located on Indian School Road. In the 1980s, the road ended near a large abandoned lot littered with derelict buildings and itinerant tumbleweeds. I did not know it at the time, but this was once Albuquerque Indian School (AIS), one of over four hundred boarding schools run by the BIA. AIS was chartered in the 1880s to assimilate Indigenous children to settler society through industrial and agricultural education.[60] Some Indian families sent their children voluntarily, but others were conned or coerced. In 1891, Congress authorized the Commissioner of Indian Affairs to suspend food rations for Native families who refused to turn over their children.[61]

Students were required to speak English, eat Western foods, learn Western trades, and adopt Western gender and behavioral norms to prepare them for a life in settler society. Generations of children lost access to traditional languages, diets, and religious practices. What the Dawes Act did to Indian land, the boarding school system did to

**Figure 1:** Schoolgirls attend sewing class at Albuquerque Indian School, Albuquerque, New Mexico, 1910. Unknown photographer. Courtesy National Archives, Records of the Bureau of Indian Affairs, NAID 292877.

the Indian body: break its ties to community and make it productive for settler society.

In the 1970s, construction of a new city park near AIS unearthed the bones of children buried in unmarked graves.[62] Deadly epidemics of scarlet fever, measles, and influenza plagued the dormitories of AIS, and up until the 1930s, administrators used a nearby lot as a makeshift cemetery for its Pueblo, Navajo, and Apache children. Some believe there are more graves to be found, underneath another city park and a nearby commercial area.[63] It is exemplary of so-called assimilation that Indigenous children were not even allowed to return home upon their death.

In the late 1800s, the United States faced two contradictory crises. The closure of the frontier shrank the amount of new land available for farming and ranching, while the rapid expansion of agriculture fed overproduction and deflated farm profits. Allotment and assimilation addressed *both* concerns. By eroding reservation lands, allotment provided a steady flow of new acres for settlement. By eroding Native self-sufficiency, assimilation forced Indians to consume settler surpluses. In the early twentieth century, these policies helped usher in the "Golden Age of American Agriculture," a period of high demand and increased farm profits.[64]

But by the 1920s, agricultural depression had returned with a vengeance, and with it a natural resource crisis of continental proportions. To weather the storm, the United States would once again change its relationship to Indian country.

## ACT 4: REORGANIZATION

For a time, growing urban and export demands bolstered confidence in the agricultural sector. Between 1897 and 1920, prices for US food and farmland increased. In response, farmers expanded their holdings and invested in new labor-saving technologies. In his description of agriculture's golden age, former USDA economist Willard Cochrane writes, "The country was settled. The world was at peace. Hard work, thrift, and 'right thinking' had indeed paid off for farmers; the good life was a reality." Of course, the world was *not* at peace—wartime chaos in Europe was keeping demand high for US farm products. The good life was *not* a reality for the millions of Native Americans forced to eat the country's leftovers. And even then, the so-called golden age was about to come to an inglorious end.

With the return of European production after World War I, demand for farm products collapsed. Between May and November 1920, wheat prices dropped from nearly three dollars to one dollar a bushel. Settlers who bought land while prices were high fell into debt as the agricultural sector skidded into recession. The general economy held on until 1929, when the banks finally went under. Droves of bankrupt farmers and farmhands hit the road in search of work as national unemployment reached 25 percent. To make matters worse, falling prices pushed farmers to increase production on marginal land and abandon soil conservation practices. By the mid-1930s, soil erosion had turned swaths of Texas, Colorado, Kansas, and Oklahoma (from the Choctaw word for "red people") into the infamous Dust Bowl, forcing a mass migration of destitute households.

To weather the depression, the United States needed to increase farm prices, invest in agricultural infrastructure, and stem the tide of unemployed workers. Settler histories describe how New Deal programs created the Soil Conservation Service, bought up agricultural commodities to support farm prices, and hired legions of unemployed workers to build roads, dam rivers, and electrify rural areas. To justify these investments and reassure an anxious public, President Franklin Roosevelt preached national unity and cooperative effort: "All American workers, brain workers and manual workers alike, and all the rest of us whose well-being depends on theirs, know that our needs are one. . . . There is no cleavage between white collar workers and manual workers, between artists and artisans, musicians and mechanics, lawyers and accountants and architects and miners."[65] But the cleavage between settlers and Indians would be preserved. By the late 1920s, the disastrous impact of US Indian policy was drawing public condemnation. The widely circulated Meriam Report blamed allotment and assimilation for rampant poverty and disease on Indian reservations.[66] In keeping with Roosevelt's brand, the new commissioner of Indian affairs, John Collier, proposed an "Indian New Deal" that would increase federal and state funding for education, infrastructure, and medical care.

The Great Depression offered other compelling reasons to make reservations more livable. Greater integration of Indigenous farmers into the settler economy would only complicate New Deal objectives: Competition from Indian farms and ranches would suppress settler farm profits, and competition from unemployed Indians would suppress settler wages. Notwithstanding Collier's good intentions, Roosevelt needed a way to keep Indians and settlers separated, and that meant preserving the very thing past administrations had steadily undermined: a stable, self-contained Indian country. An out-of-the-way elsewhere that would still consume the country's excesses.

With the Indian Reorganization Act of 1934, Roosevelt abolished the allotment program—after all, farmland was already too cheap, so there was little need to increase the supply. Tribes would be officially recognized as political entities so long as they rewrote their constitutions to be in "structural harmony" with the United States.[67] Tribal governments were reorganized around settler notions of democracy, with little acknowledgment of traditional cultural practices. And to maintain demographic erosion and reduce future welfare obligations, Congress insisted that tribal governments use blood quantum to determine membership.

Indian New Deal programs encouraged reservation agriculture, but with a twist. The BIA supported irrigation and mechanization for *cash crops*, not staples. In Navajo Nation (Diné Bikéyah), residents were encouraged to plant fruit trees, not corn that might compete with settler farms or reduce their appetite for commodity staples. Households were encouraged to raise sheep to make wool for handicrafts, not beef cattle that might compete with white ranchers for both grazing land and paying customers.[68] Meanwhile, the commodity food program was expanded. Surplus wheat flour, sugar, butter, and canned meat and vegetables flooded reservations, undermining any chance of locally sourced alternatives. In Indian boarding schools, Native children were forced to do their part for the dairy industry by drinking milk and eating cheese—despite 75 percent of them being lactose intolerant.[69]

Policies that appeared to support Indigenous agriculture invariably yielded to settler concerns. During the construction of Hoover Dam (a New Deal project), officials worried that livestock grazing on the Navajo Reservation was exacerbating soil erosion and increasing siltation of the Lake Mead reservoir. The BIA, having just encouraged growth in Navajo ranching, reversed course and ordered households to cull their herds. Those who refused were arrested. Families watched as BIA officials "killed herds on the spot, left them to rot. . . . Some herds were even driven off cliffs, while others were doused with kerosene and burned alive."[70] Livestock reductions crushed small farms on the reservation and tanked the Navajo economy. Many households stopped subsistence farming during this period and became dependent on federal food surpluses. In the end, the culling of Navajo livestock had no impact on siltation but may have increased available fodder for off-reservation ranchers. Some suspect this was the motivation all along.[71]

In Northern California and southern Oregon, six dams were built across the Klamath River and its tributaries during the New Deal period. Intended to generate electricity and irrigate settler farms, the dams blocked over 90 percent of Chinook salmon spawning grounds. Salmon is the centerpiece of the diet for Karuk, Yurok, and Klamath people. But as river levels dropped and nutrient-rich runoff disappeared, so did the fish. In response, the Department of the Interior outlawed Native fishing. Like herd culling on the Navajo reservation to "protect" Hoover Dam, it was the Native food system that paid the price for settler society's environmental mismanagement. In 2002, over 68,000 adult salmon died in the Klamath River from low water levels, the largest fish kill in US history.

Meanwhile, Karuk salmon consumption fell by 99 percent. Living in what had been the third richest salmon fishery in the West, Natives became dependent on commodity foods rich in fat and sugar. Sociologists Leontina Hormel and Kari Norgaard observe that "with the loss of the most important food source . . . the Karuk people hold the dubious honor of experiencing one of the most recent and dramatic diet shifts of any native tribe in the USA."[72] In 2010, Karuk were three times more likely to suffer type 2 diabetes and four times more likely to experience heart disease than the general population. Among the Navajo, 33 percent have type 2 diabetes or are prediabetic. The national average is 8.3 percent.[73]

The reorganization of Indian country occurred at a time when the settler economy was struggling with economic depression and wanted to limit competition from Indigenous land and labor. Natives were made productive not by eliminating the reservation system, but *by protecting and managing it.*

With the end of World War II, the US economy emerged from depression with booming industrial growth. Farms produced record crops with the help of chemical and mechanical inputs. The new economic landscape came with two new challenges: find cheap labor and dispose of agriculture's extraordinary surpluses. BIA administrators became convinced that Natives would be more useful assembling consumer products in urban factories than raising goats on the reservation. Having worked to preserve reservations through the Great Depression, the United States would reverse course once more.

## ACT 5: TERMINATION

By the late 1940s, the provisioning of food and basic services on reservation lands was in disarray. As the settler economy enjoyed a postwar boom, many reservations lacked electricity, running water, functional roads, hospitals, and schools. Life expectancy for Natives on reservations was 44 years, compared to 77 years for white people. In the winter of 1947–48, starvation was so widespread on the Navajo and Hopi reservations that the government had to airlift emergency food.[74] With workers in high demand, states saw reservation conditions as an economic opportunity. Backed by white-owned agricultural, manufacturing, and mining interests, states lobbied the federal government to end its treaty obligations and push Natives into wage labor.[75]

By the 1950s, these efforts had paid off and the elimination of reservations became national policy. The Termination Act of 1953 sought to end trust protection of Indian lands and delist tribes from federal recognition. One of the signature programs was the Indian Relocation Act of 1956. Designed by BIA commissioner Dillon S.

Myer, the program offered two hundred dollars and a one-way ticket for Natives to move from reservations to cities "where they would disappear through assimilation into the white, American mainstream."[76] Federal trust protection would end and their land would be sold off for development. Myer saw relocation as the first step in ending reservations, tribal governments, treaty obligations, and the BIA. In an echo of Richard Henry Pratt, one white BIA official explained in 1957: "I've always felt that the only real solution for the Navajo was to cease to be a Navajo—to get off the reservation and become a citizen just like everybody else, and make his living in the same way as other people. Forget that he is a Navajo, in other words."[77]

But even if he forgot, "everybody else" still remembered. Under the program, 100,000 Natives were scattered across the country into poor, working-class neighborhoods. Forced to compete for low-skilled jobs, Native transplants were isolated and targeted for racial abuse and job discrimination. Many struggled with unemployment and destitution. By 1970, half of all Indigenous people lived in cities, more than a hundred Indigenous nations had been legislatively terminated, and another million acres of Native land had been lost to non-Natives. The Indian Relocation Act swapped rural poverty, land ownership, and cultural integration for urban poverty, precarious wages, and lasting psychological wounds. The impact was not limited to transplants: A longitudinal study in 2012 found elevated rates of depression, substance abuse, and antisocial behavior among the *grandchildren* of those who were transplanted.[78]

Since 1975, Indigenous nations have won greater self-determination with increased self-management of federal programs. Termination tried to convert Indigenous people into market-dependent consumers and producers, to dissolve Native America into mainstream white America and continue the erasure of Native foodways. But in this task it largely failed. Because of termination, today's Indian Country (capital C) spans reservations, rural areas, *and* cities. "With Native people being more inclusive," notes historian Donald Fixico, "[termination] only expanded Indian

Country in a very unusual way."[79] In other words, the city-dwelling Navajo never forgot who he was.

## MAKING ELSEWHERE

Over the past 250 years, Indian country emerged as a convenient elsewhere for the United States, one that evolved to suit the changing social and economic needs of the settler food system. In the early 1800s, the United States faced a looming natural resource crisis, agrarian class conflict, and a debt-burdened treasury. Treaties and forced removals were immediate territorial solutions to these problems, all of which could be mitigated through land sales. With the Indian Removal Act, Congress separated the land it wanted from the people it didn't. "Elsewhere" was "anywhere but here," an amorphous purgatory where the human by-products of dispossession could be mothballed and kept out of sight.

By the 1850s, Congress sought territorial solutions to its impasse over slavery and growing concerns with urban class conflict. Cavalry raids, forced starvation, and homestead acts helped secure land for railroads, colleges, and urban laborers looking to escape poor working conditions. Mostly, these efforts preserved a long tradition of land speculation. Elsewhere became a prison—staked out on the worst land, legally severed from settler space, and dependent on the government for its existence.

With closure of the frontier in the late 1800s, a century of rapid farm expansion came to an end. Vast monocultures of wheat and corn produced more food than the market could consume. As agriculture looked for ways to parcel out new acres and consume its looming surpluses, reservations were mined for valuable allotments and captive consumers. Elsewhere became a land bank and a dumping ground for unhealthy foods.

As agricultural depression lingered through the 1920s and 1930s, Indian land and farm products became a threat to the settler economy. The Indian Reorganization Act preserved Indigenous nations as political entities so long as tribes adopted settler forms of governance and surrendered to federal oversight. Elsewhere became a separate space with separate rules, both walled off and micromanaged.

Finally, by 1950, revitalized industries demanded cheap labor as new agricultural technologies helped farmers fill the country's grain stockpiles. To complete the conversion of Indigenous people into wage laborers and market consumers, the BIA tried to consolidate Indian country into other elsewheres: poor urban neighborhoods where the country's underclasses competed for low-wage jobs.

## CARCERAL FIXES AND THE MAKING OF "AMERICAN" AGRICULTURE

Through twists and turns, the US food system relied on spatial fixes to solve its problems. One form that exemplified US-Indian relations is the *carceral fix* in which distinct territories are cordoned off and managed under separate rules.[80] These unique spaces provide opportunities for states to manage resources and people: Indian country variously served as a land bank, a population of captive consumers, and a reserve army of cheap labor.[81] Operating under different rules, carceral spaces can be tweaked, experimented on, and upended without disrupting "free" space. The government did not need to regulate land speculation so long as allotment rules were properly adjusted; white ranchers in the Lake Mead watershed did not need to cull their herds so long as BIA agents could force the Navajo to cull theirs.

Carceral fixes are often defined through formal "interfaces"—treaties, congressional acts, court rulings, and policies. In the United States, these interfaces defined the economic, social, and territorial relationship between Indian and settler. Early treaties bound Indigenous people to the settler economy through the need to purchase food and pay debts, but denied them economic opportunities like selling land. Later policies bound Indians to the settler food system through commodity foods but denied them the resources to grow their own. In fact, the power of US Indian policy came not from *isolating* Natives from the settler economy, but from *partially integrating* them. Land, food, and working bodies were encouraged to move across the Indian-settler interface in ways that benefited settler society.

Like other spatial fixes, carceral fixes redirect structural violence from privileged spaces into diverse elsewheres. The Indian Removal Act redirected class antagonism from the plantation-yeoman frontier to the yeoman-Indian frontier. Homestead acts and allotments redirected class tensions from urban areas into the contested countryside. And commodity food programs redirected the violence of falling food prices from settler farms into Native bodies.

Indian country was not the only place shaped by these fixes. By absorbing the structural violence of the settler food system, elsewhere helped define "here," its privileged opposite. As Indian country came to be "barren," "backward," and "poor" in the settler imagination, non-Indian agriculture became "productive," "principled," and "free." Jefferson's "cherished cultivators" appeared all the more vigorous and virtuous because the system's self-destructive tendencies were directed onto others. The country's agricultural sector was all the more profitable because its onerous social and environmental costs were borne by others. In fact, Indian country was not an encumbrance to settler agriculture, but *essential* for its meteoric growth.

As colonialism erased Indigenous food systems from the landscape, settler agriculture took its place, inverting the historical narrative once more. Monoculture replaced integrated fields of corn, beans, and squash. Cattle ranches replaced herds of plains bison. Commodity production became the new normal as once-common practices of Indigenous cooperation and self-provisioning became literal and figurative outliers. By the mid-twentieth century, it was white family farms that clung to ancestral acres in the face of land-hungry businesses and speculators. Settlers had become nouveau Natives—albeit, with somewhat less genocide.[82]

## MISSED OPPORTUNITIES

It is tempting to think that an alternate history without Andrew Jackson, Kit Carson, or Richard Henry Pratt might have been more egalitarian and humane, that Indigenous lifeways east of the Mississippi might have been preserved, that Navajo might never have been marched into starvation, or that Native children might never have been confined to boarding schools. But the removal of a few individual cogs is unlikely to have stalled the machine. Settler colonialism was itself the symptom of deep contradictions in the logic of Euro-American agriculture. After all, colonial North America was the spatial fix for class conflict across the Atlantic, where elites enclosed common land and forced peasants to produce a surplus.[83] European agriculture was organized around removal and dispossession—it is no surprise that its North American offshoot had violence in its DNA.

This does not mean that the *forms* of violence were inevitable. Genocide is still a series of choices—choices made by powerful people with familiar names. Since independence, other voices have argued for peaceful coexistence, meaningful and lasting Indigenous sovereignty, and the honoring of treaties. Indians and non-Indians alike have pushed back against settler colonialism at every turn. Ultimately, the dispossession and extermination of Native Americans was a decision to ignore such voices—to double down on Jefferson's genocidal business plan not from a *lack* of humane alternatives, but *in spite* of them.

Early land policies could have put the brakes on speculation. The government could have regulated trade with Europe, slowing growth in the plantation sector. US Indian policy could have allowed Natives to keep some portion of ancestral lands rather than prioritize removal, and it could have supported cultural land tenure rather than impose a real estate market that benefitted settlers. The BIA could have protected Native foodways rather than undermine them with unhealthy commodities. For that matter, US agriculture could have found something else to do with its surpluses than dump flour, fat, and sugar on a captive population. But to pursue any of these alternatives would have required the food system to find other solutions to agrarian class conflict, land speculation, urban unrest, economic stagnation, and chronic over-production. It would have required settler society to look within itself for a solution to structural violence, rather than redirect that violence onto others.

Without land hunger as an explanatory crutch, this odd territorial shell game makes more sense. Settler colonialism was animated not by any manifest "destiny," but by a deliberate, calculated, and never-ending commitment to export its own problems.

Indian country was the food system's first elsewhere, but not its last. The spatial fixes first employed against Indigenous people have been redeployed to create new elsewheres and prepare other populations for the system's excess violence. Distant land and labor were exploited to maintain the country's rapacious monocultures (chapter 2) and powerful monopolies (chapters 5 and 6). Japanese American farmers were dispossessed and incarcerated under the pretense of national security (chapter 3). Poor tenants and landowners were kept in debt to facilitate the extraction of land and labor (chapters 4 and 5). Infrastructure was used to undermine local economies (chapter 6), and dangerous surpluses were dumped on other markets to protect profits (chapters 4, 5, and 7).

For the past 250 years, the food system has used geography to solve its chronic crises, turning islands, countries, neighborhoods, and prison camps into new versions of Indian country, all to reinforce a predominantly white, male, and capitalist vision of *American* agriculture. Placenames like this one are a work in progress.

• • •

Placenames are also hopes. For Indigenous Americans today, Indian Country is a place of pride, resilience, and community, a reminder that settler colonialism—for all the damage it did—has yet to succeed.[84] Indian Country is symbolic of Indigenous *survivance*: an active sense of presence and continuity; a conscious rejection of tragedy and victimhood; a lively promise of rejuvenation.[85]

One expression of survivance taking root in Indian Country is "place-based" food provisioning.[86] Industrial agriculture distances consumers from the soil and supplants local diversity with one-size-fits-all monocultures. Place-based approaches do the opposite, using what is available to rebuild the web of relationships between land, people, animals, and plants. At the center of this web is food. In Diné Bikéyah, community leaders are teaching traditional farming, ranching, and cooking to a new generation of Navajo. At the annual Miss Navajo Nation pageant, each finalist slaughters a sheep and prepares a traditional meal over a wood fire. Tribal programs and nonprofits offer training and financial support for Navajo who choose to farm for subsistence. In Oneida Nation (Wisconsin), farmers raise Tuscarora white corn, a variety from ancestral lands in New York. Community members work together at harvest time, barter heirloom seeds, and relearn cultural knowledge and practice. Oneida cornbread now reclaims shelf space from snack foods at gas stations throughout the reservation. And in the Klamath basin, twenty years of work by Indigenous leaders has finally paid off. To help restore the wild salmon population, PacificCorp agreed to remove four of its hydroelectric dams, the largest dam removal effort in US history. It is hoped that future generations will witness Chinook salmon returning to their ancestral tributaries—as Yurok, Karuk, and Klamath return to theirs. From Massachusetts (Algonquin for "great hill small place") to Alaska (Inuit for "great lands"), similar efforts are reclaiming Native foodways for the twenty-first century.

But place-based food provisioning does more than feed people. At a former boarding school in Tseshaht Nation (c'išaaʔatḥ), tribal members have converted the old playground into an organic garden.[87] "Children never played in these schools," says Charlotte Coté, whose sister established the garden, "they were always waiting to see if they would be the next victims of the people who abused them." Today, the

Tseshaht Garden Project nurtures prolific carrots, squash, and kale on the same grounds where children were beaten, tortured, and experimented on.[88] Coté writes,

> So, I asked her [my sister], "Well, do you think the land can feel the pain the same way these children felt the pain?" And she says that the land, the plants, the animals, everything around here, they saw what happened. There is trauma in that land. We need to heal that land too. Believe me, I can't grow kale for the life of me although I've tried . . . [now] it's like a Jurassic Park garden. That land needed to heal so badly that it's producing all of this bounty for us and for the people, some of whom went to those schools and experienced extreme hunger.[89]

If extraction is a common theme in Indigenous-settler relations, reclamation is a throughline in Indigenous survivance. As Indigenous farming reclaims its cultural place, and settler agriculture struggles to extract more from its monocultures, the narrative of progress is inverted once again. It is *Indigenous* farmers—vigorous, independent, tied to the land, and "wedded to its liberty and interests by the most lasting bonds"—that best embody Jefferson's agrarianism, not the billion-dollar agribusinesses that control the settler food system. There is still work to be done, but today, the prospects for Indigenous food systems look bright; it is the future of settler agriculture that needs reckoning.

# 2

# SCATOLOGICAL IMPERIALISM

## *Squandering Soil and Grabbing Guano for Ravenous Monocultures*

Sometimes success is its own worst enemy. In 1492, Columbus sailed the ocean blue, claiming the Caribbean, its resources, and its people for the Spanish Crown. But in 1503, one year into his fourth voyage, Columbus's fortunes literally ran aground. With his last two vessels infested with shipworm, the explorer beached his ships on Jamaica's northern shore. For the next year, Christopher Columbus—conqueror of hemispheres—was forced to rely on the kindness of strangers. Things started out well: Jamaica's Indigenous Taíno kept him and his crew alive with gifts of food. But the shipwrecked Spaniards stuck with the game plan that had won them fame and fortune, returning the generosity of their hosts with murder, rape, and pillage. After seven months, Taíno patience wore thin, and the gifts of food ceased. Sensing correctly that he had overstayed his welcome, Columbus put some of his crew in canoes and dispatched them to Hispaniola, where he hoped the Spanish garrison could launch a rescue.

Eighty-five miles east of Jamaica, the crew came upon a small, teardrop-shaped island. Seeing no evidence of fresh water or human inhabitants, they chose not to disembark. But before sailing on to Hispaniola, they marked the tiny speck on the map and gave it the name "Navaza." Had they gone ashore, they would have encountered dangerous reefs and fifty-foot cliffs. Upon scaling those cliffs, they would have found poisonous plants, venomous scorpions, and squawking seabirds. For the treasure-hungry Spaniards, Navaza would have seemed as unremarkable up close as it did from the sea, and yet, the island contained enough wealth to fund several expeditions. In 1504, there was no way for them to know this, and even if they had known, there was no way to cash in. It would be another three hundred years before Europeans realized just how valuable a tiny, unremarkable island like Navaza could be.

For millennia, only seabirds visited Navaza, drawn by the thousands to fish in the island's teeming reefs. Colonies of red-footed booby and frigatebird ate, bred, defecated, and died on the island. When it came to defecation, Navaza's avians overachieved: Thousands of years of prolific pooping had left the island's two square miles under a million tons of guano. One notable heap, later dubbed the Mountain, was as tall as a ten-story building. By the 1700s, Haitian fishers were making infrequent visits, drawn to the island's reefs for the same purpose as the birds. But otherwise, Navaza remained uninhabited, uncharted, and uninteresting to the human world.

That would change in 1857. There are conflicting accounts of what transpired, but in September of that year, the island was claimed in the name of Baltimore businessman Edward K. Cooper.[1] Such a claim would ordinarily lack legal standing but for the recent signing of the Guano Islands Act. According to the act, any citizen of the United States could lay claim to an uninhabited (and previously unclaimed) island bearing guano deposits and—upon presidential approval—the island would become US territory. The Guano Islands Act triggered a maritime bonanza. Like homesteaders rushing to claim Western allotments, American entrepreneurs cruised the Pacific and Caribbean in search of rocky outcrops and their fecal bounties. No fewer than ninety-four islands were claimed under the act; sixty-six became official possessions, and nine remain US territories.[2]

It may seem odd that the country's first foray into overseas empire was inspired by excrement (Columbus, at least, was looking for gold).

In fact, the prize wasn't the guano itself, but the nitrates and phosphates it held. By the mid-1800s, agricultural soils in the United States were so lacking in these vital nutrients that acquiring bird guano had become a matter of national security. In 1850, President Fillmore declared guano "so desirable an article to the agricultural interest of the United States that it is the duty of the Government" to obtain it.[3] At the time, obtaining guano meant importing it from Peru, where British traders held a monopoly. But high demand (and healthy gouging) drove up prices. The United States wanted guano of its own, and with the 1856 Guano Islands Act, it crowdsourced the treasure hunt to hundreds of latter-day Columbuses.

Of course, finding a private guano island was one thing; removing the toxic excrement and getting it to the United States was another. Mining guano involved breaking apart the enormous deposits with dynamite and pickaxes, loading the pieces into sacks, and hauling the sacks to waiting boats. Omnipresent guano dust filled workers' mouths and sinus cavities. Acrid vapors burned the lungs. Those who worked below deck loading the ships could succumb in minutes to the foul smell. Conditions were so horrendous that companies had to rely on forced labor: convicts, indentured workers, and army deserters, all under armed guard. Infractions were punished by the whip, and those killed on the job were simply replaced by the next in line. As the *Christian Review* noted in 1862, companies that mined guano had perfected "the infernal art of using up human life to the very last inch."[4]

On Navassa (renamed to suit its English-speaking proprietors), the Navassa Phosphate Company slowly dismantled the island, one hand-loaded sack at a time. Former slaves were recruited from Baltimore with promises of a tropical paradise in the Caribbean. Instead, they wound up on Navassa where they were shackled, beaten, and prohibited from leaving. Eventually, workers' patience wore thin, and in 1889 they went on strike. When the white overseers threatened to shoot them, the workers fought back, and four overseers were killed. The arrival of a British warship brought the brief rebellion to an end, and the strikers were captured. Eighteen Black guano workers were taken back to Baltimore, where they were charged and convicted of murder in federal court. The case, *Jones v. United States*, would reach the Supreme Court in 1890, not over the legality of abusing former slaves, but because the court couldn't decide if Navassa was really part of the United States.[5]

*Jones v. United States* is notable for several reasons. It opened the door to later territorial acquisitions, revealed the appalling working conditions of former slaves in the Reconstruction era, and was the first Supreme Court case argued by Black attorneys. But Navassa is also a story about the food system. The Guano Islands Act would never have existed without a long-standing commitment to intensive agriculture, the repeal of Britain's Corn Laws, and growing hunger in the Southeastern United States. The act was never intended to kick-start imperial expansion (though it did), nor instigate a legal crisis over the definition of "territory" (though it did that too). The act was established to save agriculture from a self-inflicted wound. American monocultures stuck with the game plan that had supported them for decades, returning the generosity of the soil by pillaging its nitrogen and phosphorus. By the 1850s, soil fertility had worn thin, and its gifts of abundance ceased. Agriculture's productive success had become its own worst enemy.

• • •

In the mid-1800s, long before the Dust Bowl years of the 1930s and the environmental movement of the 1970s, US farms struggled through the country's first great natural resource crisis. This is the story of how agricultural policies and practices squandered the continent's fertile soil, then scoured the world for a solution. At the heart of this story is *expropriation*, the unequal exchange of one thing for another—the unequal exchange of nutrients between soil and crop, between city and countryside, and between the United States and distant elsewheres; but also, the unequal exchange of value between those who profited from guano and those coerced into extracting it. This is the story of how a basic principle of capitalist agriculture kicked off a brutal and unprecedented movement of crops, chemicals, and captives around the world, one that continues today.

## YOU CAN'T SPELL "PLANT" WITHOUT N AND P

Plant life relies on seventeen essential chemicals for growth. For most plants, only one of these chemicals is derived from the atmosphere: carbon dioxide ($CO_2$). The rest are exchanged with the soil and taken up by roots. All seventeen nutrients are necessary, but shortages of nitrogen (N) and phosphorus (P) are often the limiting factors in plant growth.[6]

Nitrogen forms the chemical bonds between amino acids and is essential for making enzymes and proteins, the basis for cellular activity. Nitrogen atoms bond with carbon in chlorophyll, the molecule upon which all plant life—and, consequently, animal life—depends. Nitrogen is easy to find, making up 79 percent of the air we breathe. But this atmospheric form ($N_2$) is inert, locked up behind triple bonds. Plants need reactive nitrogen. One way to get it is biological: Certain bacteria can unlock $N_2$ with enzymes, releasing oxides of nitrogen (NOx) or ammonia ($NH_3$) in a process called "biofixation." Leguminous plants like clover, soybeans, and peas nurture colonies of biofixing *Rhizobium* bacteria in their roots, returning more nitrogen to the soil than they use. Another strategy is to blow up the $N_2$ lock altogether, but this requires significant energy. To complicate matters, reactive nitrogen is readily dissolved in water and leached out of the soil. The upshot is that the amount of nitrogen available to plants is often more limited than other nutrients, especially in places with abundant rainfall.[7]

Phosphorus, on the other hand, makes up far less of the earth's biosphere than nitrogen but hangs around for longer. Its most abundant form, phosphate ($PO_4^{3-}$), is the mortar for living bodies, holding together the structural tissue of plants and the bones and teeth of animals. The primary energy source for cell metabolism is adenosine triphosphate (ATP), built on phosphate bonds. Unlike nitrates, which readily move between air, soil, and water, phosphorus cycles more slowly. Over time, phosphate is eroded from rocks, carried by rivers, and deposited in lowland soils. Most of the earth's phosphorus ends up in the ocean, where it is deposited in the seabed. Millions of years later, geologic uplift and volcanism create new phosphate-rich islands, and the process starts over.

Phosphate also hangs out around people. Of the phosphate we eat, 98 percent leaves the body as waste. Historically, this excreted phosphorus accumulated near places of human settlement. In fact, archaeologists use phosphorus concentrations in the soil as an indicator of past human habitation. Meanwhile, the remaining 2 percent goes into making our teeth and bones, returning to the soil when we die. Follow the trail of phosphorus and it will lead you to both the outhouse and the grave.

## METABOLISM AND RIFT

We refer to the conversion of chemical inputs into beneficial outputs in living organisms as *metabolism*: the sum of all chemical reactions that sustain life. Cells metabolize inputs to produce energy and synthesize large molecules. Plants metabolize $CO_2$, water, sunlight, and nutrients to build and nourish cells. And animals metabolize the cells in plants (or other animals) for the chemicals to feed their own cells.

But as a concept, metabolism is not limited to individual organisms. At the scale of an ecosystem, primary production and nutrient cycling ensure the survival of entire biotic communities through processes we might call "metabolic." Even social processes can be thought of as metabolic: To produce food and income, a farm metabolizes soil, crops, and livestock, but also the daily labor of the farmer.[8] A powerplant metabolizes inputs of coal or oil to produce electricity, and a factory metabolizes that electricity, raw materials, and the labor of its workers to produce commodities. Meanwhile, workers metabolize their wages to buy food, and metabolize the food to provide more energy for labor.

Every link in this metabolic chain involves both production and consumption—to produce one thing, another thing is consumed, itself the product of an earlier round of metabolism. Economists think of production and consumption as separate activities, but metabolically speaking, that which produces something must consume something, and that which consumes something must produce something, even if these "somethings" are not counted in human ledgers.

Ideally, what is produced by one entity is replenished. For example, a portion of the soil nutrients consumed by an animal are returned to the soil through its manure and the decomposition of its corpse. These metabolic circulations happen at different speeds and spatial scales: Chemical transfers between bacterial colonies take place over minutes and microns, while the geologic processes that expose phosphorus-rich bedrock can span centuries and continents. The round trip may be complex—from soil to plant, animal, ocean, and atmosphere before returning to the soil.

Sometimes, a link in the metabolic chain breaks. For example, vital nutrients are not returned to the soil in time to satisfy the needs of dependent plants. Karl Marx referred to this as *metabolic rift*, a breach in "the natural rhythms of regeneration and recycling" that limits the ability of life to reproduce itself.[9] For Marx, rift was a defining feature of capitalist agriculture, a system that requires farmers to increase production per unit of land, time, and labor to stay profitable. But metabolic rift predates capitalism. In fact, since the emergence of agriculture 12,000 years ago, farmers have not only pursued increases in production but sought better ways to manage the metabolic *consequences* of those increases.

## MANAGING AGRICULTURAL RIFT

As farming invented techniques for extracting more calories per acre, it placed greater demands on the soil and raised the specter of metabolic rift. Historically, farmers have used four strategies to mitigate this threat.[10] The most basic involves clearing land of existing plant communities for use as fields and pastures. Farms make use of the nutrients already stored in the soil to grow crops. In the mid-1800s, homesteaders struggled to break up the dense sod of North America's central plains, but the reward was some of the richest topsoil in the world—a bounty of chemical resources amassed over millennia by the interaction of roots, insects, and microbes.

When land is readily available and population pressures are low, farms simply move after exhausting soil nutrients. Some forms of cultivation, like swidden, accomplish this through systematic land rotation: After harvest, farmers let fields lie dormant while they clear and cultivate other land, only to return to the original fields after a period of time. Managed correctly, these fallow periods allow native plant and animal communities to return and replenish soil nutrients.

Another strategy involves planting crops that fix atmospheric nitrogen. The nutrient benefit from planting legumes was known to farmers in ancient Greece and China. The Indigenous people of the Americas planted corn alongside squash, root crops, and nitrogen-fixing beans in a system called *milpa*. In Native America, corn, beans, and squash—the "three sisters"—continue to play an essential role in cultural and dietary traditions.[11] In Europe, farmers planted clover, a leguminous herb that provided fodder for livestock and helped return nitrogen to the soil between grain harvests.[12] Today, a season of nitrogen-fixing soybeans or alfalfa may not replace the nitrogen lost to a single crop of wheat or corn, but it helps.

A third strategy is to return the organic by-products of production back to the soil. Farmers use crop residue, animal manure, and left-over food to fertilize fields, replacing some of the nutrients extracted during harvest. Historically, the need to maintain soil fertility forged important partnerships between farms, villages, and ranches. In East Africa, sedentary Haya farmers acquired manure to fertilize their banana gardens from neighboring Hinda cattle herders. In Edo-period Japan, merchants collected excrement from urban areas to sell to rural farmers.[13] Until recently, human excrement was a valuable and frequently traded commodity—quite the opposite of "waste."

Preindustrial farming systems employed some combination of these three strategies. These forms of cultivation are sometimes called "closed" because nitrogen and phosphorus stay (more or less) within the system. But under the economic and demographic pressures of the Industrial Revolution, these closed systems failed. As Europeans left the countryside for cities, farms shifted from growing diverse crops for local consumption to raising a handful of valuable crops for export.[14] It became increasingly difficult to keep the nutrient loop closed as rural soils lost abundant nitrogen and phosphorus to the sewers of European cities. Throughout the nineteenth century, concerns with metabolic rift preoccupied Europe's scientists, philosophers, and artists. In *Les Misérables*, Victor Hugo offered this encomium to misplaced excrement:

> There is no guano comparable in fertility with the detritus of a capital. A great city is the most mighty of dung-makers. Certain success would attend the experiment of employing the city to manure the plain. If our gold is manure, our manure, on the other hand, is gold. . . . Those fetid drippings of subterranean mire, which the pavements hide from you—do you know what they are? They are the meadow in flower, the green grass, wild thyme . . . they are perfumed hay, they are golden wheat, they are the bread on your table, they are the warm blood in your veins, they are health, they are joy, they are life.[15]

Parisian drippings may have been life, but in the mid-1800s that life was being flushed out to sea. The steady expropriation of soil nutrients during the Industrial Revolution pushed European agriculture toward a fourth nutrient-management strategy: taking nitrogen and phosphorus from other sources. These "open" agricultural systems overcame metabolic rift by taking vital nutrients from other parts of the world, effectively exporting the rift from local farms to distant phosphate quarries, sardine fisheries, and guano islands. In *Grundrisse: Foundations of the Critique of Political Economy*, Marx lamented, "Agriculture no longer finds the natural conditions of its own production within itself, naturally arising, spontaneous, and ready to hand, but these exist as an independent industry separate from it—and, with this separateness, the whole complex set of interconnections in which this industry exists is drawn into the sphere of the conditions of agricultural production."[16] Marx wrote these words in 1857, the same year the United States claimed Navassa. By then, these interconnections included foreign governments, island ecosystems, and coerced labor—all braided together to save a food system that could no longer grow its own food.

## VAMPIRES AND VANDALS

Before synthetic fertilizer it was bird dung, and before bird dung it was human bones. For centuries, butchers sold the ribs, femurs, and vertebrae of slaughtered animals to be ground into bonemeal, a rich source of phosphorus. But as agricultural demand grew, it wasn't long before bone merchants began scouring more macabre sources. Human skeletons were dug up from European battlefields and cemeteries, shipped to Britain, and ground up for fertilizer. There was plenty to go around: With the end of the Napoleonic Wars in 1818, Europe's graveyards were ripe for picking. In 1822, the *London Observer* wrote, "It is now ascertained beyond a doubt, by actual experiment upon an extensive scale, that a dead soldier is a most valuable article of commerce. . . . For aught known to the contrary, the good farmers of Yorkshire are, in great measure, indebted to the bones of their children for their daily bread."[17]

The illicit bone trade was met with outrage on the continent, not just because the English were ransacking graves and stealing dead people. More to the point, they were stealing phosphorus. The chemist Justus von Liebig protested,

> England is robbing all other countries of their fertility. Already in her eagerness for bones, she has turned up the battlefields of Leipsic, and Waterloo, and of Crimea; already from the catacombs of Sicily she has carried away skeletons of many successive generations. Annually she removes from the shores of other countries to her own the manurial equivalent of three millions and a half of men whom she takes from us the means of supporting and squanders down her sewers to the sea. Like a vampire she hangs from the neck of Europe.[18]

The vampire's thirst reached a new level in the 1840s. For twenty years, Britain's Corn Laws limited grain imports to protect English producers. At the same time, Britain enriched its wealthy landowners and fed its workers at the expense of colonial elsewheres, especially Ireland. Marx noted that by taking so much corn, beef, and wool from Ireland, England "indirectly exported the soil . . . without even allowing the cultivators the means for replacing" its nutrients.[19] But when the potato blight devastated Irish agriculture in 1845, the golden goose stopped laying eggs. Desperate for food, Parliament repealed the Corn Laws, opening the door to grain imports. Demand for fertilizer soared as English farms were forced to compete with foreign producers. Farmers had gotten their first taste of Peruvian guano in 1841, and by mid-decade hundreds of British vessels were scouring the globe for bird manure.[20]

English farmers were not the only ones looking for ways to remain competitive. The repeal of the Corn Laws influenced foreign producers as well: By the mid-1850s, a quarter of the wheat consumed in Britain was grown abroad. One of the country's biggest suppliers was the United States, another place where the poor were going hungry from depleted soils.[21]

US wheat was just the latest in a long list of North American farm exports. The first was tobacco, a New World crop that Columbus introduced to Europe. From the early 1600s, agriculture in the Mid-Atlantic colonies was organized around tobacco production. By 1775, over 100 million pounds of tobacco was being exported, most from the rich lowland soils of Maryland and Virginia. So much labor was committed to tobacco in Virginia that the colony forced its settlers to plant a portion of land in corn just to prevent starvation.[22] Tobacco is also nutrient intensive. By 1800, depleted soils in the Mid-Atlantic forced some plantation companies to move west and south, around the southern foothills of the Appalachians and into Georgia. By 1812, large coastal estates in Virginia were picking up stakes and heading for Alabama and Mississippi, leaving exhausted land behind.

Southern soils did not fare well under the new owners. Growers repeated the wasteful practices that had saved time and money in the East, planting the same acres with intensive crops until soil nutrients were depleted, then moving on. As historian George Crawford observed, "Crop rotation existed as a nice theoretical concept. Few landholders attempted to apply manures. . . . After they despoiled one tract, farmers, planters, and herdsmen moved on to another and ruined it, in turn."[23] And why not? Buying new acres was cheaper than improving existing ones and would remain so as long as the government paid its debts with land sales. Indeed, farmers in Georgia rarely invested much in homes and structures, knowing they would be moving on before long.[24]

What tobacco started, King Cotton finished. Between 1800 and 1860, US cotton exports grew from 21 million to 1.8 billion pounds. More than half was destined for British factories.[25] But this extraordinary growth came from an expansion of cotton *land*, not from increases in *yield*. In fact, by the 1840s, yields were falling. Agricultural journals reported declines of 50 to 70 percent per acre in nearly every Georgia county. Many planters blamed the South's favorite export: “It is the successive crops of cotton which have been grown on our lands that so exhausted them,” said one Baldwin County farmer. As cotton consumed land to solve its nutrient problem, it spread metabolic rift to other areas. One Floyd County farmer lamented, “The same vandal system which has desolated so large a portion of Middle Georgia—ribbed it with gullies, which the waving broomsedge vainly attempts to conceal—has invaded this beautiful section.”[26]

The “vandal” system would ultimately undermine the ecological basis for food production. Between 1850 and 1860, corn and pork consumption declined across the South, in some places falling by more than half. Livestock died from lack of fodder, and rural households endured food shortages that could last weeks. An April 1859 editorial in the *Savannah Republican* described the city as “but one stage distant from the starvation point,” warning that soon “our citizens will be compelled to send to Charleston and New York for their daily table supplies.”[27] As cotton acres pushed out other crops, town markets that used to carry a diverse assortment of locally raised meat and produce could no longer keep stock. Desperate Southern communities considered unprecedented measures. The editor of the *Sandersville Central Georgian* was asked to organize a meeting to help get food to hungry people in the county: “The plan proposed is to raise a general fund with which to purchase corn, and each one who is able taking and paying for same at cost prices, as much as he may necessarily need, those who are unable to buy being furnished gratis.” To make matters worse, farmland expansion was no longer a viable option. Some left for new lands in Texas and Arkansas, but most were stuck with their exhausted acres. Metabolic rift had a devastating impact on long-term wealth. As Crawford notes, “The future of each household was tied to permanent returns from agriculture. . . . How parents would fare in their later years depended upon the land they owned. . . . Children would get their start by drawing the resources their fathers had managed to accumulate.”[28]

By the start of the Civil War, the value of agricultural land in the South was half that of land in the North, due primarily to poor soil management.[29] To save itself and mend the rift, Southern agriculture turned increasingly to a new and seemingly miraculous cure—one the British were already raving about.

## "THE TURDS OF FOREIGN BIRDS"

Three hundred years after Columbus returned home with samples of New World corn, potatoes, and tobacco, Alexander von Humboldt returned from the Pacific coast of South America with another transformative addition to Old World agriculture. On his visit to Peru in 1802, the German explorer observed Quechua-speaking farmers spreading bird guano on their fields (in fact, the English word "guano" comes from the Quechua word for bird droppings). There seemed plenty to go around: Some of the islands off the coast were covered in layers a hundred feet thick. Curious about its chemical properties, Humboldt took a sample of guano to show his friends back home.

The Chincha Islands off the coast of Peru, where Humboldt collected his sample, offer perfect conditions for the accumulation and preservation of guano. As cold ocean currents run into the islands' underwater slopes, deepwater nutrients are pushed up into the shallows. Phytoplankton feed on the nutrients, anchoveta eat the plankton, and at the top of the pyramid, millions of cormorants, pelicans, terns, and penguins feast on the abundant fish. Bigger birds that live in large colonies tend to be the most prolific poopers—like the Guanay cormorant, affectionately dubbed Peru's "billion-dollar bird."[30] With less than an inch of rain reaching the islands each year, the prodigious excrement piles up, forming barren, pockmarked outcroppings. A visitor to the Chinchas in 1863 described the approach from the sea: "From the distance of two or three miles, the cliffs and slopes . . . appear to be rather formed of chalk, having the undulations of their summits covered with a reddish olive coloured carpet" that gives off a "perceptibly ammoniated atmosphere."[31]

Humboldt's chemistry colleagues were impressed with the Chincha guano, noting that it contained 11 to 17 percent nitrogen: up to thirty times the nitrogen content of traditional manures. It was also rich in phosphates.[32] But the discovery did not inspire a rush, not right away. It would take a few more years of falling soil fertility, grave robbing, and an Irish famine to truly light the candle on the guano craze. Once lit, however, there was no going back. The first large shipments arrived in England in 1841. Between 1850 and 1860, over three thousand ships carried guano around the world. And by 1870, Peru was exporting 700,000 tons per year, much of it bound for the United States.[33]

Guano transformed American agriculture. In 1853, the *Farmer's Journal* of North Carolina cheered guano's effect on the Upper South: "There are many rich fields in many parts of the country, which but for guano would have still been barren. . . . Wagons laden with the increased products of the grain fields, blocked the streets of market towns. Millers and forwarders were overwhelmed with business, and the banks drained to pay for wheat." The *New England Farmer* reported that "the introduction of guano into Maryland and the tidewater region of the lower States" has ensured "all branches of industry have been stimulated to new life and activity."[34] Liebig estimated that from 1841 to 1855 in Europe, the addition of Peruvian guano was sufficient to grow ten million metric tons more grain, enough to feed over twenty-six million people for a year.[35]

Guano's success in the field was translated into extraordinary profits for two private companies. The Peruvian firm F. Barreda y Hermano obtained exclusive rights to mine the islands, and the British firm of Antony Gibbs & Sons was granted a monopoly to sell the guano to the world. As demand for Peru's magic dust soared in Britain and the United States, so did monopoly-controlled prices. In 1843, a ton of guano sold in Baltimore for seven cents. In 1850, it topped seventy-six dollars.[36] As one ribald poem proclaimed,

The House of Gibbs made their dibs
Selling the turds of foreign birds.[37]

The company certainly made their dibs. Over four decades, Peru exported some fourteen million tons of guano at a value of £150 million, the equivalent of $13 billion today.[38] But mining and selling guano were not the only ways companies profited from this increasingly global trade.

## SLAVERY 2.0

For all the salubrious benefits guano provided plants, its laborious extraction did considerable harm to humans. In 1844, the *Vermont Watchman* described guano as "a beastly smelling-bottle sort of mess, looking like bad snuff mixed with rotten kittens."[39] The sharp ammonia burned the lungs, despite the hemp masks that workers wore. Those tasked with loading ships would emerge from the holds "gasping for breath, sometimes suffering nosebleeds or temporary blindness."[40] Extended exposure led to long-term respiratory and digestive ailments, not to mention daily labor under a hot sun on an island with no drinkable water. It was not uncommon for exhausted workers to fall into the loading chutes along with the guano, ending up "suffocated by the ammonia, before they could recover the shock of the fall."[41] Under these working conditions, few were willing to sign up for guano duty, even if the paycheck was substantial.

For the firms involved, guano mining was precisely the kind of business that benefited from slavery. There was only one problem: Slavery was coming to an end. Britain and the United States had banned the importation of slaves by 1808. Spain abolished slavery in all its colonies (except the Caribbean "sugar islands") by 1811. Initially, Peru used slaves and convicts to mine guano, but banned the practice in 1854 under international pressure.[42] Where was a profit-seeking enterprise going to find people desperate enough to be convinced, conned, or coerced into an early death?

The alternative to chattel slavery was debt peonage. Chinese men, contractually bound to pay for their passage across the Pacific, were shipped from East Asia as part of the so-called coolie trade.[43] In the mid-1800s, British officials promoted the coolie trade as an alternative to slavery and set up recruiting operations on China's east coast, and by the peak of the guano craze, major American trading houses like W. R. Grace were in on the act. Recruiters concocted a vision of future prosperity to convince the men to sign on. Those who refused were often kidnapped. Coolies were told that their contracts guaranteed passage to California's gold fields in exchange for a few years of hard labor. In fact, before being loaded onto ships, the men were branded with the letter C (ostensibly for "California").[44] But few ended up there, and those who did were more likely to build railroads than mine for gold.

Instead, over ninety-thousand coolies were shipped to Peru. The *New York Daily Times* described the vessels that carried them as "based upon the model of slave ships . . . [with victims] crowded down between low decks, where any other than a prone or sitting posture is out of the question."[45] Coolies were kept in shackles during the voyage, and 10 percent died enroute. On arrival, the majority were forced into plantation labor, but approximately 10,000 ended up on the guano islands. One observer described the conditions: "Besides being worked almost to death, they have neither sufficient food nor passably wholesome water. . . . Each man is compelled to clear from four to five tons of guano a day. . . . The hard lot of the coolies in the guano works was magnified by the cruelty which they suffered from hard taskmasters."[46] Contractual obligations were rarely honored by the company. As Maurice Nash, a crew member on a US guano ship, observed, "Once on the islands a Chinamen [*sic*] seldom gets off, but remains a slave, to die there."[47] In fact, the treatment of workers was so horrendous and the prospect of freedom so dim that many chose suicide. A US consul to Peru described how guards had to be stationed "around the shores . . . to prevent [coolies] from committing suicide by drowning, to which end the coolie rushes in his moments of despair."[48] Should a worker succeed in killing himself—or die from the job or harsh treatment—traders were ready with another. As late as the 1920s, you could find Chinese graveyards on every island, "where garments, detached bones, and twisted corpses of poor coolies have for many years been exposed to the merciless sun."[49]

Like so many corpses, time exposed the worst atrocities of the coolie trade. Peru eventually bowed to international pressure and signed a treaty with China regulating debt peonage. But as English bone merchants did in the 1830s, guano companies simply went back to the map. Their next source of labor was closer to home: Rapanui (Easter Island). In the 1860s, slave traders kidnapped one-third of Rapanui's population to work Peru's excrement mines. The guano trade would find its expendable workers someway, somehow, from somewhere.

### NAVASSA

Two thousand miles away, in the inhospitable quarries of the Navassa Phosphate Company, that "somewhere" was Baltimore. By the 1860s, American farmers were fed up with the British-Peruvian monopoly on guano. Antony Gibbs's firm prioritized its British clients and gouged everyone else. Guano prices were high enough that the Maryland and Virginia legislatures passed resolutions demanding the government improve its relationship with Peru.[50] In response, Presidents Millard Fillmore and Franklin Pierce promised to make guano a national priority, and in 1856 Congress passed the Guano Islands Act, which allowed privately claimed islands to "appertain" to the United States as territories.

The country's first appurtenances were in the Pacific. Wealthy financiers organized oceanic expeditions in search of reefs and rocky outcrops. In 1856, the hastily formed American Guano Company used the Guano Islands Act to claim Baker Island and Jarvis Island, fifteen hundred miles southeast of Hawai'i. The next year, Edward Cooper claimed Navassa in the Caribbean, and by 1863, the

**Figure 2:** An oil painting, circa 1870, of Navassa Island depicting a company settlement on the shoreline with guano mining operations in the hills. Unknown artist, private collection. Public domain.

United States had acquired fifty-seven more. But with new territories came new obligations—notably, to defend these outposts of coral and manure from would-be interlopers. In 1858, Haiti protested US claims to Navassa and sent soldiers to retake the island. President Buchanan dispatched the USS *Saratoga*, and the Haitians departed without a fight, but also refused to recognize US claims.[51]

With their excrement protected, the next concern for guano barons was how to remove it and transport it home. Content to let Britain and Peru exploit Chinese workers, US companies turned to Hawaiian labor on its Pacific islands and former slaves on Caribbean ones. Many of Navassa's laborers were recruited in Baltimore, where they were conned into signing fifteen-month contracts under promises of an easy tropical life. Wages were eight dollars per month, to be paid at the end of the contract. The company deducted one dollar per day for sickness and fined workers for infractions. The contract covered food, but the rations were so poor that workers had to supplement with high-priced canned goods from the company store. By the end of the contract, it was possible for a worker to receive no pay at all.[52]

Meanwhile, the company's overseers—often former slave drivers—punished offenses by "tricing workers up by their wrists for up to six hours in the tropical sun."[53] The *Kansas City Journal* did not mince words: "An American slave in Navassa has no alternative. If he does not wish to work he may starve, for all the food on the island is controlled by the company . . . the crack of the slavedriver's whip was soon heard where only the scream of the seagull had broken the silence for centuries."[54] Physical and mental abuse eventually pushed the laborers too far, and on September 14, 1889, they refused to work. When one of the overseers threatened to shoot, the men fought back, killing four. Following the "riot," one of the company's white overseers offered this racial take: "There are 137 darkies employed in the phosphate mines and they were all sent from the city of Baltimore. . . . They were a pretty hard class of men and up to all kinds of viciousness when the opportunity presented itself. If they had been unsophisticated natives of that region it might have been an easy matter to subdue them, but being colored laborers picked up in an American city they were an entirely different class to deal with."[55]

On December 2, 1889, eighteen Black workers were tried for murder in federal court. Three were convicted and sentenced to death. On appeal, the defense argued that the United States did not have legal authority over Navassa because the island was claimed by Haiti and the word "appertain" in the Guano Islands Act was vague. But the Supreme Court disagreed and upheld the sentences, affirming that Navassa—and, by extension, all other guano islands—fell under US law.

Facing public outcry over the ruling, President Benjamin Harrison commuted the death sentences to life in prison. In fact, Harrison worried that the ruling set a precedent for corporations to convert US territories into prison colonies.[56] But rather than step back from the colonial precipice, successive administrations pushed forward. Victory in the Spanish-American War in 1898 would crown a new era of agricultural imperialism for US companies in Latin America and the Philippines (see chapter 5).

## ROBBING PETER TO PAY PAUL

In capitalist agriculture, farm activity is geared toward commodity production: growing crops and raising animals for sale. Survival in this context means maximizing returns, and to do this, farms try to increase production per unit of land, time, and labor. But this intensification has limits: Only so many plants and animals can be raised on an acre of soil, only so much growth accrues to plants and animals in a season, and only so much can be expected of a worker. It is possible to exceed these limits, but only temporarily, and only at ecological and social costs. Since the sixteenth century, capitalist agriculture has involved the art and science of *expropriation*—how to take more from the system than what is returned, and how to push the ecological and social costs onto other people and places.[57]

Intensive cultivation expropriated nitrogen and phosphorus from the soil. In the United States, wheat and corn were marketed to distant cities, where human metabolism released nutrients into waterways by way of the sewer. Cotton and tobacco were exported to Europe, where their nutrients were woven into textiles or sent up in smoke. Over time, this metabolic rift deprived soils of the means to support life. In England, where land was limited, consumers came to rely on agricultural imports, exporting the rift to Britain's settler colonies. In the United States, where land expropriation dovetailed with larger political goals, planters exported metabolic rift by colonizing new lands in the South and West, leaving a trail of degradation in their wake. In both cases, agriculture looked beyond its immediate borders for a solution—an *extraterritorial fix* for metabolic rift.

Land expansion in the United States supplied more nitrogen and phosphorus for expropriation, but only temporarily. By the mid-1800s, these lands were depleted, too. Once more, agriculture looked beyond its borders to repair its nutrient rift, finding extraterritorial fixes in the metabolism of island ecosystems, debt-bound Chinese, and its own unwanted workers. Capitalist agriculture, a system antithetical to ecological and social self-sufficiency, used imperialism to stay alive.[58]

Commodities hide their sordid histories of expropriation behind seemingly equal market exchanges.[59] Extraterritorial fixes make this easier by pushing ecological and social costs beyond the horizon of awareness and concern. So, too, with the guano trade. Before it could give life to US wheat and cotton, Peru's guano drove coolies to throw themselves off Pacific cliffs. Navassa's phosphate kept free men in chains. With their purchases, farmers encouraged the next nefarious labor contract, slave ship, and crack of the whip. And though farmers may not have known the bloody provenance of their miracle cure, there were plenty who did—overseers, traders, businessowners, and diplomats. Like Columbus stranded on Jamaica, they stuck with the game plan that had proven profitable for so many years.

## MISSED OPPORTUNITIES

Not everyone wanted to maintain this commitment to expropriation. In the 1850s, some farmers warned of the dangers of vandal agriculture and pushed for reform—voices like Garnett Andrews, a Georgia planter: "If the fertility of this territory had been preserved," he told a crowd in Macon, "it would now support in comfort, ten times its present population." Or Solon Robinson, who warned against giving up livestock dung because "dung embodied the cycle of nutrients in a given place; [whereas] guano came from outside the loop."[60] Faced with a crisis of fertility, large-scale agriculture could have followed this advice and tried to "close the loop." US strategists could have limited exports of cotton and wheat, slowed or halted the opening of new agricultural lands, or created incentives to conserve soil resources.

When the stark brutality of the coolie trade became clear, the United States could have prohibited companies from participating in it. It could have repealed the Guano Islands Act and prevented the exploitation of former slaves, Hawaiians, and island ecosystems. Instead, laborers were forced to move the mountain themselves. Eighteen men on Navassa refused to accept brutal treatment. Coolies crossing the Pacific mutinied sixty-eight times between 1847 and 1874.[61] And one by one, each suicide drew attention to conditions in Peru, ultimately shortening the coolie trade one human life at a time.

These changes in course would have required policymakers to listen to the country's small farmers, poor laborers, and whistle-blowers, not just its powerful plantations, guano companies, and trading houses. It would have meant abandoning extraterritorial fixes, mending metabolic rifts, and renouncing expropriation. In essence, it would have required US agriculture to throw out large parts of the imperialist game plan it inherited from Britain. Instead, the United States followed Britain's footsteps.

The proof is in the wording of the Guano Islands Act. As Daniel Immerwahr notes, "appurtenance," the term used to describe these new territories, suggests something like an accessory, an auxiliary organ neither integrated into—nor separate from—the whole.[62] As appurtenances, what political future was available to Navassa, Baker, Jarvis, and other islands? With enough people, could they become guano *states* in the union? Guano *territories* belonging to existing states? Privately owned guano-*stans*? The ambiguity of the word served an important function: It ensured these islands were attached to the United States in a relationship of convenience; one that could be reinterpreted to address changing needs. The Indian reservation system turned Native land into a versatile, internal elsewhere; the Guano Islands Act—and its codification of "appurtenances"—turned the country's elsewhere-making aspirations outwards.

## NEW SOURCES, RIFTS, AND ELSEWHERES

By 1880, the Peruvian guano trade was in decline and sodium nitrate ($NaNO_3$) from South America's Atacama Desert was taking over. British and American financiers backed the new nitrate business and encouraged its monopolistic tendencies: By 1890, 71 percent of the ten-million-pound trade was controlled by one British company.[63] Meanwhile, others scoured the Pacific for phosphate. Multinationals like the British Phosphate Commission—a partnership between the United Kingdom, Australia, and New Zealand—removed phosphate-rich land from islands across the Pacific. In Banaba, Kiribati, the commission took 90 percent of the island's surface, creating a rift so severe that the population had to be relocated to Fiji.[64]

The emergence of synthetic fertilizers accelerated guano's decline. In 1909, German chemists Fritz Haber and Carl Bosch pioneered a process for extracting ammonia from atmospheric nitrogen. By 1914, the Haber-Bosch process was producing more reactive nitrogen per year than had been extracted during the entire guano trade. Synthetic fertilizers were so valuable after World War I that Germany was allowed to pay some of its debts with nitrogen.[65] But the new process came with a trade-off: what *Rhizobium* accomplished with enzymes, Haber-Bosch achieved with energy. Lots of it. Since 1914, the primary source of that energy has been petroleum. Today, much of the food we eat is a fossil fuel by-product: Agriculture is still a vampire, only now, the "bones" it depends on are much, much older.

To protect its industrial monocultures, the United States pushes metabolic rift into distant places. As a result, the map of agricultural elsewheres keeps changing. In the 1850s, the United States was ready to defend Pacific and Caribbean islands over guano; by the 1950s, it was destabilizing Middle Eastern governments over oil. Beyond the farm, nitrogen collects in lakes and oceans where it spawns new ecological rifts. As Persian Gulf states receive their own oil in the form of US wheat exports, the Gulf of Mexico receives it in the form of poisonous runoff.

• • •

In 2013, historian Gregory Cushman prophesized that "someday soon, excrement may again be equated with silver."[66] It didn't take long for the world to prove him right. Today, Navassa is quiet, but Peru's guano islands are not. A century-long lull in extraction has come to an end as the trade in excrement makes a comeback. This time, growing interest in guano is fueled by demand for *organic* alternatives. Rising prices and environmental concerns with synthetic fertilizer have reignited the rush for bird dung, still one of the best natural sources of nitrogen ever found.

It remains one of the most unpleasant substances to mine. Like their nineteenth-century counterparts, today's *campañeros* ("campers") cover their mouths to filter acrid dust, sleep in flimsy tents, shovel guano by hand, and carry 110-pound bags down steep paths to waiting boats. These seasonal workers make 1,200 Peruvian soles ($310) a month, less than half Peru's average income. Most *campañeros* are Indigenous Quechua. Observers have accused the state-run company of "semi-slavery," but when the semi-enslaved are poor migrants from an ethnic minority, and when the trade is lucrative, it is easy to stick with the capitalist game plan.[67] Convicts, coolies, and Quechua—same old snake, different skin.

Today, the annual haul is 12,000 tons, less than 10 percent of what was extracted each year during the craze of the 1850s. The difference is not from a lack of demand but from an emerging metabolic rift: the region's sixty-million-strong seabird population has shrunk to just four million. The waters off Peru's coast—once teeming with anchoveta—are threatened by rising temperatures and overfishing. Today, one in five industrially harvested fish come from these waters. Peru's billion-dollar bird now competes for food with billions of people.[68] It remains to be seen what new frontiers of fish and fertilizer will be exploited once Peru's disappears.

# 3

# GRAND THEFT AGRICULTURE

## *Japanese American Incarceration and California's Produce Empire*

For the Salinas Valley Vegetable Grower-Shipper Association, the date of December 7, 1941, did not "live on in infamy."[1] To the contrary, the Japanese bombing of Pearl Harbor would bring good things for the white growers of the Salinas Valley, but first they had to pull some strings.

Within hours of the attack, the association dispatched its managing secretary, Austin E. Anson, to Washington, D.C. His mission: to convince the War and Navy Departments, the attorney general, and "every congressman who would listen to him" that people of Japanese ancestry living in the Salinas Valley posed an imminent threat to the United States.[2] Over the next few months, the American Legion, the California Farm Bureau, the "quasi-fascist" Associated Farmers, and prominent California politicians joined Anson in arguing for the enforced removal of Japanese immigrants and their US citizen children from the West Coast.[3] On February 19, 1942, President Roosevelt responded to their concerns with Executive Order 9066, providing the legal basis to exile Japanese Americans from California, Washington, and Oregon. And by August, 125,284 people had been forcibly removed to concentration camps where they would be held for the next three years.[4] Nearly two-thirds were US citizens.

We consume our histories the way we approach a big meal, gorging ourselves on familiar and palatable dishes until our hunger is satisfied. Meanwhile, the unfamiliar or less palatable remains uneaten. Something may catch our eye after the meal, but by then it is too late; we have had our fill. Japanese American incarceration—a less palatable part of US history—has the misfortune of being served alongside a monumental one: World War II. This blockbuster narrative has a compelling arc and moral subtext: It is the story of a country that overcomes its differences, conquers its adversaries, and emerges as a world leader. Like the "tireless yeomanry" that drove Western expansion in the 1800s, the "greatest generation" of the war years reified the freedom-loving character of a nation through extraordinary personal sacrifice. Consuming such an aggrandizing narrative leaves little room for homegrown racism, bigotry, and greed.

Over time, survivors, journalists, and historians have called attention to the harrowing experiences of incarcerated Japanese Americans during the war years.[5] Today, this chapter in history is considered something of a black eye, one of only a few national injustices garnering a presidential apology and some form of financial remuneration for victims.[6] Despite these acknowledgments, popular histories continue to frame removal and incarceration as a tragedy born of fear and military necessity. The public, it is claimed, suffered a bought of temporary xenophobia brought on by war. Apologetic history texts still equivocate with words like "relocation" and "internment," even though the concentration camps had towers with armed guards, inmates were kept behind barbed wire, some were shot and killed, and their property was systematically stolen.[7]

There is an economic story here, too; one that starts well before the attack on Pearl Harbor and one that is deeply entangled with the food system. Lurking behind Anson's hasty mission to Washington was the disposition of some 220,000 acres of prime agricultural land in California—land owned and operated by Japanese immigrants and their descendants. Anson's grower-shipper association was supported by the farm interests of Montgomery Street in San Francisco, considered the "Wall Street of the West" and "one of the most powerful aggregations of wealthy corporations in the US."[8] Since 1900, the white farmers, transporters, wholesalers, and bankers behind Montgomery Street had watched as Japanese immigrants steadily expanded their share of California's fruit and vegetable market. The bombing of Pearl Harbor offered Anson and his white associates a golden opportunity to claim this coveted territory. Five months later, in an interview with the *Saturday Evening Post*, Anson made his revanchist motives clear: "We're charged with wanting to get rid of the Japs for selfish reasons. We might as well be honest. We do. It's a question of whether the white man lives on the Pacific Coast or the brown men. They came into this valley to work and they stayed to take over. . . . If all the Japs were removed tomorrow, we'd never miss them in two weeks, because the white farmers can take over and produce everything the Jap grows. And we don't want them back when the war ends, either."[9]

Anson's anti-Japanese sentiments were widely embraced, but in December 1941, support for forced removal was not. Two-thirds of Japanese Americans living on the West Coast worked in agriculture, and powerful figures like California Governor Culbert Olson were concerned that removing so much labor from the fields would cripple the state's economy and threaten the war effort. Montgomery Street's capitalists needed a way to convince the Roosevelt administration, state legislatures, and the public that removal was in the country's best interest. It would not be easy, but if they played their cards right, they might just capture a more complete share of California's food system at the precise moment when the country needed food the most.

• • •

After the war, Japanese Americans returned home to find their remaining property sold off, stolen, or destroyed. Large agribusinesses and members of the white-dominated grower-shipper associations received confiscated farms and farm equipment at little cost.

Tax returns that would have identified the individual and corporate beneficiaries were inexplicably destroyed by the IRS.[10] A study conducted in the 1960s found that Japanese Americans lost over $4 billion in assets, though many losses could not be given a dollar value. And Anson would get his wish: Fewer than half of the evacuees from California returned to the state following incarceration, relocating to other parts of the country for fear of racial violence.[11] Fourteen years after the last concentration camp closed, the number of Japanese American farmers was a mere 25 percent of what it had been in 1941.[12]

This is a story of how powerful actors reorganized an essential part of the US food system along white-supremacist lines. As with Indian reservations, the incarceration of Japanese Americans was a spatial fix for capital and labor, destroying nonwhite competition in the marketplace, providing a captive workforce of agricultural labor, and reallocating valuable land into privileged hands. In its wake, US agribusinesses would increasingly turn from the Pacific to Central America, and from the labor of US citizens to the exploitation of precarious noncitizens.

## ISSEI AND NISEI

Not long after California became a state (1850) and US Commodore Matthew Perry made his famous overtures to Japan (1853) the first Japanese immigrants began arriving on California's shores. These early flows were dwarfed by the influx of Chinese immigrants, many of whom were part of the same coolie trade that supplied labor to Peru's guano islands. But that imbalance wouldn't last. Economic hardship following Japan's 1904–5 war with Russia drove waves of new migrants, and by 1907, tens of thousands of Japanese were arriving each year. As with other immigrant populations, the willingness of Japanese to work for low wages helped them secure diverse forms of employment. Many first-generation immigrants (*issei*) found work as field laborers in California's burgeoning vegetable and sugar beet sectors where they quickly made a name for themselves.

Part of their success came from the popular contract labor system in which large "gangs" of issei workers were organized under Japanese "bosses."[13] Bosses secured work for their gangs, negotiated wages, and distributed pay to workers, substantially reducing paperwork for employers and ensuring a reliable supply of labor. The system offered a simple, inexpensive, and dependable source of field hands for white growers who would otherwise have had to negotiate with and pay each worker individually.

What pleased white growers, however, did not please white workers who complained that issei gangs and bosses suppressed wages. As early as 1905, anti-Japanese sentiment began to follow the trajectory of anti-Chinese sentiment decades earlier. In May of that year, the Asiatic Exclusion League was founded to bring an end to Japanese immigration, and in 1909, seventeen different anti-Japanese bills were debated in the California legislature. Four years later, the Alien Land bill became law, prohibiting Asian immigrants from owning land, leasing land for more than three years, or bequeathing land to their heirs.[14]

Restrictions on Japanese immigration were relaxed during World War I as demand for agricultural labor grew, only to return in the 1920s as farm profits collapsed. This decade saw the rise of the "Yellow Peril" propaganda machine that stoked fears of cheap Japanese labor displacing white workers throughout the US economy.[15] The campaign resulted in the passage of the Asiatic Exclusion Law of 1924 that halted nearly all Japanese immigration to the country.

By then, the next generation of Japanese Americans (*nisei*) had become key players in Western agriculture's tug-of-war. Born in the United States, nisei were full citizens with constitutional rights, including ownership of land. Issei, prohibited from owning land because of their alien status, circumvented the racist spirit of the 1913 land law by titling farms in the names of their nisei children. The California legislature made several attempts to amend the 1913 law to close this "loophole," including prohibiting US-born nisei from acquiring citizenship in the first place. But these efforts were generally unsuccessful.[16] By the 1930s, the practice was tacitly accepted even among white landowners, and by the 1940s, many nisei were full-time owners and operators.[17]

## FROM FIELD HANDS TO FARM OWNERS

The success of the contract labor system helped Japanese Americans transition from farm labor to farm operation and ownership. To secure a reliable workforce, many California growers leased their land to issei bosses. Meanwhile, Japanese American operators distinguished themselves through their willingness to pay higher rents (or shares) and make improvements to the land. In 1900, there were only 37 Japanese American farms in California. Ten years later, there were 1,816. In the 1910s, Japanese-owned farmland increased from 16,000 to 75,000 acres, while leased and shared land grew from 140,000 to 313,000 acres. And it wasn't just quantity: According to the California State Board of Control, Japanese American farmland was some of the "best in the State."[18]

Growth in ownership and tenancy was aided by Japanese farmers' cooperatives (*nogyo kumiai*) that helped issei find employers and lessors. As land prices increased, cooperatives set a maximum rental price that members were willing to pay, protecting their members from price gouging and reducing entry barriers for newcomers. Cooperatives assisted farmers in marketing their crops and obtaining inputs, helped resolve disputes between landlords and tenants, and published periodicals to educate their members on horticultural best practices.[19]

As Japanese Americans transitioned from field hands to farm owners, issei workers were steadily replaced by Mexican and Filipino immigrants. Rarely did white workers become permanent field hands. This followed a racial pecking order that had governed agriculture in California since Spanish missions enslaved Indigenous people in the eighteenth century.[20] For European colonists, each race possessed physical characteristics that made it suitable for some productive tasks and not others—a worldview that justified the enslavement of Africans for plantation work in the New World and Chinese for plantation work in Southeast Asia. The *Saturday Evening Post* article in which Anson confessed his hope for a Japanese-free California had this to say: "White men and women, largely Oklahomans, handled the trimming, icing and crating in the packing plants, but they were never able to endure the back-breaking stoop work in the fields. Only the short-legged Japs could take that."[21] For those "short-legged Japs" to move from "stoop work" to farm ownership was an audacious leap across racial boundaries. In the eyes of their white neighbors, Japanese American owner-operators were quite literally out of place.

## RESENTMENT

White growers and their Montgomery Street backers were vociferous in their complaints. First, growers accused issei of possessing specialized horticultural knowledge from their homeland, giving them an unfair advantage. To be sure, some farmers concentrated on crops they were familiar with in Japan: It was issei who founded California's rice industry by improving some of the least suitable land in the state.[22] But many more pursued agricultural work out of necessity, having been excluded from non-agricultural opportunities. Due to long-standing resentment against the Chinese and the availability of European migrants, issei were denied jobs in urban factories. They were also excluded from labor unions: California chapters of the American Federation of Labor wrote Japanese exclusion into their bylaws.[23] In the early twentieth century, farming was one activity that did not directly prohibit Japanese participation.

Second, white growers claimed that merely being Japanese gave Japanese Americans an unfair advantage. V. S. McClatchy, a representative of the racist Japanese Exclusion League, lamented, "The Japanese possess superior advantages in economic competition, partly because of racial characteristics, thrift, industry, low standards of living, willingness to work long hours without expensive pleasures, the women working as men, etc."[24] Of course, the characteristics of hard work, austerity, and egalitarianism to which McClatchy and others objected were the same qualities so admired in the white yeomanry of Thomas Jefferson and rhapsodized in the speeches of Theodore Roosevelt and FDR. At issue was not the *fact* of "superior advantages," it was the supposed possession of such advantages by *nonwhites*. Resentment over Japanese American success was fueled by the failure of white-owned capital to compete, even with the legislative deck stacked in its favor.

Third, white-owned agribusinesses complained that cooperation between Japanese American producers, distributors, wholesalers, and retailers was fundamentally anticompetitive and a violation of capitalist principles. The cooperative groups that helped farmers acquire inputs and market their crops were disparaged as "syndicates" or worse. Journalist Frank Taylor described the growing network of Japanese-operated enterprises as the "Yellow Octopus," and McClatchy framed all Japanese business activities as part of a grand and sinister imperialist plot, a "carefully planned system of 'peaceful penetration' by the Japanese to displace whites and secure economic control."[25]

Of course, these grievances emerged only after Japanese Americans became landowners. The cooperative groups so hated by white growers *after* 1920 were merely an extension of the contract labor system so appreciated by white growers *before* 1920. It was one thing for issei to organize in support of white businesses, another thing entirely to organize as competition. Furthermore, Japanese American cooperation and solidarity emerged in response to widespread public prejudice, formal exclusion from economic activity, and political pressure.[26] Racist laws and everyday discrimination helped bind issei together. In some ways, so-called Japanese syndicates followed the historical example set by white grower-shipper associations that had successfully excluded nonwhites from West Coast agriculture for decades.[27]

Despite growing discrimination, Japanese Americans worked to deepen their ties to the United States. By May 1942, 20 percent of Japanese families living on the Pacific Coast had contributed a family member to the US military. A larger number of Japanese immigrants spoke English than any immigrant group other than Scandinavians and Germans.[28] Even after being branded saboteurs and spies, Japanese Americans continued to contribute materially and symbolically to the war effort. Fisher writes, "The Japanese continued in every way possible to prove their loyalty . . . they bought war bonds, donated blood, implicitly obeyed every law and regulation."[29] But it would have little impact: For Anson, McClatchy, and Montgomery Street, the issue was never about loyalty, it was always about land.

## ON THE EVE OF WAR

In 1941, the land owned by Japanese Americans on the West Coast was worth over $72 million. In California, it comprised 220,000 acres primarily in the Sacramento River Delta, the San Joaquin Valley, and the area around Santa Maria and Santa Barbara.[30] Japanese Americans raised 40 percent of California's vegetables, including over 90 percent of the state's snap beans, celery, peppers, and strawberries; 75 percent of the state's spinach, onions, and cucumbers; and over 50 percent of the state's artichokes, cauliflower, and tomatoes. Rounding it out, Japanese Americans farmed 30,000 acres of grapes and 19,000 acres of almonds, apples, apricots, cherries, peaches, peas, plums, prunes, and walnuts, accounting for $42 million of California's $105 million commercial vegetable haul in 1941.[31] For any given vegetable eaten in the United States in 1941, there was a one in four chance it was grown by a Japanese American.[32]

Most of the 5,500 farms responsible for this bounty were small, averaging forty acres compared to the national average of two hundred.[33] But issei and nisei made up the difference by focusing on high-value crops and intensification strategies, wresting greater value per acre than their non-Japanese neighbors. In 1940, the average acre of

California farmland was valued at $38 and was under cultivation one-quarter of the year. By contrast, the average acre of Japanese American farmland was valued at $280 and under cultivation three-quarters of the year.[34] Despite concerns that they were taking over existing farms, Japanese Americans helped bring new territories into production. Routinely excluded from the best lands, issei and nisei invested heavily to improve some of the least-suitable tracts in the state, land deemed unusable by white growers.

In the West Coast's major cities, issei and nisei enterprises dominated the supply chain for fruits and vegetables. In Los Angeles's wholesale sector, "the largest metropolitan produce-distributing center west of Chicago," white operators accounted for only 10 percent of the market.[35] Japanese American businesses literally fed Los Angeles, controlling the produce trade from farms in the Santa Maria Valley to the city's food retail outlets. Removing them from the food system would not be an easy task.

## FROM "FIFTH COLUMN" TO FORCED REMOVAL

The dust had barely settled on Pearl Harbor before Japanese Americans in Hawaiʻi and on the West Coast were being accused of aiding the enemy. Rumors spread of spies using powerful radio transmitters to report the movement of US ships and saboteurs plowing giant arrows into Oahu's sugarcane fields to direct the emperor's bombers toward the target. Pearl Harbor's expanse of warships was quite possibly the most conspicuous thing on Oahu, making any reliance on plowed signposts somewhat laughable. But these and other claims stuck in the minds of a frightened public despite any concrete evidence of saboteurs or their nefarious activities. As high-level military and government officials repeated these rumors for media consumption, public fears of a "fifth column" of Japanese agents in the United States grew.[36]

In the absence of evidence, imagined threats took on a life of their own. On his trip to Washington, D.C., Austin Anson made sure to paint a clear picture of "how the Salinas Valley sloped off into Monterey Bay, a most inviting landing place for an invading army; how the valley's Japanese, if they chose to do so, might blow up bridges, disrupt traffic and sabotage defense."[37] The *Los Angeles Times* published a map designed to depict how Japanese landholdings were spaced "in a manner to permit disastrous assaults on every military objective."[38] But these fears were never realized: Despite an enormous FBI dragnet and warrantless searches, no conspiracy or subversion was ever found.

Over time, anti-Japanese propagandists changed tactics, arguing that the *absence* of nefarious activity was proof of enemy coordination: There had been no sabotage because Tokyo had yet to give the order. In "a novel perversion of the laws of evidence," Japanese Americans, as a group, became "guilty by *lack* of evidence."[39] Any person of Japanese ancestry—not just those known to the FBI—could be an agent lying in wait. In the words of General John DeWitt, head of the Western Defense Command and the official ultimately tasked with removal and incarceration, "The Japanese race is an enemy race and while many second and third generation Japanese born on United States soil, possessed of United States citizenship, have become 'Americanized,' the racial strains are undiluted."[40]

By January, politicians, celebrities, and radio hosts were calling for the forced removal of all Japanese Americans, including those with US citizenship. Congressman Leland Ford of Los Angeles demanded that "all Japanese, whether citizens or not, be placed in inland concentration camps."[41] In his widely syndicated column, Henry McLemore wrote, "I am for the immediate removal of every Japanese on the West Coast to a point deep in the interior. I don't mean a nice part of the interior either. . . . Let 'em be pinched, hurt, hungry, and dead up against it. . . . Personally I hate the Japanese. And that goes for all of them."[42] By mid-February 1942, anti-Japanese sentiment had reached a tipping point, and with Roosevelt's February 19 executive order, the path toward mass incarceration was cleared for General DeWitt. Through the summer and fall, as the Navy sent warships across the Pacific, the Army sent Japanese Americans to makeshift prisons "deep in the interior." Over time, the popular press framed this as an unavoidable consequence of war. But just six months earlier, forced removal seemed far from inevitable. In fact, before discussion of removal could get underway, Anson and his associates had to somehow convince Californians they should be more frightened of the Japanese than of a food shortage.

## FOOD OR FEAR?

On the morning of December 8, 1941, as President Roosevelt prepared to ask Congress for a declaration war, the Treasury Department was busy freezing issei and nisei assets on the West Coast. Japanese American farmers, shippers, wholesalers, and retailers found they could no longer access their bank accounts or process payments, effectively suspending the distribution and sale of all Japanese-grown vegetables. In Los Angeles, where issei-owned enterprises were deeply integrated into the food system, three major produce markets shut down. After an emergency meeting of market operators, issei merchants handed over their vegetables to non-Japanese merchants to prevent the produce from going to waste.[43]

It took five days—and pressure from the USDA—for the Treasury Department to reverse course. But by then, the consequences of the asset freeze had already rippled through the supply chain. Some issei, wary of additional freezes, stopped participating in the market, while others feared that *not* participating would be seen as sabotage.[44] From Los Angeles to Seattle, urban residents began to worry about food shortages and price increases if Japanese Americans were removed from the food system. From the outset, the public understood that forced removal would impact the economy of the Western United States. But the treasury fiasco put the potential suffering literally on the table. Issei and nisei planted, watered, picked, shipped, and sold the fruits and vegetables that fed millions of people, both within and beyond California.

For the next few weeks, claims of military necessity ran headlong into a groundswell of concern for the food system. In response to Henry McLemore's fiery essay, the *Bainbridge Island Review* wrote, "Just who would grow our fruits and vegetables if Mr. McLemore's advice is followed? . . . It certainly is an apparent fact that the bulk of our produce and, we dare say, the bulk of the produce grown for our Army and Navy encampments on the West Coast come from Japanese gardens. The economy of one-third of the nation would be thrown into utter confusion if all Japanese were herded into the interior."[45]

In fact, the USDA was deeply concerned with how to meet the needs of a country preparing for war. Through an array of food-related programs—Fighting Foods, Health-for-Victory, and Food-for-Freedom—agency officials sought to increase food production even as agricultural workers were leaving for the front lines or for military industries. As Secretary of Agriculture Claude Wickard explained, "Not only must we keep our own soldiers strong physically by producing the food they need, but we must back them up in another way by keeping workers in our war factories well fed."[46] With industry gearing up, agriculture was already bleeding workers, and that was *before* removing Japanese Americans. To solve its military crisis, the country had to solve a farm labor crisis.

The impasse between the USDA, the US Army, and anti-Japanese sentiment inspired some officials to explore alternatives. One option was to remove only issei, leaving nisei to continue farming. Nisei, it was argued, had closer ties to the United States than their alien parents, making them less of a security threat. This option was popular with Justice Department officials who wanted to avoid any constitutional challenges with the wartime incarceration of US citizens (a keen bit of foresight). P. A. Minges, a specialist with the California Extension Service of the USDA, suggested that white growers temporarily take over Japanese American farms and hire back the owners to work as laborers. Minges's proposal had the advantage of keeping labor in place where it could support the state economy. For the US

Army, Minges's proposal was a nonstarter because it kept Japanese Americans close to the military installations they were accused of targeting.

Politically, the most promising alternative to complete removal was what historian Roger Daniels called the "California plan," the brainchild of Governor Culbert Olson.[47] Olson wanted to remove California's issei and nisei to other locations within the state and away from vulnerable targets, ensuring that their productive value continued to benefit California's economy and the war effort. Olson reassured the public that Japanese Americans would be kept under "surveillance and protection" and reminded his opponents that to "lose the benefit of this Japanese labor in agricultural production would be a serious loss to our war economy."[48] In late December 1941, twenty-five California counties told the USDA they were interested in receiving Japanese Americans as agricultural laborers should evacuation take place.[49]

But none of these alternatives were acceptable to the Salinas Valley Grower-Shipper Association, their Farm Bureau allies, or their Montgomery Street partners bent on dispossessing Japanese Americans of their valuable land. Removing issei but keeping nisei in place would not help: The land coveted by white growers was already titled to nisei. Minges's proposal to temporarily take over Japanese American farms would complicate plans for dispossession, especially with owners and operators working their own fields. And under Olson's plan, it was not clear what property protections would apply to evacuees—evacuees who would be close enough to check in on their farms, oversee their tenants, and call their attorneys. White growers needed a solution that would justify the redistribution of *all* Japanese American land at the same time, a solution that could not be undone later. The state already had market-based mechanisms for the sale of real estate. All they needed was for Japanese Americans to be excluded en masse from that market.

In the meantime, public fears of food shortages had to be assuaged. The California Farm Bureau, Associated Farmers, and various grower-shipper associations launched media campaigns to downplay the importance of Japanese American agriculture and push for volunteer labor programs (like "Victory Vacations") to backfill losses.[50] There would be no crisis, they claimed: White growers would take over Japanese farms, and migrant laborers would pick up the slack. To help recruit that new army of fieldworkers, powerful agricultural interests turned to an old source of labor and an unexpected ally in Los Angeles.

## THE LABOR ROTISSERIE

In the decade before World War II, the United States deported over a million people of Mexican origin. Sixty percent were US citizens—many had never been to Mexico. At the time, the Mexican workforce that supported US agriculture was seen as a threat to the moving mass of white laborers displaced by the Great Depression. Officers raided farms, businesses, neighborhoods, schools, and even hospitals, rounding up individuals and families for deportation. Major companies laid off Mexican workers, and local governments passed laws prohibiting Mexican employment.[51] During this period, the Los Angeles Chamber of Commerce became a powerful champion and business partner of Japanese American enterprises, encouraging the expansion of issei and nisei labor to undermine the well-organized Mexican farmworker movement.[52]

But by December 1941, the economic trajectory for Los Angeles had changed. The city was anticipating rapid growth in war-related industries and high demand for manufacturing jobs. Meanwhile, the agricultural sector—where Japanese American businesses dominated—was no longer a key concern for planners. So long as the immediate threat of food shortages could be addressed, Los Angeles might just evolve from an agricultural marketplace to an industrial hub. In this context, Mexican field hands posed less of a threat than deeply rooted Japanese Americans.[53] In fact, at the highest levels, preparations were being made to supplant the once-essential Japanese with Mexican labor through a cross-border partnership known as the Bracero Program.

Through vocal representatives like Thomas Drake, the chamber helped keep California's labor rotisserie turning, lobbying for the War Department to get "immediate control over enemy aliens and their families."[54] With replacement field hands in the works, it was no longer necessary to keep Japanese Americans employed within the state. By March 1942, Governor Olson had given up on his California plan, and by July, he was excited to welcome back the same people California had spent the past decade removing.

## THE GREAT FIRE SALE

On March 18, 1942, Roosevelt set the final piece in place with Executive Order 9102 establishing the War Relocation Authority (WRA) to oversee the evacuation of over 110,000 Japanese Americans in the largest forced relocation in US history.[55] The man Roosevelt put in charge of the WRA was Dillon S. Myer, a New Deal–era administrator with the USDA. Myer made a name for himself managing this unprecedented operation under tight time frames and with limited resources. In fact, his name may sound familiar: After the war, Myer was promoted to director of the Bureau of Indian Affairs (BIA). Twelve years after managing the removal of Japanese Americans, Myer was responsible for relocating 100,000 Indigenous people under the Indian Relocation Act of 1956 (see chapter 1). If you needed an inconvenient racial group moved, Myer was your guy.

As the WRA began constructing permanent concentration camps (officially "relocation centers"), orders were given to evacuate Japanese Americans to numerous assembly centers for temporary confinement. The first compulsory evacuation orders were delivered on March 24 to the issei and nisei of Bainbridge Island, Washington. Evacuees were given six days "to take care of all business matters and dispose of those possessions that they could not physically carry."[56] Over the following weeks, similar orders were given across the

Western states. Depending on the location, Japanese Americans were given between forty-eight hours and two weeks to prepare. Homeowners had to sell their homes and belongings, business owners had to dispose of their inventory, and farmers had to sell or lease their land and dispose of their farm equipment in a hostile marketplace, receiving a tiny fraction of what their property was worth.[57] As the final day approached, desperate sellers took whatever price they could get. A witness to evacuation in Los Angeles testified,

> What I remember most was my father who had just purchased a Fordson Tractor for about $750 a few months prior to the notice. Imagine his delight, after a lifetime of farming with nothing but a horse, plow, shovel and his bare hands, to finally be able to use such a device. He finally had begun to achieve some success. A dream was really coming true. He had much to look forward to. Then came the notice. And his prize tractor was sold for a measly $75.[58]

For farmers, one of the most distressing consequences was the loss of income from their current crop. The spring and summer harvest in 1942 was setting up to be a bumper crop, but evacuation orders forced many farmers to abandon their fields before it could be taken in. The secretary of agriculture insisted that Japanese Americans continue farming until the very last day, even though owners and operators would see little or no income from their labor. On March 10, the Western Defense Command issued a statement encouraging continued cultivation: "There can be no doubt that all persons who wish to show their loyalty to this country should continue farming operation to the fullest extent." Three days later, the request became a threat: "Any failure to [maintain crops] might be considered as sabotage."[59] A sympathetic employee assigned to the Federal Reserve recounted the story of a man who wanted to harvest his small crop of strawberries, but the deadline for evacuation was fast approaching. The man requested an extra twenty-four hours but was denied, and in frustration, he plowed his crop under. The next day, he was picked up by the FBI and thrown in jail for sabotage.[60]

The Farm Service Administration (FSA) was tasked with overseeing the disposition of farms owned or leased by Japanese Americans. The FSA's primary mandate was to maintain production on the land while evacuees were incarcerated—not to protect evacuee claims. For the 232,000 acres under FSA authority, this meant facilitating the transfer of leases, the appointment of new operators, or the sale of owned land. To that end, FSA agents sought out and recruited potential buyers, helping them secure loans and file the necessary paperwork to take over evacuee leases and farms. In some cases, the FSA turned farms over to corporate agribusinesses connected with grower-shipper associations. FSA regional director Laurence I. Hewes Jr. referred to this enormous change of fortune as "probably one of the most dramatic events in the agricultural history of the United States."[61]

Many evacuees who chose to hold onto their farms merely delayed their losses. New tenants refused to pay rent, failed to maintain the property, or simply abandoned it. Warehouses were vandalized and looted. Japanese Americans found it challenging to enforce contracts or make insurance, tax, and mortgage payments while incarcerated. Bankruptcies led to property confiscation and forced sale. In the past, the Japanese American community weathered economic and social upheaval by looking after each other's property. But mass incarceration took that away: There were no issei or nisei left. Even orphans were taken.[62]

By August 1942, millions of dollars in property had already switched from Japanese to non-Japanese hands, and as foreclosures mounted, that number swelled. Adon Poli, a USDA agricultural economist, estimated that by the end of the war, Japanese American farmers would control no more than 23 percent of their prewar holdings.[63] Austin Anson's white-supremacist wish was coming true.

But the dollar value of lost or destroyed property paled in comparison to the loss of family heirlooms, journals and photographs, wedding gifts, childhood homes, meticulously maintained client bases and business reputations, and painstakingly improved farmland. Forced removal destroyed the vibrant matter that had anchored so many lives and accreted so many precious meanings. It poisoned economic potential and terminated promising careers. But these violences were only the beginning.

## GROWING FOR THE STATE

Names like Heart Mountain, Topaz, and Manzanar ("apple orchard" in Spanish) fail to evoke the grueling conditions facing inmates at these inland concentration camps. From March through August 1942, Japanese American men, women, and children were forcibly relocated from urban assembly centers to ten permanent (but hastily constructed) facilities spread across seven states. The rural landscapes selected by the WRA were not parklands or scenic byways but desolate and isolated elsewheres. The camp newspaper at Minidoka described its siting as "a vast stretch of sagebrush stubble and shifting, swirling sand—a dreary, forbidding, flat expanse of arid wilderness."[64] Shoddy construction offered little protection from wind, dust storms, extreme heat, and bitter cold. Buildings needed constant maintenance. Regardless of location, many inmates spent their years of incarceration toiling to maintain basic necessities: fresh water, sanitation, shelter, and heat in winter.

Concentration camp sites were selected for their agricultural potential, but in the summer of 1942, these were not working farms—not yet.[65] Former WRA director and USDA employee Milton Eisenhower envisioned each site as a farm in the making where Japanese Americans would use their expertise to improve the land for agriculture and through this labor prove their loyalty.[66] Existing water supplies would be developed for irrigation; in fact, three of the concentration camps already had Bureau of Reclamation irrigation schemes in the works. As soon as each site was populated, inmates were put to work leveling land and building irrigation ditches. Fields

**Figure 3:** Johnny Fukazawa, an evacuee of Japanese ancestry, weeds a cornfield at the Manzanar internment camp, California, June 30, 1942. Records of the War Relocation Authority. Photograph by Dorothea Lange. Courtesy National Archives NAID 538048.

were plowed, crops planted, and stockyards erected. A primary objective of WRA administrators was to contribute to the USDA's wartime production quotas, mandating that camps be "self-sufficient within one season with surpluses going to the war effort."[67] By late 1942, Tule Lake, Manzanar, and Gila River had already produced over a million pounds of food and feed crops. By 1943, not only were the remaining camp farms online, many were shipping out surpluses.[68]

Another WRA objective was the improvement of land for private settlement after the war. Regional director Era Reeseman spoke of his intention to use Topaz to "resettle non-Japanese farmers now farming on 'submarginal' land." The *Millard County Chronicle* referred to the Topaz site as a "self-liquidating project."[69] In fact, at nearly every site, land was being prepared for postwar occupation by white homesteaders, with priority given to veterans. The government stood to make a sizeable profit in the land market: At Topaz, where some of the site's 20,000 acres had been privately owned, the WRA purchased unimproved land for a dollar per acre. With irrigation canals and functioning fields, that same land was valued at up to thirty dollars per acre.[70] As they had done for decades in California's San Joaquin Valley, Japanese Americans set to work improving agricultural lands; and yet again, it was non-Japanese who reaped the benefits.

For the most part, locals did not welcome the concentration camps and their inmates. State governments strongly opposed WRA plans to incarcerate Japanese Americans in their territory, citing anti-Japanese sentiments. Gila River and Poston were located on Indian reservations in Arizona, and in both cases tribal governments voted against the camps. At Poston, the Tribal Council of the Colorado River Indian Reservation wanted "no part in inflicting injustices on Japanese Americans similar to what they had suffered."[71] Construction proceeded despite these objections, and over time some communities saw a boost in the local economy. Each camp was designed to house 8,000 to 20,000 people: a small city of roads, barracks, offices, mess halls, schools, hospitals, and warehouses. Building and operating these facilities offered a lifeline for communities still recovering from the Great Depression.

## "PINCHED, HURT, AND HUNGRY"

WRA concentration camps may have had roads, schools, and hospitals, but that is where any resemblance to comfortable living ended. Inmates at Heart Mountain saw winter temperatures fall to –28°F, summer temperatures at Gila River and Poston hit 120°F, and at Topaz temps could vary as much as 50°F in a single day and 150°F over a year.[72] Barracks were poorly constructed and initially uninsulated. As snow blew in through unsealed doors and windows, inmates layered up in World War I army surplus clothing and gathered around coal stoves. This was no guarantee of comfort: Minidoka ran out of coal in the winter of 1942 due to wartime demand. The stoves, like those who huddled around them, were at the mercy of distant decision-makers.

Rations were bleak despite the variety of fresh produce grown on camp farms. Lucille Hitomi, incarcerated at Tule Lake with her family, recounted how her brother took advantage of his job at the camp slaughterhouse to supplement the daily grind: "I don't know if this was legal . . . but sometimes [my brother] would bring bits of meat home. My mother brought to camp a hot plate and a frying pan, and she'd cook the meat in the barrack. . . . I guess it was more like home."[73] Over time, administrators allowed inmates to grow daikon, mung beans, nappa, and habucha (for tea), ingredients used in Japanese American food preparation. But the supply of healthy food was inconsistent. Poor diet and sanitation led to intestinal disorders. The fresh water supply at Topaz kept breaking down due to the use of inappropriate pipes. Arid conditions led to dehydration, heatstroke, and silicosis from inhalation of dust. And some camps experienced outbreaks of tuberculosis and infantile paralysis.[74]

## INCARCERATED LABOR

With its deep structural ties to the USDA, the WRA saw its concentration camps—first and foremost—as agricultural stations.[75] The plan was for each camp to grow enough food and silage to sustain its human and livestock populations, trade goods with other camps, and contribute its surplus to support the war effort. To improve the land and maintain a high level of production, the WRA expected that 50 percent of working-age adults in each camp would be available as field labor. This proved a gross overestimate.

Low wages made fieldwork in the camps undesirable. Inmates were paid between $12 and $19 a month for full-time fieldwork, depending on the camp. Meanwhile, outside the camps, the going rate for fieldwork was between $132 and $164 a month.[76] In an effort to benefit private farms and improve relations with local communities, the WRA allowed some inmates to request seasonal leave for agricultural labor outside the camp. In places like Grenada, Colorado, the result was an exodus of Japanese Americans into surrounding farms where labor was in demand.

The seasonal leave system offered important advantages for both growers and laborers. Growers could tap into a ready supply of field hands willing to work for less. Meanwhile, inmates could earn more outside the fences than they could inside. Private growers throughout the West came to view incarcerated Japanese Americans as a solution to wartime labor shortages. As early as 1942, sugar beet growers in Idaho, Montana, Utah, Wyoming, and Colorado were posting ads in camp newspapers looking to recruit Japanese Americans. That year, 80,000 acres were harvested by evacuees, and by the end of the war, Japanese Americans were responsible for 20 percent of the Intermountain West's sugar beet crop.[77]

In places where seasonal leave was limited or prohibited, low wages and dangerous working conditions led to worker frustration, strikes, and physical violence. In October 1943, a truck carrying thirty field workers at Tule Lake was involved in an accident and one person died. The inmates went on strike, demanding safer working conditions. In response, the camp administrator called in strikebreakers from other concentration camps to bring in the harvest. Each was paid a dollar an hour—ten times the rate of a Tule Lake inmate. As tensions mounted, strikers blockaded trucks and warehouses, leading the camp administrator to declare martial law. Tanks were deployed, MPs searched barracks, strike leaders were confined to the stockade for months, and fines were imposed on those who refused to work.[78]

At Topaz, the killing of James Wakasa by a sentry near the perimeter fence outraged inmates.[79] Fearing an insurrection, the camp's MP commander armed his men with machine guns and teargas. A general strike was called at Poston in November 1942 due to a shortage of heating stoves and the arrest of two popular community members. Inmates went on strike at Minidoka in 1944 in response to low wages and the appointment of a white maintenance foreman. And tensions at Tule Lake were further enflamed on May 24, 1944, when Shoichi James Okamoto was shot and killed by a guard after obstructing a truck at the main gate. Following the investigation, the guard was acquitted and fined one dollar—the cost of the bullet. His charge: "Unauthorized use of government property."[80]

Despite labor actions, WRA concentration camps still managed to produce an extraordinary haul, achieving forty-three million pounds of harvested vegetables in 1943 alone. Surpluses went to support wartime food programs like Lend Lease or were sold on the open market. The Tule Lake camp—where strikes were most frequent—still delivered over two million pounds of vegetables for public sale. The WRA learned what the penitentiary system had known all along: Prison agriculture is profitable.

## THE END OF INCARCERATION

Public Proclamation 21 ended the West Coast Exclusion Order on December 17, 1944, and by early January, inmates were beginning to leave the camps. But there was no mad rush for home. Fearing violent reprisals, many Japanese Americans—especially those caring for children and elderly—remained in the relative shelter of the camps. By August 1945, there were still 44,000 Japanese Americans living at WRA camps, many of whom were scared to leave.

They had good reason. On January 5, the Doi family was the first to return home to Placer County, California. Three days later, shots were fired into their home and their packing shed was destroyed with dynamite. A group of men and women confessed to the crime, but all were acquitted.[81] Similar incidents greeted returning evacuees across the country. Of the few who held onto their property, many came home to find their farms abandoned, their land use changed, their farm equipment stolen, and their improvements in disrepair.

Others—having lost their farms to foreclosure or been forced to sell—were starting from scratch.

Who were the new owners recruited by the FSA during the chaos of 1942? The Commission on Wartime Relocation and Internment of Civilians noted that white-owned agribusinesses "did very well . . . in the wake of evacuation," expanding their holdings and receiving valuable farm equipment at bargain prices.[82] Many individual owners were "naturalized European immigrants, or Americans from the Dust Bowl region."[83] In other words, the same poor white laborers so despised by growers and shippers during the Great Depression. As the WRA rounded up Japanese Americans, many of the once-hated "Okies" took their place as owners and operators. Previously deported Mexicans, welcomed back under the Bracero Program, arrived to work the fields. In an irony often repeated in US agricultural history, the most recent enemies of capital became its new saviors. Before long, they would become enemies once again.

## REAPING THE REWARDS OF A NATIONAL EMERGENCY

The mass incarceration of Japanese Americans reorganized the food system in four ways. First, it allowed white-owned capital to claim some of California's most valuable agricultural land, effectively expropriating decades of hard labor and careful investment. This theft—what geographer David Harvey calls *accumulation through dispossession*—was achieved not only by the operation of a "free" market but through government intervention.[84] The WRA defined the terms of evacuation with tight deadlines and strict rules on the disposition of property. With mere hours or days before evacuation, desperate sellers offered steep discounts. From there, prices followed supply and demand as Japanese Americans sold everything and buyers offered next to nothing. Under the mandates of the Roosevelt administration, the USDA, and the FSA, the free market's "invisible hand" became a fist.

Second, incarceration gave the USDA a chance to improve private and public lands for agriculture. By 1946, the WRA had begun to sell off the now-productive camp farms to homesteaders. The value of this land came from the expropriated labor of underpaid inmates, none of whom were allowed to benefit from their work after the war.

Third, incarceration dramatically altered the playing field for white-owned businesses, dissolving the most successful enterprises in California's lucrative vegetable market at *precisely* the moment when those vegetables were in peak demand. As the USDA called on farmers to ramp up production, it was white businesses—flush with new territory and new clients—that capitalized on the wartime rush. After the war, many Japanese American families returned to operate food-related businesses, but by then the damage had been done. Japanese American enterprises that survived the war were severely disadvantaged by losses of land, clients, and assets.

Finally, incarceration—a spatial fix for labor and capital—justified other fixes along the way. Seasonal leave programs offered private growers a supply of cheap workers, a fix for regional labor shortages. The removal of Japanese Americans from West Coast farms justified the importation of cheap labor under the Bracero Program, in effect a government subsidy for large agribusinesses. And removal justified FSA assistance for newly recruited white owners and operators, a selective form of welfare. In the end, removal allowed labor and capital to find new and diverse relationships, ones in which a greater share of the benefits would accrue to white growers and the state.

## CARCERAL FIXES FOR RACIAL CAPITALISM

"I dissent, therefore, from this legalization of racism. . . . Racial discrimination in any form and in any degree has no justifiable part whatever in our democratic way of life," wrote Justice Frank Murphy in his dissenting opinion for *Korematsu v. United States*, the 1944 Supreme Court case that legalized forced removal. According to Murphy, by excluding all people of Japanese ancestry from the Pacific

Coast, the government had exceeded constitutional authority and fallen into "the ugly abyss of racism."[85] Murphy was right to call it racism: Japanese Americans and Asian Americans more broadly had been targets of racial violence in the United States since the mid-1800s through exclusionary acts, sanitary laws, job discrimination, public harassment, property destruction, and mass killings.[86] Some of these violences continue today.

But Murphy's characterization was incomplete. This was not just racial "discrimination." Yellow Peril propaganda converged with powerful agribusiness interests to remove Japanese Americans from their farms and into a segregated part of the food system—one where their past, present, and future labor could be more effectively expropriated. Issei and nisei were robbed, locked up, and forced to work not because of racism per se, but because of racial *capitalism*.[87]

To facilitate accumulation, capitalism takes advantage of inequalities: differences in the relative availability of raw materials and capital, and differences in demand for goods and services. To extract the greatest value from labor, capitalism actively constructs differences in the valuation of human *life*, inequalities historically produced through ideologies of race.[88] Indigenous people were constructed as lazy savages to encourage the theft of their land, Black and Chinese men were constructed as disposable workers to justify their forced labor, and Japanese Americans were constructed as devious enemies to encourage their incarceration. In later chapters, the food system will deny Black sharecroppers access to federal support to keep them hungry, classify meatpackers into racial categories to keep them from unionizing, and use redlining to keep Black grocers from competing with white-owned supermarkets. As these cases demonstrate, capitalism uses diverse elsewheres to reinforce the racial inequalities it needs: Concentration camps like Heart Mountain, Topaz, and Manzanar were just some of many forms. Racism may have "no justifiable part" in a democratic way of life, as Justice Murphy argued, but it is the cornerstone of a capitalist food system.

## MISSED OPPORTUNITIES

In the early months of World War II, the US food system weathered a series of labor and production crises. Rather than distribute the resulting financial, material, and bodily harms across its producers and consumers, the system focused these violences onto an unwanted and easily scapegoated population. Powerful actors in the food system did this to win a long-coveted prize: the best agricultural land on the West Coast. Japanese Americans were condemned to concentration camps because of specific choices about who should benefit and who should suffer from a national emergency.

The Roosevelt administration could have worked to counter anti-Japanese sentiment and protect Japanese American businesses in the weeks following Pearl Harbor. Alternatives to incarceration like the California plan might have assuaged public paranoia and kept Japanese American businesses intact. Long before Pearl Harbor, California's legislators could have rejected Yellow Peril logic and voted down the exclusionary laws that underpinned legal and economic discrimination. Doing so might have improved agribusiness relations with other workers—Chinese, Mexican, Filipino, and poor whites. California could have reined in its powerful agribusinesses and grower-shipper associations. Instead, Sacramento lawmakers let them loose.

Even then, advocates of incarceration faced a daunting challenge. It was one thing to convince a frightened public of a "foreign" enemy in its ranks; it was another thing to physically remove a population responsible for 25 percent of the country's produce. Japanese American workers and business owners were essential to the US food system: Eliminating them would exacerbate the crisis, not relieve it. To succeed, the architects of this revolution had to strategically undermine more practical alternatives, convince the public that the food system was safe, and mend fences with once-vilified laborers. In the end, being essential did not prevent Japanese Americans from being dispensable. Two hundred thousand acres of prime farmland was simply too much for Montgomery Street to ignore.

• • •

Executive Order 9066, which cleared the path for forced removal, signaled the beginning of Japanese American incarceration. But their systematic isolation and exploitation began much earlier, with policies that restricted opportunities for issei, facilitated the expropriation of their labor, and denied their children rights as US citizens. The historical success of Japanese American businesses during this period is a testament to hard work, resilience, and community-building in a system designed to limit their potential. By the time families were herded into waiting buses, Japanese Americans had been living and laboring in a specially constructed and racially defined elsewhere for over half a century.

Likewise, Public Proclamation 21 may have formally ended Japanese American incarceration, but it did little to end the structural violence experienced by evacuees. Issei and nisei were free to return home, restart old ventures, and set out on new paths, but exclusion and discrimination stayed the same. Evacuees simply exchanged one elsewhere for another.

Meanwhile, for white growers like Austin Anson, December 7, 1941, was the beginning of great things. In fact, the bombing of Pearl Harbor kick-started not one but two imperialist land grabs: the conquest of Europe's Southeast Asian colonies by Japan and the conquest of US vegetable production by white-owned businesses. Four years on, Japan's holdings across Southeast Asia and the Pacific had fallen by the wayside, but the rejuvenated white empire spreading across California's fertile valleys was about to hit its stride.

# 4

# GRAIN, GRAIN EVERYWHERE

*Technology, Policy, and the Making of Excess*

January 31, 1966, was a big day for Lyndon Johnson. In the morning, the president announced that peace talks with North Vietnam had failed and the United States would expand its aerial bombardment of the Southeast Asian country. Over the next eight years, US aircraft would release millions of tons of ordinance over North Vietnam, targeting not only military and strategic assets, but dams, dykes, and rice fields to demoralize the rural population. Johnson's decision to go "all in" on the Vietnam War is remembered as a defining moment of his presidency.

Far less is remembered about another pressing concern that day: the storming of Greenville Air Force Base. The United States was fighting real and potential threats from Southeast Asia to North Africa, and its constellation of military bases was always vulnerable to attack. But this was not some far-flung outpost surrounded by communists. Greenville AFB was in the Mississippi Delta, surrounded by cotton fields and hungry people.

Greenville AFB had been recently decommissioned, making its "capture" more of an orderly dispossession than a military defeat. In fact, the seventy Black sharecroppers who took control of it were armed only with blankets, stoves, mattresses, mops, and brooms.[1]

They offered no resistance when a swarm of military police and federal agents evicted them the next day. After all, they were never planning to wage a war, they simply wanted Johnson to take his *own* war seriously: his self-declared War on Poverty, that is, not the war on Vietnam. The first item on their list of demands was food, for themselves and for tens of thousands of starving sharecroppers like them: "We are here because we are hungry. Our children can't be taught in school because they are hungry. They can't even get the food in school because they have to buy it and don't have the money."[2]

The plan to occupy Greenville AFB—organized by local civil rights leaders the week before—capped off decades of sharecropper frustration. Since the Great Depression, agricultural policies had paid farmers to fallow land. White landowners in the Mississippi Delta took advantage of these programs to displace their sharecroppers, 90 percent of whom were Black. With federal subsidies in hand, landowners evicted their tenants, demoted them to day laborers, and invested in tractors. From 1950 to 1960, the number of Delta sharecroppers fell by 70 percent as the amount of machine-picked cotton increased sevenfold.[3] Displaced farmworkers who could afford to leave joined the Great Migration for Northern cities. Those who stayed behind competed for dwindling jobs. Many became dependent on federal food aid.

But food aid was often out of reach. Powerful whites, intent on maintaining the Delta's racial hierarchy, controlled the local distribution of commodity foods and food stamps. Some Mississippi welfare boards refused to dispense food stamps unless a recipient's white employer personally vouched for their need. In 1962, Leflore County deliberately suspended food aid to discourage Blacks from supporting voter registration. And in the winter of 1966–67, Washington County's Board of Supervisors withheld food aid to punish striking sharecroppers.[4] Access to food aid was a central concern of the Greenville protestors. As one elderly man explained, "We need to go there [Greenville AFB] and set us up one of those refugee camps, 'cause that's what we are. And being right by that air strip the government can fly in surplus commodities right to our door and we can give 'em out without them expensive middle men. They say the buildings don't have heat or lights or running water. Well, just as long as it don't leak it'll still be a damn sight better than the shack I been living in."[5]

In the Delta and across the South, white supremacy and agricultural change conspired to starve legions of poor farmworkers. In the words of a pediatrician who visited the Delta in 1967, "Slow starvation has become part of the Southern way of life."[6] Surrounded by the most productive agricultural system on Earth, mothers and fathers suffered from chronic anemia while their children died from marasmus, a severe nutritional deficiency more commonly found in famine zones. Children with marasmus look like elderly adults. Wrinkled skin hangs off their bodies. Their hair thins and falls out. Cheek fat is so depleted in marasmic infants that they cannot nurse or suck on a bottle. Lying in a state of perpetual exhaustion, many can no longer summon the energy to cry. Without medical care, parents can do little but wait for death: In Washington County in 1967, one in sixteen Black children died before their first birthday. Infants had a better chance of survival in war-torn Vietnam, even with Lyndon Johnson raining bombs on them.[7]

As sharecroppers struggled with scarcity, the government struggled with abundance. Technological revolutions brought increased productivity to agriculture through chemical fertilizers, pesticides, and tractors. Farms produced far more than the market could consume, even with growing demand for commodity-intensive meat, dairy, and processed foods. A burgeoning middle class enjoyed a multitude of new products at local supermarkets, but there was still excess food. To protect farmers from falling prices, the government bought up and stored the surplus. Large farm operations flourished, cranking out wheat and corn with the assurance that the government would buy what could not be sold.

Through the 1950s, government stockpiles quickly outgrew available storage spaces, forcing the USDA to rent silos, warehouses, and other makeshift structures from private companies at extraordinary cost. A congressional investigation revealed that surplus grain was being stored in empty oil tanks, railroad roundhouses, aircraft hangers, coal bunkers, tents, and, in some instances, abandoned movie theaters.[8] The USDA even requisitioned ships from the Maritime Commission's reserve fleet to squirrel away its agricultural excess. At the peak of the "surplus crisis," 384 freighters carrying 88 million bushels of wheat were tied up in ports from Oregon to Virginia.[9] Holding grain for the government was a lucrative business: Cargill and Archer Daniels Midland (ADM), the two largest agribusinesses in the world in 2024, made tens of millions of dollars as grain hoarders for the government—all at taxpayer expense.[10] In 1958, surplus storage was the third largest item on the federal budget after defense spending and interest on the national debt.[11]

By then, the food system resembled a theater of the absurd. As agriculture looked for containers to hold its leftovers, some farmers starved for want of scraps. As white suburbanites enjoyed a cornucopia of supermarket offerings, Black sharecroppers fought welfare boards for the right to bread and cheese. How did the food system get into this mess?

• • •

In *Powers of Freedom: Reframing Political Thought*, sociologist Nikolas Rose writes, "If policies, arguments, analyses and prescriptions purport to provide answers, they do so only in relation to a set of questions. Their very status as answers is dependent upon the existence of such questions."[12] If child starvation, mass migration, farm aggregation, and surplus storage were answers (or consequences of answers) to questions facing the food system, *what were the questions*? What problems motivated *these* solutions and discouraged others? For better or worse, the food system we see today reflects the long history of challenges it has faced and the solutions it has embraced. Many of its formative challenges emerged in the mid-twentieth century as the United States struggled with economic depression, World War II, the Cold War, civil rights battles, and technological revolution. In this context, policymakers questioned how the food system might boost agricultural profits, support public health, address widespread poverty, and bolster the country's vision for the world. This is the story of how powerful actors in the food system answered some of these questions and chose to ignore others.

The case of starving sharecroppers is usually told as a story of Southern political interference: Powerful white leaders obstructed federal programs that would otherwise have kept people alive.[13] In this narrative, the weaponization of food to oppose civil rights reforms seems to be a historical and geographical one-off, a distinctly Southern and mid-twentieth-century event. But at a larger scale, a new question emerges. How did broader changes in the food system conspire to support white supremacy? Here, the key concern is not how Southern governments kept food away from sharecroppers, but how *abundance*, and the federal response to it, made sharecroppers vulnerable to begin with.

In fact, the Delta's malnourished children were not one-offs. They were exemplary of a larger contradiction baked into the food system, one still expressed in shrinking farm populations and rural hunger, and one that spans racial divisions. In the words of Willard Cochrane, former chief economist at the USDA, "Two things about American agriculture seem never to change: first, its capacity to produce abundantly and, second, the inability of our social and economic system to make effective use of that abundance." This is the story of how the country's social and economic system chooses to deflect the violence of abundance into forgettable elsewheres *within its own territory.*

## FREE MARKET FARMING

Struggling farms and farmers have been a fixture of North America's rural landscape since the first European settlements. Early colonists struggled to adapt Old World cultivation to New World environments. Frontier families competed against expanding plantations for access to markets. Homesteaders in the late 1800s endured dust storms, fought the railroads, and labored to break up the prairie. And since the farm crisis of the 1980s, campaigning politicians, social critics, and national hagiographers have mourned the demise of small farms and the rural communities that support them. Osha Gray Davidson writes that with the loss of family farms, small towns "are being transformed into rural ghettos—pockets of poverty, unemployment, violence, and despair. . . . 'The most American part of America' is fast becoming 'America's Third World.'"[14]

The narrative of rural struggle fits into a broader parable of adversity: Small farmers are locked in a constant battle to eke out a living against forces beyond their control, a battle that can be won only through grit and determination. In his radio address to the nation on September 6, 1936, Franklin Roosevelt reprised this morality play with colorful prose: "No cracked earth, no blistering sun, no burning wind, no grasshoppers, are a permanent match for the indomitable American farmers and stockmen and their wives and children who have carried on through desperate days, and inspire us with their self-reliance, their tenacity and their courage."[15] To farm is to suffer, but through the crucible of suffering, true American character is forged. It is no wonder the state motto of Kansas is *Ad Astra per Aspera*, literally, "to the stars by hard ways."

But if earth, sun, wind, and grasshoppers were no match for family farms, the free market would be. For two hundred years, a common throughline of farm life has been the struggle to remain profitable in fickle markets. As agriculture evolved through changes in technology, price policy, and trade, each generation found itself challenged by rising input costs and falling food prices—what farmers call the "cost price squeeze." In the end, the story of failing farms and starving sharecroppers begins with the question of how farmers make money in the first place.

## THE FIXED-STOMACH PROBLEM

In a free market, two things influence the price of a product: how much interest there is in buying that product (demand) and the quantity of that product offered for sale (supply). If the demand is greater than the supply—for example, if people want more corn than farmers are producing—the price will rise. Rising prices will encourage farmers to produce more and discourage some consumers from buying, until an equilibrium is reached. On the other hand, if the supply is greater than the demand, the price will fall, discouraging farmers from growing more and motivating consumers to buy up the surplus. As is true of most commodities, it is the relationship between aggregate demand and aggregate supply that determines the price.

But food is not like other commodities. If the price of food goes up, people may eat a little less, and if it goes down, a little more, but changes in the price do not dramatically alter the demand. Bryan McDonald refers to this as the "fixed-stomach problem": Whether the price of food is high or low, we tend to eat until we are satisfied, and then we stop.[16] In economic terms, demand for food is *price inelastic*, that is, big changes in price result in only small changes in demand. For example, in the 1950s, the USDA estimated that the price elasticity of food was 0.3. This means that a 10 percent decrease in price would result in only a 3 percent increase in consumption.[17]

Price inelasticity has serious implications for farmers. Assuming no government intervention, had farms in the 1950s produced 3 percent more than what the market was willing to buy, retail prices would have dropped by 10 percent and prices received by *farmers* would have dropped 20 percent. Historically, relatively small fluctuations in supply or demand have had dramatic impacts on agricultural livelihoods and rural economies. As Cochrane notes, the price inelasticity of food ensures that farmers are always "at the crack end of the whip."[18]

The other major influence on price is supply. A sudden drop in production due to drought or war can raise food prices, while a bumper crop can send them falling. Due to price inelasticity, even small shifts in supply can have a substantial impact on the prices that farmers receive. Historically, farmers have planned for these fluctuations by diversifying production and holding assets in reserve. But there is also a chronic, systemic stressor: Farms consistently *overproduce*. During periods of rapid population growth or war (for example, 1900–1920 and 1940–1950), demand for food outstripped supply and farmers enjoyed high prices. But apart from these periods, farms have historically grown more food and raised more animals than the market is capable of consuming. As a result, low prices have been endemic to US agriculture.[19] Through it all, farmers have consistently pursued improvements in production, even when the inevitable growth in supply has negatively impacted prices. The question is why.

## THE AGRICULTURAL TREADMILL

As individual businesses, most farms do not produce enough to influence market prices, particularly in mass-produced commodities like wheat, corn, and cotton. Farmers are considered "price takers" because they must accept whatever price the market gives them for their products. For this reason, farmers tend to focus on two things they *can* influence: how much they produce and how efficiently they can produce it.

So long as the price per unit is greater than the cost to produce each unit, it makes sense to produce as many units as possible. This may involve expanding the number of acres and laborers, or increasing how much food is grown per acre or laborer. Prior to 1900, growth in agricultural production came primarily from the former: extensification of landholdings and growth in the workforce. After 1900, improvements in the *efficiency* of land and labor drove agricultural growth. Tractors could plow faster than horses. New fertilizers enabled denser planting. Chemical pesticides reduced crop losses to insects and the need for labor-intensive weeding. And hybrid varieties increased the amount of grain per acre. For a time, farms that adopted these and other efficiency-improving technologies reaped an advantage over their competitors.

But every time, the advantage was short-lived. An individual farm may be a price taker, but put together, the production decisions of millions of farms can have a profound impact on the price. As more farms increase production, the aggregate supply of food goes up. Without a corresponding increase in demand, prices fall, and they do not fall gently: The inelasticity of demand means that even a small increase in supply relative to demand will send farmgate prices plummeting.

This paradox is the wellspring of overproduction: The best way to make a profit is to produce more, but producing more undermines profits, which in turn motivates greater production. In 1925, the economist Edwin Nourse quipped, "The outlook for agricultural production is so good that the outlook for agricultural prosperity is distinctly bad."[20]

To make matters worse, as more farms invest in new technologies, the prices of these inputs rise in response to growing demand. Before long, farmers are once again stuck between rising costs and falling prices, and the advantage of adopting the technology in the first place has disappeared. Since the 1930s, the farms that survived the cost price squeeze were those that *quickly* adopted efficiency-improving technologies, reaped temporary returns, and invested those returns in even more efficient technologies. Meanwhile, those farms that hesitated got trapped. As assets dwindled, many of these farms were left with no way to catch up. "The average farmer is on a treadmill with respect to technological advance," wrote Willard Cochrane in 1958.[21] Those who can keep up reap the rewards of higher production and accelerate the pace for everyone else. Between 1935 and 1975, US agriculture doubled its yields for wheat and nearly quadrupled its yields for corn on the back of extraordinary technological advances.

But it also pushed two-thirds (5.3 million) of its farms off the back of the treadmill.[22]

## THE SHARECROPPING TREADMILL

The agricultural treadmill and its "adopt or die" imperative pushed farmers to pursue efficiency improving technologies. But the diffusion of new tractors, seeds, and pesticides was neither consistent nor even. Large-scale Western grain producers were the first to adopt new mechanical technologies, followed by Midwestern farms. Harrows, grain drills, and McCormick's reaper were all available by 1855, and a century later, over 80 percent of Midwestern and Western farms were using tractors to plow, plant, and harvest wheat and corn.[23] But in the South where cotton was king, horses, mules, and laborers continued to work the fields well into the twentieth century. As late as 1958, a scant two percent of Alabama's cotton crop was harvested by machine. The rest was picked by hand in a routine unchanged since the Civil War.[24]

Part of the reason was the crop. The edible parts of corn and wheat emerge high above the ground and tolerate rough handling, simplifying the mechanization of harvest. Vast Western monocultures were ideal for mechanical cultivation. But cotton bolls are fragile and often grow low to the ground. In the South, mechanical cotton pickers were impractical until planters introduced new varieties of cotton with higher and sturdier bolls. Early harvesters were easily fouled by weeds, and machine-harvested cotton often received a lower grade.[25] It took time for machines to match—let alone surpass—the efficiency of a picker's hand.

But this does not fully explain why manual labor lasted well into the twentieth century, long after biological and technical challenges had been overcome and industrialized cotton production was possible. Cotton planters' adoption of new technologies languished, in part, because the field workers whom machines would have replaced were more than just cheap labor: They were also a source of income.

In the aftermath of the Civil War, former slaves sought ways to make a living in the South's cash-strapped economy as plantation owners looked for new sources of labor. Government proposals to solve the crisis died on the vine: Land reforms to redistribute plantation acres to freemen were unpopular with planters, and the contract labor system, designed to negotiate labor agreements between freemen and landowners, was unpopular with former slaves. These failures led to growth in sharecropping, a system of peonage in which landless households (croppers) worked a planter's land in exchange for a portion of the harvested crop. Planters received a guaranteed workforce that needed to be paid only when the harvest came in. Depending on the arrangement, landowners allowed croppers to live on the land and sow a portion of it in food crops for subsistence (known as "corn acres"). In theory, croppers were free to leave for other employment and use the income from their share to buy land of their own. But in practice, few earned enough to do so, and those who left sharecropping usually did so from desperation, not choice.[26]

Various sharecropping arrangements emerged after Reconstruction, but in each case, the system afforded more opportunities for planters than for their tenants. The most notorious arrangement required croppers to buy seed, fertilizer, tools, clothes, and other household necessities from the plantation store or, in the case of smaller operations, designated merchants.[27] Without cash for most of the year, croppers relied on credit advanced by the landlord to afford these purchases. At harvest time, the running tab was totaled and deducted from the cropper's portion of the harvest. Interest on these loans sometimes reached 200 percent. If the share of the harvest was insufficient to pay the debt, landlords forced croppers to sign over their *next* harvest, effectively binding them to the planter. Over time, sharecropping families became trapped between growing debt and the cost of inputs. Their only option was to work harder and hope for fair consideration. Meanwhile, planters supplemented their profits by manipulating the interest rate, charging more for supplies, and, if needed, falsifying the cropper's share at harvest time.[28]

**Figure 4:** A Mississippi Delta plantation owner near Clarksdale, Mississippi in 1936. Photograph by Dorothea Lange. Courtesy Library of Congress LC-USF34- 009599-C.

The sharecropping system was so profitable for planters that it continued to expand even as cotton collapsed. Between 1920 and 1932, the price of cotton plummeted from 77.8 to 11.6 cents per pound. Cotton production initially fell, but quickly rebounded, defying market logic. By 1932—with prices at historic lows—the South churned out 13 million bales, matching its output in 1920 when prices were at their peak.[29] Southern landowners continued to grow a commodity that produced little, if any, return. But it hardly mattered; they were making money another way. Over that same period, with no increase in the number of landowners or cultivated area, the number of croppers grew nearly 40 percent.[30] Unable to make money in cotton, landowners were investing in human debtors.

To make matters worse, landlords had the final say in what crops would be grown, how much land would be cultivated, and how much fertilizer and pesticide would be applied. Owners often allocated the worst land for household subsistence crops, reserving the best land for cotton. The South could have grown more food to support its workforce. Instead, it imported corn from the Midwest, kept its workers dependent on landowners, and doubled down on cotton. In other parts of the country, the market put the cost price squeeze on the farm owner—in the South, the farm owner squeezed the tenant. Cotton may have been the cultivar, but the sharecropper was the crop.

From the 1930s to 1960s, the food system lost more than half of its 6.5 million farms and nearly all of its 776,000 sharecroppers. By the time protestors took over Greenville AFB, sharecropping was limited to regional pockets like the Mississippi Delta. Those who could afford to leave pursued jobs in urban centers. Those who could not were hungry, destitute, and angry. This transition might have been slower and less painful, but for the manner in which new technologies were allowed to take over.

## THE STRUCTURE OF AGRICULTURAL REVOLUTION

All great revolutions in agriculture have two things in common. First, each begins with a cross-pollination between farming and something *non*-farming: something outside of agriculture that inspires new ideas, irritates the status quo, or imposes new demands. External catalysts have taken the form of foreign cultivars, industrial innovations, and surplus capital with no place to go.

Second, every revolution leaves an excess in its wake: something made unnecessary, valueless, or obsolete by new forms of production. Between 1920 and 1970, the cross-pollination of agriculture and industry generated unprecedented excesses of capital, food, and labor. It was from these excesses that the violence of farm failure, unemployment, and starvation emerged.

## CROSS-POLLINATION IS THE MOTHER OF INVENTION

History tends to celebrate inventors as individual heroes who—perhaps tired of working long hours for little gain—fertilized their imaginations with the right questions and reaped inspiration.[31] In the case of agricultural advancements like gasoline tractors, chemical pesticides, and hybrid seeds, we assume the underlying questions reflected problems on the farm: how to increase production, reduce risk, lower costs, and maintain profits. We assume that agricultural necessity was the mother of agricultural invention.

But this is a narrow view of innovation. Every new machine is built upon a long history of small improvements; many successful experiments, many failures, and the bygone labor of thousands of innovative (and often unrecognized) predecessors. No technology can ever be isolated from the social and economic context in which it emerged: the public and private funding that supported it, the engineers who designed and built it, the people who opposed it (and therein shaped it), and the myriad expectations placed upon it.[32] Every innovation involves a *cross-pollination of ideas and influences*, often from vastly different sources. And it answers questions and addresses unmet needs far beyond the farm; for example, how to grow the economy, support industry, and feed the world.

So, too, with the innovations that drove agricultural change in the United States. Long before self-driving tractors and self-terminating genes, cross-pollination inspired the first revolution in settler agriculture. English, Dutch, French, and Spanish colonies in the early 1600s faced an uncertain future. European farming practices were ill-suited to New World environments, and all along the East Coast, colonial beachheads struggled to survive the challenges of unfamiliar soils, harsh winters, and new pests. In these early years, Indigenous knowledge and trade in food and seed were key to colonial survival.[33] Food historian Jennifer Wallach notes that colonists "learned to forage for wild greens as they traveled, to fish when they came upon water, and to hunt for buffalo and antelope" to maintain their survival.[34] Over time, this exchange permanently altered how settler farming was conducted.

By the mid-1800s, cast steel plows, harvesters, and steam tractors were beginning to ply the country's waves of grain, leveraging earlier advancements in sawmilling, mining, and transportation. By 1900, the advent of internal combustion enabled the rise of the first small-scale gasoline tractors: the Bull, Farmall, and Fordson. Later designs quickly adopted innovations from the automotive industry, like rubber tires that allowed tractors to travel greater distances off-field. Tractor production did more than increase labor efficiency on the farm. It gave birth to new industries for parts and implements, creating factory jobs for a growing urban population. By 1939, tractor assembly lines employed the majority of workers in the agricultural equipment industry.[35]

After World War II, manufacturers that had invested in wartime production sought new markets. Factories were retooled to make heavy equipment for farmers, many of whom were committed to the agricultural treadmill and eager for new advancements. In the decade following the war, the number of tractors doubled. US manufacturers built half a million new tractors in 1951 alone.[36] Mechanical innovations impacted every aspect of farming as new towed implements—plows, harrows, cultivators, and sprayers—multiplied what a farmer could do with a basic tractor.

Perhaps the best-known example of industrial and agricultural cross-pollination was the conversion of munitions factories to fertilizer production. Plants and bombs have one thing in common: They both need nitrogen. On the upside, nitrogen is everywhere, making up 78 percent of the air we breathe. On the downside, it is notoriously difficult to "fix" nitrogen into reactive compounds that can help plants grow and make bombs explode. In fertile soil, bacteria do this work, turning nitrogen into nitrate that can be assimilated by plant roots. But if intensive cultivation has depleted the soil and deprived nitrogen fixing bacteria of a place to live, crops will suffer (see chapter 2).

Enter chemical fertilizers. In the early 1900s, discoveries by German chemists Fritz Haber and Carl Bosch offered the possibility of artificially fixing nitrogen into ammonia that could be sprayed on fields. As a bonus, the same industrial process could produce abundant nitrates for ammunition, thereby combining two long-standing national preoccupations—food and war. But fixing nitrogen in this way requires abundant capital in the form of chemical plants and power generation, regardless of whether the nitrates are used for weapons or wheat. With the onset of World War I, the government invested $100 million in nitrogen fixation plants to make ammunition. And it never looked back: One enthusiastic War Department report would later proclaim, "Man has passed through the stone age, the bronze age, and the iron age, and now is in the nitrogen age."[37]

But as the war wound down, industry officials warned about the future of these capital-intensive operations.[38] Chemical companies, flush with wartime profits, wanted government help to secure a steady supply of peacetime customers. As part of the National Defense Act of 1916, Congress appropriated $20 million to convert munitions plants into fertilizer factories after the war. Section 124 of the act authorized the president to "determine the best, cheapest, and most available means for the production of nitrates and other products for munitions of war and useful in the manufacture of fertilizers."[39] Under this mandate, President Wilson established the Fixed Nitrogen Research Laboratory (FNRL), tasked with funneling new discoveries, patents, and experimental catalysts to private companies.[40] The government dammed the Tennessee River and built a power plant when the American Cyanamid company wanted a nitrogen fixation plant there. And as new fertilizers hit the market, publicly funded extension programs taught farmers how to use them.

This cozy relationship between the government and chemical companies would propel a national fertilizer boom thirty years later, as surplus industrial capital looked for new markets after World War II. By this measure, some consider the 1916 National Defense Act "one of the most important pieces of agricultural legislation in the nation's history."[41]

The cross-pollination of chemistry and agriculture was not limited to fertilizer. Between 1947 and 1949, pesticide manufacturers invested $3.8 billion to expand production. Dichlorodiphenyltrichloroethane, better known as DDT, was a broad-spectrum insecticide developed for use during World War II. After the war, chemical companies marketed it to farmers. Sales of DDT increased 1,100 percent between 1944 and 1951, with nearly all growth in the agricultural sector. To speed the application of new chemical inputs, farming took to the air. Crop dusting—the aerial application of chemical pesticides or fertilizers—hit its stride after World War II when thousands of surplus Piper Cubs used to train military pilots became the backbone of agriculture's burgeoning crop-dusting fleet. Some of the veterans who trained on these aircraft during the war ended up working as "dusters" in the fight against weeds and weevils.[42]

Over time, chemical companies became seed companies, breeding new varieties to work with proprietary herbicide and fertilizer formulations. Here, innovation involved the literal cross-pollination of emerging biotechnologies with an assortment of domestic and foreign plant strains—a genetic library meticulously curated by hundreds of generations of farmers. From this bounty, researchers bred new strains to produce more fruit, use less water, and respond to nitrate-rich fertilizers. As seeds became corporate property, the right to use and share them became a matter of licenses and legal precedents.[43]

If tractors, chemicals, and novel seeds were the answers, what were the questions? As the technological revolution was busy addressing important agricultural challenges, it was also solving problems far beyond the farm. In the first half of the twentieth century, industry was stuck on a treadmill of its own, weathering wartime boom and bust. Where could manufacturers and chemical companies find new markets after the war? Where could industry find new sources of cheap labor? It was agriculture that answered the call. On the farm, industry found a conspirator addicted to technological improvements—one that would pay dividends in the form of surplus workers.

## THE MAKING OF EXCESS

Popular narratives of agricultural revolution tend to celebrate what each technology contributed to the food system: greater yields, labor productivity, and consumer choice. New devices, chemicals, and cultivars are usually presented as improvements over the status quo. By such accounts, the great technological advancements of the past 250 years consistently helped agriculture do more and do it better.

But what technology *adds* is only half the story. A new machine demands conformity from all that it touches. A host of living and nonliving things must adjust to its rhythms and embrace its limitations. Land must be cleared of all but what the machine expects; plants and animals must be altered so it can work efficiently; factories must be retooled to build it; distant landscapes must be mined to provide its raw materials and supply its fuel; and workers must learn to operate it, tend to its needs, and aid its proliferation. All these things must make way for each new tractor, novel molecule, or hybrid seed to "work."[44]

Over time, some of these things become increasingly necessary, like petroleum, bank loans, and chemical engineers. Other things become increasingly obsolete. New capabilities plow under old practices. Once-crucial plants and animals, factories, landscapes, and workers either adapt or get out of the way. *Every agricultural revolution leaves an excess in its wake*, and it is from this excess that violence emerges—particularly when that excess includes people.[45]

In European North America, the agricultural revolution with the most violent repercussions was the first. By integrating Indigenous cultivars and practices into European farming, settlers became less dependent on the cooperation of Indigenous communities. As colonists switched from subsistence to cash crop production, trade increasingly crossed the Atlantic and not the frontier. Over time, Euro-American farmers chose to see Native producers as competitors for limited markets, and by the early 1700s, any possibility of a lasting and mutually beneficial relationship between settler and Indigenous food systems had evaporated. British colonies actively obstructed Native trade to weaken their European enemies, suppress Native resistance to expansion, and support non-Native growers.[46] This first agricultural revolution in European North America turned once-indispensable Indigenous people into an impediment.

The mechanical revolution would do the same for horses, mules, and field workers. Between 1924 and 1960, the number of tractors increased 1,800 percent. Over that same period, farmers gave up 90 percent of their draft animals. By World War II, tractors had eliminated an estimated 1.7 billion man hours of annual labor, the equivalent of 850,000 workers. By 1960, that figure had risen to 1.7 million workers.[47] During this period, so many tenant farmers lost their livelihoods to machines that losing one's farm job was popularly known as "getting tractored out."

But it wasn't just tractors. Chemical and biological advancements—especially herbicides—further reduced demand for labor. The weed killer 2,4-D, developed in the late 1940s, effectively eliminated the need to manually weed corn and wheat fields. In concert with mechanization, chemical fertilizers and pesticides substantially reduced the human labor needed to produce a calorie of food. In 1820, one agricultural worker could feed five people. A century later, one worker could feed nine, and by 1955, over twenty.[48] In economic parlance, these overlapping revolutions vastly improved labor efficiency, reducing the fraction of people needed to feed the country's growing population. In 1820, 72 percent of the population worked in agriculture. By 1955, only 5 percent labored on farms.[49]

These figures reveal the aggregate impact of technological revolution on agriculture during this period, but they hide geographic and social differences. As tenants and croppers in some parts of the country left to compete for urban jobs, sharecroppers in the Mississippi Delta stayed in place—until they couldn't. What structural conditions protected sharecropping? And what changes triggered its violent collapse? The answer has to do with how government policies handled the extraordinary excesses brought on by technological revolution.

## MANAGING EXCESS

From the 1920s to the 1970s—with a brief remission during World War II—the US economy struggled to manage two interconnected and economically dangerous excesses: surplus food and surplus industrial capital. Agricultural overproduction had been a chronic condition since the early 1800s. Managing it was, as one economist put it, "the basic problem in American agriculture."[50] The problem became more complex in the early 1900s as agricultural and industrial surpluses combined to create a third kind of excess: *farmers*. Between 1929 and 1932, farm prices fell 50 percent and the net income of farm operators fell by two-thirds.[51] As unemployed workers hit the road in search of work, the Roosevelt administration made choices that would profoundly impact the future of large agribusinesses, small farms, and tenants. Its most violent impacts would not be realized until the 1950s and 1960s as Depression-era reforms dismantled the last stronghold of field labor: sharecropping.

### PUMPING THE BRAKES

Desperate to limit agricultural production during the Great Depression, Congress enacted a series of acreage-reduction programs that paid farmers to voluntarily idle some of their land. These incentives were quite popular: Between 1937 and 1944, acreage under wheat dropped 30 percent and acreage under cotton dropped by half.[52] But reducing cultivated *area* did not reduce *output*. Over this same period, US farms increased total production in every key commodity as the adoption of tractors, chemical inputs, and hybrid seeds more than compensated for any loss in area. As a means to control overproduction, acreage reductions generally failed.

But they succeeded in transforming Southern tenancy. Now paid to idle their land, planters prioritized their own acres over those of tenants. Between 1930 and 1940, the amount of land harvested by tenants fell from 82 percent to 58 percent. As their own acres disappeared, sharecroppers spent more time working on the landowner's fields, reducing their shares and making it increasingly difficult to pay their debts. Acres allocated to subsistence farming were often the first to be idled, forcing croppers to find other sources of food. Many were forced off their land only to be hired back as hourly laborers without any guarantee of work or the security of a home.[53]

Meanwhile, government payouts for acreage reductions allowed Southern planters to invest in tractors, accelerating the loss of croppers. Long spurned by the South in favor of indebted laborers, the technological revolution was finally embraced. During the 1950s, the number of Delta sharecroppers fell by 70 percent as the amount of cotton picked by machine increased 700 percent. By the time protestors occupied Greenville AFB in 1966, 95 percent of Delta cotton was being picked by machine.[54]

Of course, the human pickers didn't simply vanish. Those who could afford to leave became rural-to-urban migrants, joining a steady flow of labor headed for assembly lines in Northern cities. Those who stayed endured unemployment, destitution, and—for Black croppers—institutionalized white supremacy.

## DISPOSING OF THE EXCESS

With new yield-improving technologies in hand, farms continued to overproduce and suppress prices. To help consume the excess, USDA programs encouraged consumers to eat more meat, dairy, and processed foods to boost demand for grain and other "built-in" farm products.[55] Chemurgy, the use of agricultural products as industrial raw materials, took off in 1938 with the creation of four government laboratories tasked with finding non-food uses for farm products. By 1951, the USDA's *Yearbook of Agriculture* advocated turning corn starch into asphalt sealer, corn fiber into fabric, peanut meal into fire extinguisher fluid, and buckwheat into potential pharmaceuticals.[56] Today, the agricultural treadmill continues to prop up industrial capital as biorefining companies convert one-third of the country's corn into ethanol.[57]

The most contentious and costly remedy for overproduction was to buy up and store the surplus. Beginning with the Agricultural Adjustment Act (AAA) of 1933, the government established limits on how much of each commodity could be sold on the market, and through the Commodity Credit Corporation (CCC), purchased and stored the excess. The price paid for these purchases was calculated to make up a fraction of the difference between current market prices and what farmers received between 1909 and 1914 (when prices were high). This system, known as parity pricing, was intended as an emergency measure to support farmers through the Great Depression, but quickly became a fixture of US agriculture. Over time, to maintain its parity obligations, the USDA was forced to significantly ramp up its surplus storage program. By 1960, one-third of the country's corn crop and over half of its wheat crop were being held in government and privately owned facilities.[58] Secretaries of agriculture to Eisenhower, Kennedy, and Johnson found themselves caught between congressional demands to reduce surpluses and the farm lobby's demands to maintain parity. The "basic problem in American agriculture" took center stage in presidential debates of the period.[59]

One way to reduce grain stockpiles was to find more mouths to feed. Following World War II, the United States sent millions of tons of surplus food overseas, first as aid under the Marshall Plan and Public Law 480, then through increased exports. (The violent impact of US abundance on foreign countries is the subject of chapter 5.) The United States also ramped up domestic food distribution. Under New Deal legislation, commodity food programs distributed canned goods, flour, cheese, and other staples free of charge to people in need. The program was feeding millions of people by the 1960s, but its success depended on cooperation from state and county welfare boards. Many local governments refused to participate, arguing that free food cut into sales at local stores.

In 1964, President Johnson signed the Food Stamp Program into law. Unlike commodity foods, food stamps reimbursed local stores and allowed recipients some degree of choice. But food stamps weren't exactly free. In Mississippi, a recipient might have to pay $30 to get $80 worth of stamps. The actual cost was based on the needs of the household using a formula set by county welfare boards—boards often run by powerful white landowners. Decisions about who would get federal aid were used to punish Black workers for organizing strikes or supporting voter registration drives.[60] In the end, the cost of food stamps was simply out of reach for many poor Black households. In her testimony before Congress, Marian Wright, a lawyer with the NAACP, described the case of a mother in Washington County, Mississippi: "[She] is a welfare recipient, who makes approximately $60 a month, who has seven children, who has to pay rent, light, water, and who cannot meet the minimal requirements of food stamps. In asking several of the people in the State what responses they are going to make to this kind of situation the only thing they can say is 'I guess they will have to starve.'"[61]

Welfare initiatives like food stamps and commodity foods may have slowed the growth of agricultural surpluses, but at a cost. Sharecropping was built on an implicit agreement that landowners would ensure the bodily survival of their tenants—the allocation of acres for food crops was part of that long-standing social contract. But with the expansion of federal aid, planters enthusiastically relinquished this obligation to the government. Some planters even made a tidy profit: Excluded from free commodity programs, some of the poorest sharecroppers were forced to take out loans from their landlords so they could buy food stamps.

## THE COST OF DOING NOTHING

As surpluses mounted, economists, politicians, and government officials debated what to do. For free-market proponents, the solution was simple: Do nothing and let market forces reorganize farming. This was the position of Theodore Schultz, chair of the University of Chicago economics department. Schultz and others argued that if the government would only leave agriculture alone, prices would fall, supply and demand would move toward equilibrium, and farms would stop overproducing corn, wheat, and cotton because it would no longer be profitable to do so. In theory, farms would switch to growing alternative crops where a profit could still be made.

But in practice, most farmers lacked good alternatives. Few could claim the appropriate soil, climate, and acreage for anything but a handful of commodities. Few had the capital to buy additional implements, chemicals, and seed whenever the market demanded. And without alternatives, farmers would keep growing what they had always grown. Letting the market determine the price might eliminate overproduction, but only after eliminating the farmer. Cochrane referred to this solution as the "long, hard 'let things alone' approach," one that would reduce excess only through the "painful process of low prices, low incomes, financial distress, reduced levels of living, and business failure."[62] This bleak outcome was not lost on Schultz, who remarked that "if earnings are too low then there are too many people and some must get out [of farming]."[63]

The USDA was more circumspect, committing itself to acreage reductions and price supports to protect midsized farms from the violence of overproduction. This approach still benefited large operations over smaller ones, but it slowed the loss of family farms and tenants. In the aggregate, the USDA's approach succeeded in mitigating the most severe impacts of technological revolution. What it *failed* to do was distribute those impacts evenly.

## SURPLUS FARMWORKERS AND THE MINIMUM WAGE

Between 1930 and 1970, technological improvements and agricultural policies conspired to eliminate the Southern sharecropper. Acreage-reduction programs paid landowners to squeeze their tenants, government subsidies for industrial inputs created cheap alternatives to human workers, and food welfare programs alleviated landowners' responsibilities to feed their croppers. These efforts to manage surplus food and capital created a *new* excess: underemployed farmworkers. In 1967, the government's attempt to alleviate this surplus would be the final nail in the coffin for Black field labor in the Delta.

By the mid-1960s, most poor farmworkers in Mississippi were former sharecroppers who had been tractored out. Paid an hourly rate to weed the fields and bring in the harvest, a farmworker could expect $3.50 for the occasional twelve-hour workday. In 1967, hoping to address the plight of poor farmworkers, Congress raised the minimum wage to one dollar an hour. The result was catastrophic. Overnight, a human weed puller became twelve times more expensive than chemical weedkiller.[64] An estimated 25,000 people in the Delta lost their jobs as landowners switched to chemical alternatives. Those who weren't fired received wages below the minimum from "outright underpayment or of vague deductions for rent or for a 'Christmas fund.'"[65]

With cheap labor out of the way, chemically intensive agriculture took over. Before Congress raised the minimum wage, planters sprayed 5.5 million pounds of herbicide on cotton per year. In 1968, that figure jumped to 13.6 million pounds.[66] Decades after wheat and corn, industrialized agriculture had finally arrived in the Cotton Kingdom as planters switched from paying their workers to paying large input suppliers. Chemical companies couldn't have asked for a better government subsidy than Lyndon Johnson's War on Poverty.[67]

## WEAVING SCARCITY INTO ABUNDANCE

If starving children and overflowing granaries were the answers to questions facing the food system, what were the questions? What problems on and beyond the farm motivated these solutions and discouraged others? And how did these choices perpetuate violence? The paradox of scarcity and abundance from the 1940s to the 1960s resulted from the way the food system responded to four interconnected challenges: how to keep industry profitable, keep farm prices high, manage agricultural excesses, and manage surplus labor.

Flush with capital, industry sought new markets to relieve its own chronic overproduction. Meanwhile, trapped between high costs and low prices, farmers looked for ways to increase productivity before the next round of falling prices. Technological innovations in agriculture offered a spatial fix for surplus industrial capital, redirecting the violence of factory closures onto farmers stuck on the agricultural treadmill. To support this transfer, federal and state governments invested in new chemical plants, research facilities, and agricultural extensions.

These efforts made the challenge of falling farm prices more complex. To relieve the cost price squeeze, the government enacted acreage reductions and marketing quotas. Acreage reductions provided the resources for landowners to invest in new technologies at the expense of tenants and farmworkers, intensifying the impact of the agricultural treadmill and creating a mosaic of producing and nonproducing spaces. In the South, this mosaic broke up along racial and class lines. Meanwhile, marketing quotas redirected and delayed the violence of falling prices into an increasingly complex surplus storage system.

Production controls aggravated a third challenge: how to manage agriculture's increasingly costly excesses. Foreign disposal and domestic food programs provided a spatial fix for overproduction, redirecting the economic impact of stored commodities into remote elsewheres where negative consequences might not be noticed (or serve an ulterior motive). The introduction of commodity foods and food stamps did, in fact, benefit millions of struggling households. But it also accelerated the loss of tenancy and the transfer of wealth from farmworkers to landowners and corporations.

Finally, as farmers fell off the treadmill and tenants were evicted, rural-to-urban migration became a spatial fix for surplus labor. Government support for mechanization coincided with programs to reskill farmworkers in non-agricultural trades, completing the conversion of food producers into consumers, and farms into factories.

## DOMESTIC REARRANGEMENTS AND THE DESTRUCTIVE CREATION OF PLACE

These spatial fixes are a defining feature of *domestic rearrangements*, geographic responses to overproduction that transform the way existing spaces are used within the food system. Domestic rearrangements serve to redirect structural violence from privileged regions, economic sectors, or social classes to less privileged ones. Agricultural industrialization (of which this sharecropper story is a part) is exemplary of such rearrangements as rural economies lose out to urban ones, field workers become factory workers, family farms consume industrial surpluses, and their products are turned into cheap industrial inputs.

Whereas carceral fixes create new spaces into which excesses can be dumped and managed, domestic rearrangements repurpose existing ones. This tends to pull nearby places in opposite directions, creating a patchwork landscape of abundance and scarcity. Shantytowns of farmworkers coalesce near thriving cotton plantations. Global agribusinesses colonize counties where family farms are struggling. In places where money and jobs are scarce, a surplus of cheap labor takes hold. Nearby concentrations of capital maintain this lucrative paradox until more profitable alternatives are offered up. Over time, this uneven landscape begins to look normal and natural, when in fact it was heavily engineered by policy, social inertia, and calculated ambivalence.[68]

Domestic rearrangements tend to create friction. After the Civil War, farmer coalitions like the Granger movement fought government subsidized railroad monopolies. Wheat-, corn-, and cotton-producing blocs battled each other over who would get the most favorable treatment from the USDA.[69] And from South Asia to Latin America, the last half century has seen the violent conversion of subsistence farmers into migrant workers. Historian Allan Kulikoff writes, "Capitalist transformation is not an automatic process but one fraught with conflict and violence, as capitalists attempt to reduce independent farmers to dependent wage workers."[70] Kulikoff was writing about the United States in the nineteenth century, but his observation applies equally well to more recent changes in India, Ghana, Mexico, and elsewhere. By the late 1960s, the United States was exporting its blueprint for domestic rearrangement to the world—even as its own sharecroppers starved.

In his 1942 book *Capitalism, Socialism, and Democracy*, economist Joseph Schumpeter coined the phrase "creative destruction" to describe how new innovations make existing technologies obsolete.[71] For Schumpeter, creative destruction is endemic to capitalism and drives economic growth. As industrial innovations colonized agriculture, USDA economists searched for policies that would maximize creative benefits and minimize destructive costs. But they asked the wrong question. The key question was not *whether* a new policy was destructive, but *where and to whom*. From 1930 to 1970, domestic rearrangements ensured that those who benefitted from creation were different from those who suffered from destruction. "I guess they will have to starve" was a satisfactory answer so long as "they" were poor, rural, and Black.

**MISSED OPPORTUNITIES**

Domestic rearrangements are not inherently violent. In fact, other, less harmful rearrangements were widely discussed. During Reconstruction, proposals designed to divide up plantation land among freedmen were a form of domestic rearrangement that would have impeded the emergence of sharecropping. During the economic recession of the late 1800s, there were opportunities to restructure Southern cotton to redistribute wealth more equitably. Instead, the sharecropping system was encouraged to dig in, exploiting tenants for another fifty years. Eventually weakened by New Deal domestic rearrangements, sharecropping collapsed on its most vulnerable dependents.

The USDA was well aware of what these policies were doing to the rural poor. By the time Delta children were dying of malnutrition, the social consequences of acreage reduction had been documented for decades in the department's own almanacs. In fact, the architects of New Deal legislation had predicted these consequences in 1933: Early drafts of the AAA stipulated that landowners must share their acreage-reduction payments with tenants. If croppers were going to be kicked off the land, at least they would receive something in exchange.

But this stipulation was removed to appease Southern congressmen, a decision that benefited not only white supremacists, but also manufacturing, the chemical industry, and large cotton plantations.[72] Courtesy of the US government, planters enjoyed cheap labor, subsidized inputs, and a paycheck for not farming. Meanwhile, Black workers had to beg white welfare boards for food. In its handling of rural poverty in Mississippi, Marian Wright described the federal government as "the single, largest supporter of segregation."[73]

Another proposal involved regulating the adoption of new technologies in agriculture.[74] Had yield-increasing inputs been disseminated more slowly, late adopters might not have been punished as severely by falling prices. Had access to these technologies favored small farms and tenants, croppers might have gained leverage with landlords.[75] Had farmers been allowed to share in industry's profits, the dissemination of new technologies might have mitigated the cost price squeeze. But these proposals were never seriously considered. Industrial capital and input markets were so deeply entangled that placing limits on large manufacturers and chemical companies was politically impossible. Meanwhile, government intervention to help large agribusinesses received broad support.

By the 1940s, Thomas Jefferson's vision of a rural nation filled with hardworking family farms was evolving into an urban one increasingly filled with the landless. Those farmers who survived the treadmill spent their days toiling to support DuPont, Dow Chemical, and International Harvester. As industrial giants captured more of agriculture's inputs and processors claimed more of its output, farmers were left holding the riskiest and least profitable slice of the food system. Farming had become the economy's ugly stepchild—a far cry from Jefferson's cherished cultivators and Roosevelt's indomitable yeomanry.[76]

• • •

From the 1930s to the 1970s, the food system displaced its small farmers and sharecroppers in response to interlocking crises of agricultural and industrial overproduction. In agriculture, the government pursued policies of acreage reduction and surplus storage. In industry, public funding for research fueled mechanical, chemical, and biological advancements. These contradictory goals ushered in both the golden age of agricultural research and the golden age of commodity surpluses as industry found a coconspirator in large farm operations.[77] Across the United States, small farms were squeezed out as food prices fell and costly inputs became necessary. In the South, the government's efforts to offload excess food on the hungry only accelerated the collapse of sharecropping. The violence of falling prices and changing markets was redirected from large producers onto small ones, effectively removing poor producers from the productive landscape.

There were alternatives to violent displacement and child starvation. But to consider them would have required powerful actors to strike a different balance between agriculture and industry, between large and small farms, and between laborers and landowners. In the South, a nonviolent alternative would have required a commitment to undoing structural legacies dating back to Reconstruction. Most importantly, it would have involved getting farmers off the agricultural treadmill, and that would have meant valuing something other than ever-increasing production and wealth concentration.

On January 31, 1966, speaking slowly into the TV camera, President Johnson announced the resumption of aerial bombardment and decried the violence of an ongoing communist revolution in Vietnam. Meanwhile, seventy sharecroppers at Greenville AFB decried the violence of an ongoing *capitalist* revolution in the United States. North Vietnamese farmers and Southern sharecroppers were on radically different trajectories, but in that moment, they shared a common bond. From the Sông Hồng Delta to the Mississippi Delta, planes rained ammonium nitrate, infants perished, and workers abandoned their fields for cities as industrial capital found new and old elsewheres to dump its excesses.

# 5

# INVADE, AID, TRADE

*Making the World Safe for Profit*

The secret communique from Frederick Chapin, US ambassador to Guatemala, was unusually long and detailed. The day before, November 9, 1982, five Indigenous refugees had contacted embassy officials claiming they were victims of atrocities committed by the Guatemalan army. Too frightened to come to the embassy, the refugees met with US officials in the relative safety of a private apartment. They had reason to be afraid: General Efraín Ríos Montt, Guatemala's military dictator, was ruthlessly pursuing suspected communist insurgents and had specifically targeted the Indigenous Maya communities to which the refugees belonged.

One of the refugees was a seven-year-old girl found wandering alone near the village of La Estancia. At first, she refused to speak, but as the others told their stories, the girl slowly opened up. In the communique, Chapin reports,

> The young girl told EMBOFF [embassy officials] that on September 30 she and a group of women and children were hiding from the [Guatemalan] army in the countryside of Chimaltenango. A baby cried out from the bushes they were hiding in and they were discovered by the army. The soldiers, she said, tossed grenades

> into their hiding spot, killing many people instantly. The young girl was wounded in the wrist, her mother was too hurt to get up and run. A soldier came by and robbed the mother of money and jewelry. A short while later another soldier returned and ordered the mother to rise; she was unable to do so. The soldier then shot her to death. . . . Although she [the girl] was able to relate her story coherently, she still spoke of her mother in the present tense.[1]

The refugees described other atrocities near La Estancia. Seven men were taken by soldiers and never seen again. A young man walking alone near his village was shot by an army patrol and "his head was then ripped open with machetes." The next day, his family was too frightened to claim the body. Seventy men, women, and children were cut down "while fleeing along the Pixcaya River." They were "tired of running from the army and had resigned themselves to death or pardon."[2]

Ríos Montt's campaign of terror was the latest chapter in a civil war that had run twenty-two years and would continue for another fourteen. When he came to power in 1982, Ríos Montt was convinced that Maya families were supporting communist rebels. His counter-insurgency turned its attention from student activists and union leaders in the cities to peasant farmers in the countryside. The horrors visited on Maya villages make for grim reading. One month after Chapin's communique, 300 people—including 114 children—were tortured, raped, and murdered over the span of three days by Guatemalan special forces in the town of Dos Erres. The first to be thrown in the village well was a live baby; the rest of the village soon followed. For the next few days, passersby did not know what had happened to Dos Erres's residents, noting only that the well had been covered over. No one dared look inside.[3]

As a snapshot of Ríos Montt's scorched earth policy, Chapin's communique revealed a great deal: endemic fear, distrust, rage, and resignation in a society experiencing genocide. What was *not* revealed was a fact already known to everyone on the distribution list: The perpetrators of these massacres were trained and armed by the United States. CIA agents were supplying weapons and ammunition, designing and directing assassinations against suspected insurgents, and instructing Ríos Montt's death squads on the finer points of counterinsurgency: how to deprive an enemy of local support, confine potential threats in concentration camps, and torture prisoners. The CIA certainly knew a thing or two about rebel movements: It was actively supporting one in Nicaragua and had defeated ones in neighboring El Salvador and Honduras. Under the guise of fighting communism, US support kept Guatemala's civil war burning for over three decades, leaving 200,000 dead, 50,000 disappeared, and 1.5 million displaced from their homes.

Ríos Montt would stand trial for genocide against the Maya in 2012, but the tribunal could not account for all those responsible for violence. Nor could it address the structural causes of Guatemala's civil war. Ruthless as he was, Ríos Montt was still a product of history and geography. Were you to trace the fallen dominoes back in time and space—beyond burned villages in the highlands and sprawling mansions in the capital—you would find the United Fruit Company and its banana monopoly in the 1930s, a US-led invasion to preserve that monopoly in the 1950s, a botched food aid program in the 1970s, and a relentless commitment to agricultural development better known as the Green Revolution.

Between the 1980s and the 2020s, as rural livelihoods collapsed, waves of Guatemalans headed north looking for work. Despite fleeing state-sponsored violence, the United States refused to recognize them as refugees—doing so would have forced the government to acknowledge it was supporting such violence.[4] Instead, Guatemalans entered the country as undocumented migrants. Farms growing labor-intensive fruits and vegetables took advantage of these vulnerable workers as free-trade agreements dumped US corn on Guatemala, perpetuating the cycle of displacement. The seven-year-old orphan who spoke to embassy officials in 1982 and the migrants who risked crossing the Rio Grande forty years later are linked together through the US food system and the unrelenting treadmill at its heart.

• • •

This is the story of how powerful actors in the food system pursued their interests and solved their problems by remaking distant places. It is also the story of how diplomats used food to achieve and justify their geopolitical goals. At the turn of the twentieth century, the United States pursued these interests the colonial way, through annexation, invasion, and corporate conquest. From the 1950s to the 1970s, food aid was a versatile tool to dispose of dangerous surpluses, capture international markets, and undermine foreign competition. Since then, the dissemination of agricultural technologies has subsidized domestic industries, propped up foreign elites, and pushed smallholders into precarious labor—all in the name of export-oriented development.

Through it all, the United States touted its First World resources as the solution to Third World problems, when in fact the opposite was more consistently true: It was *Third* World resources—land, labor, dependent economies, and expendable people—that were busy solving problems for the First.[5]

This chapter begins and ends in Guatemala, but the full extent of US food imperialism is not confined to this country or its Central American neighbors. Since the late 1800s, dozens of countries and territories have experienced similar misadventures.[6] Some were invaded to secure corporate and national interests, others were subdued through aid, and still others were hollowed out to make space for US imports and investment. A vast body of work documents these histories, making this story a necessarily cursory one. Lively debate continues over the impact (and ethics) of US actions.[7] But few will argue that US companies benefited from the way the country put its agricultural and industrial bounty to work beyond its borders. This is the story of how the United States variously invaded, aided, and traded its way to the top of the world's food pyramid.

## INVASION: CONQUEST BY ARMS

Strategic obsessions change. Before communism, it was bananas, and before bananas, it was sugar. In the late 1800s, global trade introduced US consumers to an array of imported commodities. Urban populations began to demand Central American coffee and cacao, East Asian tea, and Caribbean fruits—crops that could not be grown on continental farmland. Overproduction of wheat and corn kept food prices low for a growing urban workforce, but also suppressed profits for farmers and processors. Once-powerful agribusiness and railroad monopolies, harassed by antitrust actions, faced increasing competition. Growth in agriculture had come from frontier expansion, but the continental frontier was closing. By 1896, the sector was in full-blown depression. How could agricultural capital still generate a profit? More to the point, *where*?

War would answer the question. In 1898, the United States soundly defeated Spain, vastly expanding its national map and raising the curtain on a new, tropical frontier. Among the spoils was sugar. In a single seven-month period, the United States claimed the Philippines, Puerto Rico, Cuba, Guam, and Hawai'i (annexed separately)—accounting for nearly one-fifth of global sugarcane production.[8] Americans celebrated their new territories by throwing "sugar parties." Across the United States, victory celebrations featured candies and confections. Some public festivities included candy replicas of the USS *Maine*, the battleship destroyed in Havana harbor at the start of the war.[9] Demand for the white gold was at an all-time high: Middle- and upper-class households added it to coffee and tea, bakers folded it into new dessert recipes, and manufacturers used it to enhance and preserve a growing multitude of processed foods.[10]

As the government consolidated control over its new territories, large sugar interests raced to capture a share of the prize. There was little to stop them: The American Sugar Refining Company, known as the Sugar Trust, controlled 98 percent of all US sugar refining. President William McKinley, who claimed divine inspiration when he annexed the Philippines, was a prominent Sugar Trust lobbyist and received healthy campaign donations from the trust in 1896.[11]

But the new tropical frontier, like the old continental one, was not free for the taking. Nationalist movements demanded independence, not only from colonial oppression but from the landowning elites who had protected Spanish authority for centuries. Nowhere was this more obvious than in the Philippines. To beat Spain, the United States had cooperated with a Philippine independence movement led by Emilio Aguinaldo. For years, Aguinaldo's forces had fought the Spanish and paid the price in lives. But when it came time to determine the archipelago's new leadership, McKinley betrayed Aguinaldo and negotiated directly with Spain. On February 4, 1899, Congress declared the Philippines (and its coveted plantations) property of the United States. To manage its affairs and maintain order, the United States put Philippine oligarchs in charge. Aguinaldo demanded independence and fought back, switching from conventional warfare to guerrilla tactics. In response, US troops burned villages and granaries, confined peasants to concentration camps, tortured prisoners, and—as a matter of policy—did not differentiate between combatants and innocent bystanders.[12] By early 1900, the United States was behaving an awful lot like the Spanish Crown it had recently replaced.

Many on the mainland felt queasy about this metamorphosis from liberator to oppressor, and "the Philippine question"—whether to grant independence or continue the war—became a topic of heated debate in Congress. Senator Albert Jeremiah Beveridge, a staunch supporter of President McKinley and Vice President Theodore Roosevelt, backed the conquest of the Philippines as part of the "divine mission of America," claiming that God had "marked the American people as His chosen nation." The spoils of such a mission included "all the profit, all the glory, all the happiness possible to man."[13]

For Beveridge, "profit" and "glory" were not metaphors. His speeches were gilt with religious exceptionalism, paternalism, and white supremacy, but his vision for the Philippines was grounded in practical economics and geopolitics. His backers included financial and shipping interests, naval strategists, and powerful sugar companies. Crushing Philippine nationalists would open greater markets for US businesses. As Beveridge explained, "More and more, Europe will manufacture the most it needs, secure from its colonies the most it consumes. Where shall we turn for consumers of our surplus? *Geography answers the question*. . . . The Philippines give us a base at the door of all the East . . . within hail of India, shoulder to shoulder with China, richer in its own resources than any equal body of land on the entire globe."[14]

Beveridge's words exemplified how leaders wove together geopolitical, economic, and ideological objectives into a mandate to reform the world. Ten years later, these braided interests were reiterated by President Taft: "While our foreign policy should not be turned a hair's breadth from the straight path of justice, it may be well made to include active intervention to secure for our merchandise and our capitalists opportunity for profitable investment."[15] In fact, for the next century, Beveridge's understanding of geography would "answer the question" of where US corporations would find cheap labor and resources, where the economy would dump its surpluses, and where the military would be deployed. It would answer the question of which governments deserved assistance, which would be left alone, and which would be actively undermined. By 1906, Beveridge's understanding of geography had also killed a quarter of a million Filipinos, most through starvation and disease.

## LATIFUNDIOS

Aguinaldo's supporters fought for independence not only from Spain and (later) the United States, but from a centuries-old social structure that funneled the colony's wealth into the pockets of an elite ruling class. The central motive behind European colonialism was the extraction of value from each territory, "getting the colonies to 'pay their way' and generate profits" for Europe.[16] In the 1500s, New World gold and silver propped up European treasuries. In the 1600s, Indigenous genocide cleared the way for sugar plantations built on the spatial fix of enslaved African labor. Within the agrarian societies of Central America and the Philippines, extracting value entailed gaining control over land and labor. Invariably, this meant that precolonial forms of subsistence had to be broken and reset in ways that enabled peasants to be more easily exploited.[17]

To accomplish this, Spain introduced a modified form of European feudalism. The colonial state granted settlers large tracts of land (*latifundios*) along with the right to demand tribute and levy labor from the Indigenous inhabitants. These estates produced commodities for market—sometimes food crops, but usually cash crops for export: coffee, sugar, cotton, coconut, sisal, and tobacco depending on the location. Indigenous subsistence farming was confined to smallholdings (*minifundios*), and households were forced to pay the landowner with crops, labor, or a combination of both. Arrangements varied between locales, but the basic structure was maintained: The landowning class (European and mestizo) controlled the Indigenous population's access to land and managed the allocation of labor.

For the colonial state, the landowning class offered a means to control territory, population, and production. For landowners, the colonial state guaranteed property rights, provided marketing support, and deployed troops to quell peasant unrest. When Spain's Central American colonies achieved independence in the early 1800s, it was landowning elites who assumed control of state institutions. When the United States annexed the Philippines in 1898, it guaranteed the property rights of the landed class and demanded the return of lands confiscated by peasants during the revolution.[18] The *latifundia* system and its plantation monocultures were preserved. Meanwhile, *latifundistas* invented new ways to extract a greater share of

Indigenous labor through sharecropping, cash tenancy, debt bondage, and wage labor.

Spain's defeat in 1898 severely weakened its influence, not just in the territories it lost to the United States, but in all its former Central and South American colonies. In Guatemala, US banks and agribusinesses moved to fill the vacuum.[19] One of the companies that took advantage of Spain's weakening grip was United Fruit, the result of a recent merger between plantation and railroad interests. The company would permanently alter the political and social trajectory of Latin America, not by breaking and resetting Spain's colonial land and labor system, but by doubling down on it.

## THE OCTOPUS

By the 1930s, you could make the two-thousand-mile journey from Guatemala City to New Orleans without ever leaving the property of the United Fruit Company. The trains from Guatemala City to Puerto Barrios were owned by International Railways of Central America, a subsidiary of United Fruit. The port—the only one on Guatemala's Caribbean coast—was also owned by United Fruit. And the steamers bound for New Orleans were part of United Fruit's "Great White Fleet." Wherever the company's reach ended, other familiar brand names took over. The country's communication network was owned by International Telephone and Telegraph of New York City, and the country's power grid belonged to General Electric. US corporations did not invest in Guatemalan infrastructure, they simply bought out the local companies that did. In this regard, Guatemala was not unique: Throughout Latin America, United Fruit monopolized "electric light, the mails, telegraph and telephone, and—a no less important public service—politics."[20] It was this deep integration into everyday life that earned United Fruit the moniker El Pulpo, "the Octopus."

In making this trip, you would have joined a tremendous current of commodities and capital flowing between Central America and the United States; in this case, between the labor of Guatemalans and the corporate interests of Wall Street. This grand circulation helped financiers and industrialists weather the Great Depression. It also helped long-standing aristocracies hold sway in Central America. Dictators like Guatemala's Jorge Ubico—nicknamed the "Napoleon of the tropics"—enjoyed broad US support for his iron-fisted suppression of dissent. In exchange, he granted United Fruit unfettered access to the country's land and labor. Under Ubico, United Fruit held a monopoly on banana production, employed Guatemalans under brutal working conditions, and exported its products tax-free. The company came to own almost half the map of Guatemala, much of it acquired by forcing Indigenous, Black, and poor mestizo inhabitants off their land and hiring them back at starvation wages.[21] Ubico passed laws that allowed plantation owners to arm themselves and enforce order. He indemnified them from liability for the deaths of their workers. And to force Indigenous Guatemalans into wage labor, he passed vagrancy laws eerily similar to the Black Codes of the Reconstruction-era South.[22] When workers rebelled, the company counted on Ubico's soldiers to crush dissent. United Fruit made a killing by letting Ubico do the killing.

Through it all, Guatemala's oligarchy enjoyed political protection from the United States as a matter of policy. In a 1927 State Department memorandum, Under Secretary Robert Olds explained, "We do control the destinies of Central America and we do so now for the simple reason that the national interest absolutely dictates such a course. . . . Until now Central America has always understood that governments which we recognize and support stay in power, while those we do not recognize and support fall."[23]

In practice, failure to recognize and support US business interests could occasion American boots marching up the beach. The Marines had become a fixture in and around the Caribbean by the 1920s. Smedley Butler, a decorated Marine Corps general, described his career in this way: "I helped make Mexico . . . safe for American oil interests in 1914. I helped make Haiti and Cuba a decent place for the

**Figure 5:** Map of United Fruit Company steamship routes, 1924. Courtesy of Smithsonian Institute Archives SIA2012-7659.

National City Bank boys to collect revenues in. . . . I helped purify Nicaragua for the international banking house of Brown Brothers in 1909–1912. I brought light to the Dominican Republic for American sugar interests in 1916. I helped make Honduras 'right' for American fruit companies in 1903."[24] Ubico had a front-row seat to these spectacles, no doubt influencing his choice to overlook the concerns of poor Guatemalans in favor of the foreign agribusiness that ran much of his country.

Which is why, when Guatemalans took to the streets in 1944 and overthrew Ubico, elites in Washington, D.C., and New Orleans

(United Fruit headquarters) took notice. For the next decade, democratically elected presidents—the first in Guatemala's history—began to untangle the Octopus. President José Arevalo passed laws to end racial discrimination, guarantee a minimum wage, and decriminalize labor unions. His successor, the popular Jacobo Árbenz, built new railroads and ports to end United Fruit's monopoly on transportation. In 1952, Árbenz passed agrarian reforms to return large tracts of unused fruit company land to peasants. United Fruit could have spared a few acres—it used a mere 8 to 15 percent of its enormous holdings in Guatemala. Árbenz even offered to compensate United Fruit for what was taken, at the per-acre price the company declared for tax purposes.[25] But United Fruit's explosive response to the new law suggested there was more at stake than lost land: The company was worried about losing control of the country.

The Octopus launched a blistering smear campaign to paint Árbenz as a communist. Former president of United Fruit Sam Zemurray hired a PR firm to convince Congress that unless Árbenz was deposed, Guatemala would become a vassal of the Soviet Union. Zemurray knew something about coups, having overthrown Honduran President Miguel Dávila in 1911 with a small army of mercenaries (Dávila had dared to increase taxes on banana exports). This time around, Zemurray needed more help. His PR firm produced a 235-page "Report on Guatemala" and delivered a copy to every member of Congress.[26] In it, Árbenz's reforms were described as not only a menace to United Fruit, but an imminent threat to the free world.

It wasn't a hard sell. Not only was President Eisenhower taking an aggressive line on communism, but his closest advisors were Fruit Company men. Secretary of State John Foster Dulles and his brother, CIA director Allen Dulles, were partners at a law firm representing United Fruit. The director of international security affairs at the State Department was a former CEO of United Fruit, and Eisenhower's personal secretary was married to United Fruit's public relations director. Headstrong from overthrowing Iran's democratically elected leader in 1953, the CIA was ready to take its coup d'état roadshow to Central America. Agents secretly recruited and trained thousands of Guatemalan exiles in neighboring Honduras and Nicaragua, then set up an unprecedented propaganda campaign. Radio broadcasts threatened a US invasion unless the "communist" government was deposed. The CIA approached Carlos Castillo Armas, an exiled Guatemalan colonel and outspoken opponent of Árbenz, and offered him the chance to lead the invasion. Through it all, United Fruit executives worked closely with the CIA, even discussing post-coup plans for company operations.[27] Meanwhile, embassy officials stonewalled Árbenz's attempts to discuss his concerns about the company.[28]

In June 1954, Armas invaded his own country with a CIA-equipped and -financed army. US air support "provided the clincher": F-47Ns bombed bridges, trains, and ships, destroyed gasoline supply lines, and targeted government reinforcements. Air strikes had a demoralizing effect, "adding to the myth that the [rebels] were an organized, unbeatable force."[29] Fearing that a full-scale US invasion was inevitable, many of Árbenz's military units refused to fight. On June 27, Árbenz resigned and a military junta led by Armas took over.

Armas quickly unraveled Árbenz's reforms, returning land to United Fruit and terminating worker protections. Unionists and student activists went to ground as the junta rooted out leftists. For the next thirty-six years, a line of military dictators clung to power through an obsession with anticommunism and the support of Guatemala's influential northern neighbor. The CIA helped the government set up intelligence agencies as Green Berets trained governmental and paramilitary forces. This close relationship would continue through the duration of the civil war as kidnappings and disappearances—perpetuated by both sides—became daily fare.

For many Guatemalans, the country's "Ten Years of Spring" under Arevalo and Árbenz (1944–54) validated the promise of independence, that democracy could undo the structural legacies of colonialism and break the stranglehold of latifundistas.[30] But as bombers streaked overhead, it was clear that the United States would defend its corporate interests over its democratic ideals, and shed blood to do so. In that moment, reform-minded Latin Americans learned a hard lesson: The anticolonial struggle would have to be an armed one.

One eyewitness to the Guatemalan coup was a young physician named Ernesto "Che" Guevara. In 1959, Che and Fidel Castro launched the Cuban Revolution, deposed Cuba's US-backed dictator, and nationalized foreign plantations, including those owned by United Fruit. Predictably, the CIA moved to overthrow Castro in 1961, reprising many parts of the Guatemala playbook. But this time, the invasion was snuffed out by Cuban forces in the swampy Bay of Pigs. As for the doomed soldiers, they had sailed to Cuba on United Fruit Company ships.[31]

United Fruit was called the Octopus for its entanglements with Central American governments. But it was the company's influence *within* the United States that is perhaps most notable. As Austin Anson of the Salinas Valley Growers Association learned in 1942, Congress and the White House could be moved with the right message. For Anson, it was the threat of Japanese invasion; for Zemurray and United Fruit, it was communism. Creating a mobilizing narrative would become a key concern for transnational agribusinesses in years to come, as hunger gradually replaced communism as the global boogeyman.

## THE NEW LATIFUNDIOS

Long before Detroit automakers moved factories to Mexico and before electronics companies offshored production to East Asia, agribusinesses and financial institutions made a profit expropriating foreign land and labor. In some places, like the Philippines, this involved protecting the landowning class, buying its sugar, and financing loans. In others, like Guatemala, companies bought up or leased large estates, then dispossessed smallholders with government help. Either way, US business interests benefited by maintaining colonial-era relationships between rural peasants and the oligarchy. United Fruit profited not by introducing new ways of doing business, but by becoming Latin America's newest and biggest latifundista. And a truly transnational one at that: Not only could it call on the Guatemalan government to secure its interests, it could call on the United States to overthrow that government when democracy didn't go its way.[32]

## AID: CONQUEST BY BREAD

Invading the Philippines and grabbing Central American land opened a new tropical frontier right as the continental one was closing, helping address economic stagnation in the agricultural sector. But by the second half of the twentieth century, the landscape of capital, food, and influence in the United States was changing. There were new problems to address at home and abroad, problems that demanded different solutions.

After World War II, as the CIA worked with big businesses to hunt leftists and overthrow foreign governments, the USDA struggled to solve an increasingly contentious *domestic* problem: too much food. Subsidized by the conversion of wartime munitions plants, chemical companies churned out millions of tons of synthetic fertilizer. US monocultures, trapped on the agricultural treadmill, soaked it up. As yields of wheat and corn broke new records, farms produced far more than the country or its trading partners could consume. To prevent prices from falling, the government bought up what could not be sold and stored it at taxpayer expense, fueling the surplus crisis of the 1950s and 1960s (see chapter 4). The country needed to dispose of its excess grain—the billion-dollar question was *where*.

In Europe and East Asia where the war had damaged agricultural infrastructure, the United States provided emergency food aid to friends and former enemies. In the five years following the war, US wheat exports increased tenfold, 90 percent in the form of aid.[33] The late 1940s through the 1960s also saw anticolonial conflicts erupt throughout South and Southeast Asia, the Middle East, and Africa. Population growth in newly independent countries put colonial-era agricultural systems under stress. Economies that had been organized to satisfy European demand struggled to become self-sufficient in food. To complicate matters, the new global playing field was divided between capitalist and communist blocs. Picking a side could have serious implications for access to export markets and food imports—notwithstanding the ever-present threat of invasion or coup. As newly minted governments struggled to navigate postwar geopolitics and internal conflicts, the poor struggled with hunger.

## THREE PROBLEMS, ONE SOLUTION

The word "aid" implies assistance, help, and generosity. It assumes a degree of compassion on behalf of a donor for a recipient. Between 1955 and 1975, some $24 billion worth of farm products shipped from the United States went by this name. But not all of it reflected donor compassion, and certainly not all of it was beneficial for its recipients. The reason has to do with how strategists balanced competing motivations for aid.

In the early years of the Cold War, strategists saw an opportunity to address three problems at once: global hunger, the threat of communism, and the burden of surpluses. "The seeds of totalitarian regimes are nurtured by misery and want," Truman told Congress in 1947.[34] Hubert Humphrey echoed this in 1953 when he declared that "communism has no greater ally than hunger; and democracy and freedom no greater ally than an abundance of food."[35] In policy debates, political campaigns, and the popular press, abundant food was the new American brand, a "vision of prosperity and consumerism" that became an essential part of the country's Cold War propaganda.[36] Feeding the hungry was seen as not only a national obligation but as essential for realizing the country's vision for the world. Best of all, the nation's well-fertilized monocultures were ideally suited to this task. John F. Kennedy promised to "turn our great agricultural abundance into a blessing—for ourselves and for all the world. . . . The productive genius of the American farmer is one of the most potent tools in our effort for world peace."[37] At last, the curse of chronic overproduction could be put to good use.

In 1954, Eisenhower signed Public Law 480, the Agricultural Trade Development and Assistance Act. The act permitted the export of surplus grain "to expand international trade between the United States and friendly nations, to promote the national welfare and economic stability of agriculture in America, and increase consumption of US commodities in foreign countries to further the foreign policy of the United States."[38] Through three mechanisms (or "titles") PL-480 authorized commodity surpluses to be used as food aid. Title I allowed surpluses to be sold in other countries for foreign currency. The United States could use this currency to purchase local resources or accept it as payment for additional goods and services. Title II allowed surpluses to be given freely as "emergency assistance" (for example, in the case of natural disasters), and Title III authorized the president to use surpluses for domestic disaster or emergency relief.

All three mechanisms helped dispose of surpluses, but Title I aid was also tied to future purchases from the recipient country: for example, raw materials, contractor salaries, or strategic exports. Using food aid as an entry point, Title I allowed US capital to burrow more deeply into the economies of hungry countries. By 1969, nearly $14 billion worth of aid had been distributed through the program, 83 percent through Title I.[39]

Food aid came with stipulations. Communist countries could not receive PL-480 aid. Any country receiving Title I aid was prohibited from trading with communist countries or altering their trading patterns with friendly countries.[40] Food had to be shipped on US vessels, thereby benefiting American shipping companies. And aid had to be provided "in-kind": that is, it had to be actual *food*, grown and stockpiled in the United States. PL-480 aid could not take the form of monetary loans or grants the recipient could use to buy food.[41]

After all, the program was designed to relieve material surpluses, and that meant moving grain, not dollars.

For the next two decades, PL-480 was the official mechanism for the delivery of US food aid. It gave the State Department a carrot and stick to influence the course of hungry nations and provided the USDA with a pressure valve to relieve dangerous surpluses. By the 1970s, PL-480 aid had reached fifty-four countries and substantially reduced domestic stockpiles. It also fed millions of people, but not always consistently, and often with long-term consequences.

## THE AID TRAP FOR RECIPIENTS

The humanitarian intent behind food aid is to alleviate hunger. But over time, aid can have the opposite effect. Thousands of tons of grain can create a *production disincentive* for local farmers. In the same way that oversupply suppresses prices in the United States, sudden influxes of aid can suppress food prices and farm incomes in the recipient country. The connection between aid and production disincentives remains a subject of contentious debate, with aid proponents arguing that on balance, the economic benefit of aid has historically offset its negative impacts.[42] That said, PL-480 aid consistently made its way into places where food was plentiful. Disincentive effects were blamed for suppressing wheat prices in Egypt and impoverishing smallholders in Mozambique. In his investigation of aid excesses, Graham Hancock reported seeing sacks of PL-480 food in "almost every marketplace" in Haiti.[43]

Disincentive effects were most obvious when aid was overestimated or mistimed. Nearly all the Title I aid to Somalia in 1985–86 arrived during the country's grain harvest, when food was abundant. Grain prices fell 40 percent and many farmers stopped growing staple crops. A USAID audit revealed that Title I aid had exceeded demand by nearly ten thousand tons. When the auditors found the food, it was rotting in government warehouses. The Somali government lost $1.5 million, lost productive farmers, and even lost the food. Later that year, famine struck the country, in part because of a collapse in local production.[44] Of course, eight thousand miles away, US stockpiles were a bit lighter. Every dark aid cloud had a silver lining for the USDA.

Agricultural self-sufficiency was also hindered by the short duration of PL-480 contracts. In principle, food aid allowed countries to divert resources to improve agricultural infrastructure. But governments struggled to take advantage of this opportunity when contracts were too short. In fact, the United States often used short-term commitments to bargain for political and trade concessions—what Lyndon Johnson referred to as a "short-tether policy."

Aid also created incentives for leaders to ignore food provisioning and co-opt public funds. As Hancock explains,

> Food aid . . . is eminently fungible because it frees the recipient government from the tiresome necessity of ensuring that its own people do not starve. While well-meaning foreigners feed the hungry, the leaders of a country . . . can spend other funds at their disposal on whatever they like: they can buy advanced weapons with them, they can overpay their civil servants, or they can make some more hefty deposits into their Swiss or California bank accounts.[45]

Finally, aid altered local food culture. Most of the food distributed under PL-480 was wheat, the commodity with the largest surplus. But for many recipient populations, wheat was not part of the traditional diet. Over time, the abundance of refined wheat flour supplanted traditional staples, altering diets and changing tastes. PL-480 hooked consumer markets on a crop that could be grown only overseas. By 1989, African populations that traditionally grew and ate sorghum, millet, cassava, beans, and yams were spending $2 billion a year importing wheat.[46] The erasure of traditional farming, food preparation, and diets is reminiscent of North American Indian reservations where the surplus-burning commodity food program eroded Indigenous foodways.

## THE SPOILS OF FOOD AID

For PL-480 recipients, the effects of aid—production disincentives, corruption, and changing tastes—created dependency on food imports. Recipients often became US agriculture's biggest customers, a fact celebrated by USAID. In 1986, seven of the top ten importers of US farm products were PL-480 recipients. Countries that received substantial food aid expanded their agricultural imports from the United States 30 percent faster than from other sources. By 1966, PL-480 aid had contributed no less than $5.1 billion to *commercial* sales, all while fighting communism and burning through surpluses.[47] Getting countries hooked on aid had profitable repercussions long after the aid train ended.

Not only did food aid eliminate competition for US farmers through production disincentives, it mandated the trade structure of its recipients. PL-480 countries were prohibited from competing with US producers in other, non-aid, commodities. Before contracts could be signed, recipients had to agree to not export corn and rice, protecting US producers and undermining local ones.

Food aid was used to pry open stubborn markets. For example, in 2002, the United States agreed to supply aid—in the form of corn—to famine-stricken countries in Southern Africa. To the State Department's surprise, the governments of Malawi, Mozambique, Zambia, and Zimbabwe refused it. At issue was a concern that stockpiles in the United States were not segregated between genetically modified (GMO) and non-GMO corn. If hungry farmers receiving the aid accidentally planted the corn, GMO markers might enter the national supply chain, potentially jeopardizing corn exports to countries or trading blocs with GMO bans. One such bloc was the European Union (EU), which purchased the bulk of Southern Africa's agricultural exports. To protect future trade, the countries requested that the United States mill the corn before shipment, thereby preventing farmers from planting it as seed. But the State Department refused, hoping to pressure the EU to relax its GMO ban.[48] US agriculture wanted access to European customers and was willing to let Africans starve to get it.

But the biggest benefit of food aid for the United States was the elimination of surpluses. PL-480 bought time for Congress to roll out modest adjustments to price controls, preserving the agricultural treadmill and farming's lucrative relationship with industrial capital. Had mounting surpluses forced the United States to end price controls, farm losses would have become an epidemic. Instead, farm losses to the treadmill could be passed off as normal attrition.[49]

## AID MISADVENTURES IN GUATEMALA

By the mid-1970s, Guatemala's twenty-year civil war was at a stalemate. Widespread hunger and loss of land pushed some rural households to support leftist groups like the Guerilla Army of the Poor (Ejército Guerrillero de Los Pobres, or EGP). But many Maya refused to join, fearing reprisals by the state. Their fears were justified: The military government conducted rural "pacification" campaigns and murdered or kidnapped suspected communists. Rural displacement reduced agricultural production and threatened food security. Following Humphrey's 1953 principle that communism has no greater ally than hunger, the United States funneled food aid into the country to nourish bodies and fight an ideological war.

For Maya, aid made matters worse. PL-480 corn was sold for less than it cost to grow it, undercutting the poorest producers. As rural incomes dwindled, farmers had to rely on the very aid that was impoverishing them. Larger farms with diversified markets were less affected by falling prices, but many Maya had to give up their land as aid undermined its productive value.[50] To survive, smallholders turned to wage labor. By 1972, three hundred thousand Maya were migrating seasonally to work on elite-owned coffee, cotton, and banana plantations.

Had USAID purchased corn and beans from local farms and used this to feed the hungry, the injection of cash could have benefited rural economies. But by law, PL-480 aid had to be *in-kind*—food grown on US farms and occupying government warehouses.

On February 4, 1976, a magnitude 7.5 earthquake struck Guatemala, devastating Indigenous areas in the interior. The death toll topped 23,000, with 76,000 injured. Over a million lost their homes. The United States responded with 27,000 tons of PL-480 Title II aid to support quake survivors, but it took three months for the aid to arrive. USAID determined that because of the delay, the aid offered minimal benefit, serving only 20 percent of the hungry.[51] But it had a significant negative impact on small farmers. The earthquake had left rural food production largely intact; in fact, farmers had just harvested a bumper crop. The sudden influx of staple foods—especially black beans—crushed local production at the precise moment when recovery was most vulnerable. The director of Catholic Relief Services in Guatemala lamented, "The general effect was that we knocked the bottom out of the grain market in the country for 9 to 12 months."[52] Robert Gersony, a USAID researcher, placed the blame more broadly: "I think there is a strong case to be made for stopping all P.L. 480 food. . . . Even outside disaster situations, food handouts weaken minimal social structures and they discourage farmers from growing their own food."[53]

In the aftermath of the 1976 quake, rural communities across the highlands self-organized to form "reconstruction committees": collectives for sharing knowledge, resources, and food.[54] These collectives helped fill the gap left by a negligent government and exacerbated by poorly targeted aid. But military leaders saw these grassroots networks as evidence of a resurgent communist threat and "seed beds for social revolution."[55] Death squads increasingly targeted Maya villages for destruction. In fact, those Indigenous areas worst hit by the earthquake—and most likely to form rural collectives—saw a hundredfold increase in human rights violations over the next six years.[56] By 1982, massacres like the one at Dos Erres, and those detailed in Chapin's communique, were commonplace.

In the decades after World War II, US food aid was defined around three broad objectives: to achieve diplomatic and economic goals, provide humanitarian relief, and relieve the burden of surpluses. But these objectives did not receive equal weight. Diplomatic and economic goals shaped *where and when* food aid would be directed. Humanitarian concerns shaped how aid would be *justified* and *how much* would be provided. But it was the pressure to dispose of surpluses that determined *whether aid would be given in the first place.*

How do we know? Because food aid had to be given in-kind. PL-480 may have offered geopolitical and humanitarian benefits, but it was a surplus-burning machine by design. As surpluses dwindled, the State Department began cutting aid programs due to insufficient stockpiles. In fact, in 1989, Congress and the USDA discussed maintaining a surplus simply to guarantee multiyear aid programs.[57] But long before the United States emptied its warehouses, in-kind aid was hurting the very people it was designed to help. It is difficult to know how many people suffered because the United States insisted on in-kind aid—not because victims can't be found, but because their bodies are buried at the ends of long and complex causal chains. Or at the bottom of village wells that few have the stomach to look into.

## TRADE: CONQUEST BY CAPITAL

Through most of the twentieth century, US fertilizer use steadily increased as large monocultures—trapped on the agricultural treadmill—became chemically dependent. But that upward trend did not last. Between 1950 and 1970, domestic consumption grew 360 percent; over the next two decades, just 27 percent. With little unfertilized land left to conquer, demand for fertilizer began to plateau and, with it, profits. In 1971, the industry's return on investment was a mere 1.1 percent. Where could fertilizer and petroleum companies (often one and the same) find new markets? Once again, geography would answer the question. US agriculture may have achieved "peak nitrogen," but the rest of the world had not.

Seed companies and state-funded research programs experimented with high-yielding varieties of crops (HYVs). HYVs were bred to respond to specific formulations of synthetic fertilizers, not the organic composts perfected by generations of farmers. It is no surprise that some of the largest seed breeders—Bayer, Syngenta, and Dow—were chemical companies first.[58] Together, HYVs, chemicals, tractors, and improved irrigation were increasingly marketed as the formula for "modern" agriculture, a package of inputs known as the Green Revolution.

At the other end, large processors and manufacturers captured a growing share of farm outputs. Vertical integration allowed companies like General Mills, Tyson Foods, and PepsiCo (Frito-Lay) to efficiently mill, slaughter, prepare, package, and market farm produce. Big food processors pushed out or bought up smaller ones, steadily taking over national and global supply chains.

By the 1970s, the key to growth in the agribusiness sector had shifted from control over land and labor—the United Fruit model—to control over the inputs and outputs of farming. As agrichemical companies drove up input prices and processors drove down farmgate prices, farmers were forced to operate with increasingly narrow margins. Small and midsized domestic producers were already being put out of business by the cost price squeeze. Squeezing them further would have been politically untenable. Where could agribusinesses find new farmers to put on the treadmill?

## OLD PROBLEMS, NEW PROBLEMS, AND A NEW SOLUTION

Beyond the United States, developing countries still struggled to feed their populations. A global food crisis exacerbated by population growth, climate shocks, and rising oil prices began in 1972. For some countries, a legacy of productivity-sapping aid contributed to low food production. Over 250 million people suffered from hunger through the late 1970s and 1980s, a period punctuated by a string of high-profile famines in sub-Saharan Africa.[59] Strategists still wove together threats of hunger and communism, but dumping mountains of food aid would no longer work: The USDA had achieved its goal of eliminating the national grain surplus, and with it, the stockpile that supported PL-480.

To justify new solutions, a "developmentalist" narrative emerged: Hunger was caused by *poverty*, not an absolute shortage of food. Poverty explained why households had trouble acquiring food even when it was plentiful. As such, hunger became an *economic* problem, one that required changes to how people made a living, not just giant deliveries of wheat.[60]

What, then, was the cause of poverty? From a developmentalist perspective, smallholders were poor because their farms were inefficient. The solution for agricultural inefficiency was three things the United States had in abundance: Green Revolution inputs, staple crops, and demand for cash crops. Farmers could buy US-made inputs and grow staple crops for local consumption or cash crops for export. Greater yields would lead to greater prosperity, and prosperity would steer smallholders away from the abyss of communism. Advanced technologies and free trade promised to accomplish all this while *preserving class hierarchies*—in other words, without resorting to the land reforms Árbenz's had tried in Guatemala. This new developmentalist model could feed the population, suppress dissent, and push economies down the path to liberal capitalism. It would also enrich the powerful. Elite landowners could remain elite and smallholders could remain, well, small.[61]

The Third World peasant—up to now, little more than cheap plantation labor and a threat to law and order—became the lynchpin in the fight against hunger, leftist revolution, and stagnating domestic capital.

On the technological front, the Rockefeller and Ford Foundations established a constellation of research institutes to breed HYVs and promote agricultural best practices. HYVs were developed by crossbreeding local cultivars—seeds carefully curated by Indigenous farmers for centuries. The first institutes included the International Rice Research Institute (IRRI) in the Philippines, the International

Maize and Wheat Improvement Center (CIMMYT) in Mexico, and the International Center for Tropical Agriculture (CIAT) in Colombia. Supported by public funds and chemical company investments, these research centers developed input packages of HYVs, fertilizers, and pesticides for farmers in the developing world.

On the trade front, multinational institutions like the World Bank and International Monetary Fund (IMF)—heavily influenced by the United States—pushed developing countries to relax restrictions on trade and cut back support for farmers. The former opened the door for the importation of staples and inputs, and the export of cheap cash crops. The latter encouraged governments to get out of the business of agriculture, a business that foreign capital was about to take over.

On balance, Green Revolution technologies did increase food availability, particularly in South Asia. But even there, its positive outcomes were offset by persistent hunger, inequality, and environmental degradation.[62] Elsewhere, rural development interventions were disastrous for recipient countries and their peasant populations. The question is why.

## THE DEVELOPMENT TREADMILL

Consider a hypothetical country in Latin America, sub-Saharan Africa, or Southeast Asia. It's the 1970s, and most of its population is rural and dependent on agriculture. Staple crops are grown by smallholders using traditional varieties and techniques. By Western standards, yields and incomes are low: Farmers make enough to feed their families but little more. By a variety of economic measures, the country is considered "poor" and is not meeting its prescribed milestones for economic growth. Critically, it cannot afford to buy goods from the United States. According to development experts, the prescription for this disease is to improve agricultural efficiency.

How can the country improve agricultural efficiency? With scientific and industrial inputs (HYVs and synthetic fertilizers), modern practices (water- and pesticide-intensive monocultures), and new machines (tractors and harvesters). Most of these things cannot be acquired locally and must be imported from the United States. Imports require foreign exchange, and to get foreign exchange, the country must export something the United States is interested in buying.

How can the country increase exports? It could sell its industrial output, but like most developing countries during this period, it lacks an industrial sector that can compete with US manufacturers. Instead, economists advise the country to leverage its "comparative advantage"—in this case, cheap land, cheap labor, and environmental conditions suitable for cash crops. Coffee, bananas, sisal, sugar, cotton, and coconut can be exported to the United States, where demand is high.

How can the country expand cash crop production? By committing land and labor. With respect to land, the country might clear its forests or convert existing staple cultivation to cash crops. (Of course, HYVs and fertilizer could increase cash crop production *without* requiring additional land, but those are the same inputs the country lacks.) With respect to labor, converting land from staple crops to cash crops offers the chance to convert *producers* as well, either by displacing them from their land and pushing them onto plantations, or by keeping producers in place and getting them to grow for export. Displacing producers offers a secondary benefit: a pool of cheap labor to feed industrial growth.[63]

But there is a downside. Less land and labor will be growing staple crops for domestic consumption, reducing the supply of food. As the food supply shrinks, prices will rise. To prevent starvation, wages must rise as well, but higher wages will increase the cost of the country's exports and jeopardize its competitiveness on the world market. Paradoxically, the country needs cheap food to improve agricultural efficiency, which will lead to cheap food.

How can the country get cheap food? It could import it from the United States, but that would require foreign exchange. It could request PL-480 aid, but by the 1970s, that option is increasingly limited. It could increase domestic food production, but without new inputs, it will have to transfer land and labor from *cash* crops, the opposite of what it is trying to achieve.

At this point, the country may try to escape the trap by reaching out to the IMF, the global agency that greenlights large loans to foreign governments. Loans could pay for purchases of HYVs and chemical inputs, irrigation infrastructure, rural electrification, new roads, and port facilities. But loans come with interest—and preconditions. During the 1980s and 1990s, the IMF required recipient countries to abolish unions, suppress wages, reduce spending on social programs, and eliminate protective tariffs on imported goods, a package of policies known as "structural adjustment." To be eligible for a loan, your hypothetical country would have to integrate itself into global markets and expose its producers to greater competition, price volatility, and the influence of foreign capital, all while dissolving its social safety nets.

Countries like this are stuck on a treadmill. Like the agricultural treadmill that pushed small farms in the United States out of business, the *development treadmill* elevates the fortunes of some countries as it crushes others. Early success means everything. The first countries to export enough goods to achieve a favorable balance of trade can afford to buy imported technologies. The resulting improvements in efficiency allow these countries to export cash crops at competitive prices and claim a greater share of the market. Meanwhile, foreign exchange from these sales can be spent on new technological advancements (improving efficiency) and imported food (backfilling the loss of domestic staples).

But countries that are slow to transition face an uphill battle. By the time their cash crops reach the global marketplace, competition from other developing countries has suppressed prices, and greater international demand for inputs has driven up the price of fertilizer and pesticide. These two factors—less foreign exchange and higher input costs—limit the adoption of new technologies, and agricultural productivity stagnates. In response, some countries suppress agricultural wages to protect businesses. But low wages suppress domestic consumption. Low domestic consumption suppresses economic growth, and before long, the country must appeal to the IMF for more loans. To encourage loan repayment, the IMF demands growth in exports, and the death spiral continues.

Of course, there are many reasons why a country might be slow to transition. In the case of Guatemala, these included having an oligarchy intent on preserving an exploitative class structure, a foreign fruit company intent on monopolizing the country's primary export, and a powerful northern neighbor intent on rooting out whatever it could label "communist."

The biggest losers are the rural poor. If the country succeeds in acquiring new technologies, smallholders now face the agricultural treadmill that pits them against their neighbors in a race to increase yields. Should enough farms fail, food availability may follow. In the 1970s, the Philippines invested heavily in capital-intensive technologies for rice production. Large farms benefited, but many smallholders struggled: Between 1976 and 1979, the average income for rice farmers fell by half. By 1985, the national rice harvest was 50 percent under target and the country was forced to import 600,000 tons to meet its needs.[64]

Farmers who are forced to seek work elsewhere are understandably frustrated and may unionize, strike, and protest. Some may join revolutionary movements, and the state—fearful that foreign investors and donors will pull out—may respond with violent repression. In 2008, a South Korean company attempted to lease 1.3 million acres of Madagascar. When news of the deal became public, angry demonstrators took to the streets (as they had in Guatemala in 1944 to protest another land-grabbing corporation). The state responded with force, but in the end, the public groundswell brought down the government. As punishment for the coup, the United States withdrew financial support from Madagascar, discouraged investment, and banished it from trade privileges for the next five years.[65]

## TRANSFORMING HOUSEHOLDS

Development prescriptions altered the everyday dynamics of rural households. Traditionally, women have provided most field labor and taken responsibility for decisions about farming. But development strategists, blind to (or biased against) cultural norms, promoted idealized Western gender roles. Women were expected to run the home and manage food preparation, while men were encouraged to take charge of farming, especially cash crop production. Women were often excluded from participation in agricultural development programs. New technologies were targeted to men, even when the impacted activities were dominated by women. When tractors became available, men were trained to operate them. When milling equipment was introduced, it was men who were put in charge—not the women who had threshed, winnowed, and pounded grain for centuries. With newfound responsibility for production, men gained control over household financial decisions.[66]

In the early 1980s, USAID launched a demonstration program to increase rice production in the Republic of the Gambia. Traditionally, women oversaw rain-fed rice cultivation as a communal crop, provided most of the field labor, and determined how much rice would be allocated for consumption and sale. Seeking to increase market integration, USAID drained swampland for new rice fields and installed pump irrigation. Despite 97 percent of women participating in rice production, the program was rolled out without women being consulted. The new irrigated fields were given to men, who assumed control of production and allocation decisions. Three years later, deprived of their traditional roles, three-quarters of the women in the project area had switched from growing rain-fed rice to laboring in their husbands' irrigated fields.[67]

Agricultural "modernization" altered the balance of women's public and private activities to mirror the gendered landscape of US farm households and agricultural institutions. Domestically, USDA trained its female executives in home economics. Overseas, USAID encouraged agricultural ministries to create home economics and handicraft programs for women, and educational programs to prepare men for the market and institutional leadership.[68] Sociologist Carolyn Sachs notes that the consequences for women included "increased work load, loss of existing employment, changes in the reward structure of their work, and loss of control of land."[69] Excluded from valued production, women's labor became less relevant to economists and less visible to policymakers. As men took control of export production, foreign agriculture became increasingly credible to the Western patriarchies that backed it.

## THE SPOILS OF DEVELOPMENT

The development treadmill was exceptionally profitable for US businesses. Investment in research and agricultural education created new forms of dependency, opening markets for US companies. To be sure, the capital-intensive farming that emerged from research centers like IRRI was life-changing for those who could afford its prescriptions. But many could not. From the outset, IRRI's new strains were designed around imported, petroleum-based fertilizers, not local organic ones. This wasn't by chance: The institute's founding contributors included Chevron, Shell, Esso, the Potash Institute of North America, Ciba-Geigy, Stauffer Chemical, and Monsanto.[70] And these companies got their money's worth: Green Revolution technologies helped jumpstart a US fertilizer industry that had saturated its domestic demand. Between 1971 and 1974, fertilizer consumption in the United States grew a scant 2 percent. Over that same period, consumption grew 31 percent in the Philippines and 66 percent in Guatemala.[71] In 1971, the fertilizer industry's annual return on investment was 1.1 percent—three years later it topped 40 percent.[72]

As the agricultural treadmill took hold, smallholders took out loans to afford the new inputs, loans underwritten by elite-owned banks and foreign lenders. Masagana 99, the official credit program to support small rice farmers in the Philippines, was launched by J. Walter Thompson Advertising, a Madison Avenue firm hired by Shell Chemical Company. Fertilizers and pesticides purchased with Masagana 99 loans were supplied by Planter's Products, formerly Esso Fertilizer and Agricultural Chemicals. As bankrupt smallholders abandoned farming, their land was sold or leased out to grow pineapples for export through the produce giants Dole and Del Monte.[73]

For the United States, such conquest by debt was muscle memory. In 1802, Thomas Jefferson promoted the use of debt to force Indigenous farmers off their land (see chapter 1).[74] One hundred eighty years later and half a world away, his political descendants struck it rich deploying the same strategy against a rice farming peasantry.

Official government aid was particularly lucrative. By the 1980s, 80 percent of all official aid to developing countries was being spent in *donor* countries to finance consultants, contractors, and staff salaries.[75] In that decade, agribusinesses received $1.8 billion from USAID procurement orders each year. Interest on loans was another fruitful source of income. By the early 2000s, economic growth from Mexico to Argentina had stagnated as the burden of interest payments siphoned away the productive value of Latin America's workforce. Developing countries were trapped by the very structures that purported to help them, but not because these structures were broken: As engines of accumulation, they were functioning as designed.

## "DEVELOPING" GUATEMALA

By the time Chapin sent his communique in 1982, the poorest 88 percent of Guatemala's farms controlled only 16 percent of the country's arable land. Unable to expand their holdings and harassed by government death squads, Maya households struggled to grow sufficient food. To afford food purchases, hundreds of thousands migrated for seasonal work at coastal coffee plantations, where they endured horrendous working conditions and starvation wages. Only 10 percent of coffee workers on Guatemala's plantations received a nutritionally sufficient diet.[76] As Western consumers and coffee traders benefited from cheap Mayan labor, pro-export policies opened the door for waves of profitable intervention.

IMF's structural adjustment policies mandated the removal of trade barriers and dismantled state programs designed to strengthen domestic food production. With the government no longer able to support its agricultural sector, USAID took its place. Operating as a "parallel state," the agency funded irrigation projects, provided technical expertise, and offered credit for farmers to buy imported agrichemicals and seeds.[77] But USAID's programs were geared toward *exports*: Farmers could receive loans so long as they chose to grow cash crops—snow peas, strawberries, zucchini, broccoli, and blackberries—not maize, rice, and beans.[78] As aid workers taught smallholders to spread synthetic fertilizer, agrochemical companies benefitted from a new market for nitrogen. As farmers harvested food they would never eat, shippers and processors benefited from cheap exports.

But more hectares for export meant fewer hectares for staples, impairing Guatemala's ability to feed itself. In 1961, the country produced 96 percent of the maize it needed to survive. By 1990, it produced less than half. Domestic staple production took another hit in 2006 with the signing of the Dominican Republic–Central America Free Trade Agreement (DR-CAFTA), which eliminated trade barriers on US maize and rice. As DR-CAFTA undercut Guatemalan smallholders, large Midwestern and Southern farms reaped the rewards: Between 2006 and 2020, US maize exports to Guatemala increased 97 percent and rice exports grew 63 percent.[79]

At first, rural households benefitted from the switch to cash crops and the introduction of new inputs. A study in 1989 found that returns per hectare of snow peas exceeded maize by 1,500 percent.[80] Farm incomes increased, and fewer Maya migrated to plantations in search of work. But these gains were short-lived. As competition from other exporting countries mounted, returns for Guatemalan farmers dwindled. Rural purchasing power declined and hunger spread.[81]

For many Indigenous households today, the dream of "modern" farming has become a nightmare: Decades of synthetic fertilizer use have depleted soil fertility, and yields are in decline. Some vegetables can no longer be grown *without* fertilizer, putting these crops out of reach for households that cannot afford the inputs. Heavy pesticide use has killed off beneficial insects, leading to a collapse in natural predation. Pest outbreaks have spread to staple crops, exacerbating food insecurity. And there are serious human health impacts: The Guatemalan government reports an average of 1,000 deaths from pesticide poisoning each year, though observers estimate the actual number is between five and forty times higher.[82] It is not surprising that among the Kaqchikel Maya, agrochemicals are called *itzel aq'om*, the term for "poison."[83]

## INVENTING A HUNGRY AND DYSFUNCTIONAL WORLD

Food has played an important role in US foreign relations since the late 1800s, but over that time its purpose has changed. With the closing of the frontier and growing urban demand for tropical cash crops, agribusinesses looked beyond the border for opportunities to control land and labor. As the world entered American dining rooms through sugar, coffee, and bananas, foreign farming became a *resource* of national interest, one worthy of conquest.

As domestic surpluses mounted and foreign populations grew, policymakers promoted narratives of global *scarcity*. Food shortage was accepted as the essential challenge facing former colonies.[84] The world entered American living rooms through images of starving children and communist ration lines. Apparent shortages justified aid in the form of surplus commodities, and this aid—its target, quantity, and conditions placed on it—helped steer foreign governments and prepare their economies for future dependence.

As domestic surpluses shrank and industrial surpluses grew, policymakers blamed foreign governments for food *dependence* and agricultural inefficiency. The world entered American living rooms through reports of rampant poverty, economic instability, and civil war—scenes of violence seemingly unrelated to the domestic food system. Meanwhile, dependence justified greater sales of fertilizer and wheat.

Each twist in the story assumed that the world beyond the US border was dysfunctional—a place filled with backward Filipinos, drought-afflicted Somalis, and collective-minded Guatemalans. In the geographic imagination of politicians, diplomats, and investors, the border came to divide the world between modern, rational food systems and feral ones. Beyond the line was a place where principles of democracy and self-sufficiency were nonexistent, and therefore need not be respected, a place where the violent excesses of US abundance could be disposed of and weaponized against ideological foes.

## EXTRATERRITORIAL FIXES FOR DOMESTIC OVERPRODUCTION

In the 1950s, grain surpluses were dangerous for farm profits and costly for taxpayers. By moving these surpluses into other countries' economies, USDA and the State Department created temporary surpluses elsewhere. To the extent that these new gluts robbed local producers of demand, food aid redirected the violence of overproduction from US farms onto distant others. As local production languished and tastes changed from traditional staples to wheat, foreign food systems were remade into vassals of Midwestern monocultures, robbed of self-sufficient futures. At the same time, wheat and corn producers could count on decades of foreign demand.

The food aid program is rightly credited with keeping people alive, and rightly criticized for undermining foreign self-sufficiency.

But what gets overlooked is how it kept US agriculture from collapsing under the weight of its own success, how it preserved the agricultural treadmill and, in so doing, guaranteed a profitable future for the agrichemical industry. As food aid eased surpluses, it soothed political tensions between free-market proponents and price-support advocates within and beyond the USDA.[85]

These *domestic* victories could have been achieved only by reaching beyond the borders of the country into places where racial and cultural differences impaired public compassion, where long chains of causation impaired culpability, and where selfish motives could still (occasionally) harmonize with humanitarian ones. These elsewheres served as economic and social experiments, ones in which success *and* failure benefited the United States. In fact, failure created new and profitable problems for capital to solve.

Rural development policies and Green Revolution technologies were pitched as the solution to agricultural inefficiency. Sometimes these improvements were designed for cash crops, sometimes traditional staples. Either way, the solution created demand for US-made inputs and provided cheap products for US processors. The Green Revolution is rightly credited with improving overall food availability, and rightly criticized for increasing rural inequality. What gets overlooked is how it offered a new future for US input manufacturers facing a decline in domestic demand, and processing companies facing higher domestic prices.

As with food aid, these victories could have been achieved only by reaching beyond the border into countries where yields and labor costs were low, where elites felt threatened by their peasantries, and where hunger could still justify intervention. Critically, these elsewheres offered virgin soil for synthetic fertilizers and export crops—ideal conditions for new technologies to inspire competition between farms. On its face, the United States exported HYVs, agrochemicals, and staple foods. But its most lucrative export was the agricultural treadmill.

## MISSED OPPORTUNITIES

The USDA could have addressed the surplus crisis in the 1950s and 1960s by changing domestic policy, rather than dumping grain on foreign food systems. Even if aid had included surplus grain, aid programs could have been designed to promote self-sufficiency rather than future dependence on US exports. Food aid could have been based solely on need, not on a complex geopolitical calculus that masqueraded as humane. USAID could have purchased food from Guatemalan smallholders rather than dump cheap corn and beans on their customers. It could have milled the corn bound for Southern Africa rather than use the starving as pawns.

The United States could have supported Indigenous farmers, rather than steal their agroecological knowledge, exploit their land and labor, and promote their cultural extinction. Even if Green Revolution technologies had a role to play in feeding the world, there were alternatives to the development treadmill. Indigenous farmers could have been compensated for generations of careful seed breeding. Loans could have been offered without conditions. And research institutes could have focused on inputs already available to small farmers, rather than fertilizers made by US petroleum companies.

Disingenuous development programs squandered the genuinely humane motivations of foreign service officers, agronomists, educators, and civil servants in the United States and abroad. As geographer Jamey Essex notes, in its attempt to reconcile "numerous, conflicting objectives and priorities," USAID became "an easy target of critics hostile to aid and development."[86] Aid workers sowed seeds of self-sufficiency in rural communities only to watch *Fortune* 500 companies reap the harvest. Short-term gains became long-term losses as geopolitical and financial interests changed.[87] The United States could have supported its civic-minded ambassadors and leveraged the goodwill its aid often generated. Instead, it gambled this dividend on dictators and chemical companies.

Finally, the United States could have supported foreign self-determination. It could have granted independence to the Philippines in 1898 and backed a democratic Guatemala in the 1950s. It could have supported newly independent countries working to untangle the structural legacies of colonialism—a democratic and anti-imperialist mandate consistent with its own founding principles. Instead, the United States protected colonial relationships of exploitation—*relationships that would otherwise have changed.*

What if United Fruit had been forced to pay its taxes, or if Árbenz had succeeded in redistributing land? What if Guatemalan labor had been allowed to benefit Guatemalans, rather than being held hostage to US economic and political needs? What would Guatemala be like today if colonialism had been laid to rest in the 1950s, rather than emerge in a new skin every few decades? Would the girl in Chapin's communique still be an orphan? Of all the missed opportunities, this one looms the largest.

The United States became the world's food hegemon on a narrative of feeding a starving planet. It retained its title by claiming to modernize the world's backward farmers. Today, persistent hunger and inefficient agriculture are still used to justify foreign intervention, especially in sub-Saharan Africa, where agrochemical companies promote capital-intensive farming despite its negative impacts on long-term food security.[88] The marketplace may be more competitive now as Europe, Canada, China, and Japan claim a larger share of global exports. But the gradient of power between the world's poorest farmers and its largest agribusinesses—a gradient established by the United States after World War II—remains the same. In 1986, political scientist Susan George quipped that "hunger and malnutrition could be more easily eradicated through a redistribution of *power* than of *food*."[89] That was true for Guatemala in the 1930s, 1950s, and 1980s, and is no less true today.

• • •

On December 5, 1982, just days after the communique detailing Ríos Montt's most recent massacres, President Ronald Reagan hailed the dictator as "a man of great integrity" and "totally dedicated to democracy."[90] The next day, Ríos Montt's soldiers arrived in Dos Erres and began throwing bodies in the village well. Reagan would expand official aid to Guatemala's dictators from $11 million in 1980 to $104 million by 1986. He circumvented a congressional ban on military assistance by using USAID to launder aid: As death squads rounded up Maya into concentration camps, USAID supplied the military with "food and other material . . . to establish control."[91] Beholden to the Reagan administration's vision of a liberal capitalist world order, USAID became a jack-of-all-trades for making profitable elsewheres.

Much has changed since then. In 1991, United Fruit was renamed Chiquita; in 1996, a peace accord ended the civil war; and in 2018, Efraín Ríos Montt died. But much is still the same. The richest 2.5 percent of Guatemala's farmers still control two-thirds of the arable land, and the average smallholder owns a mere 0.8 hectares. After four decades of rural development, extreme poverty continues to expand, growing 50 percent between 2000 and 2014. In 2022, 4.6 million people—nearly a quarter of the country—experienced food shortages. Guatemalan children have the highest rate of stunting in Latin America and the seventh highest in the world.[92]

Not surprisingly, US businesses still benefit from the power gradient, and not just through exports of corn and nitrogen. With hunger on the rise, rural households in Guatemala have reprised an old standby: labor migration. Over the past two decades, many Maya have abandoned the interior for plantations in the lowlands or in southern Mexico. Others have joined the current headed for the United States.[93] Of those who survive the precarious journey, many receive starvation wages picking fruits and vegetables in Florida, Texas, and California—this time for gringo latifundistas.

# 6

# OF MEAT AND MIGRANTS

*Domesticating Humans and Animals Through Economies of Scale*

Sleep deprivation tastes like chicken. Literally.

The billions of broiler hens that get turned into nuggets, drumsticks, and buffalo wings each year spend their lives in a state of chronic sleep deprivation. Not because they doom-scroll the Internet or raise young chicks that refuse to go to bed. They get four hours of sleep each night because the lights in their coops stay on for twenty.[1] As most humans know, losing sleep is stressful. It makes us grumpy, confused, and physically sick. But stress also builds fat, and a fat chicken is a valuable chicken. They say turning off your lights can save you money; in Big Poultry, keeping them on *makes* you money.

But sleep deprivation has a downside, beyond the obvious one from the chicken's perspective. Unsurprisingly, if you cut their sleep time in half, chickens are more likely to die. As artificial daytime grows, so does the mortality rate. This might not seem like a problem; after all, death is the raison d'être for a broiler hen: Killing is an essential part of transubstantiating animals into food. But USDA regulations (and common sense) still require those bleary-eyed birds to make it to the slaughterhouse alive. If death *before* the knife is a loss—of feed, kilowatts, and profits—why do it? Because forcing chickens into a twenty-hour "workday" may increase mortality by 4 percent, but it increases weight by 10.[2]

Those chickens will encounter other sleep-deprived and stressed-out creatures on the way to your dinner table: the contracted grower struggling to break even; the catchers working twelve-hour night shifts without overtime pay; the meatpackers standing shoulder to shoulder like crated livestock performing repetitive tasks for hours without a break.[3] In its treatment of animals and humans, the meat industry has run headlong into the seemingly immutable limits of what can be extracted from living things—then found ways to exceed those limits.[4]

Of course, getting chickens to participate in the meat industry is no simple task. Nature has blessed the chicken with legs to run, claws to scratch, a beak to peck, and wings to fly (or, at least, flap). Like other animals, the chicken has a metabolic *joie de vivre* that leads it to seek more life and less suffering. It does not go gentle into that good night. For one thing, chickens do not like being handled or packed together, fighting anyone who ruffles their feathers. To keep the birds from mauling each other, some operations cut off their beaks. To keep the birds from mauling humans, catchers work at night and with the lights off; ironically, the same lights that make chickens fat also make them feisty.

Growing chickens is a balancing act: The birds need to be awake but docile, packed together but not suffocated, fattened to the brink of cardiac arrest but not actually dead. They need to be "natural" enough to be edible, and "artificial" enough to conform to the industrial processes that will dismantle them. To tread this fine line, poultry houses are designed to constrain the movement of chickens but facilitate the movement of fresh air. Contract farms are located close enough to processing plants to ensure that fewer birds die in transport, but far enough from each other to ensure that fewer neighbors complain of the smell. The geography of the chicken is a reflection not just of Big Poultry's quest for greater throughput, but also of the ways that chickens (their behavior, biology, and ecology) *resist* the demands of the system.[5]

So, too, with the chicken's human undertakers. Contracted growers go bankrupt, sue, and abandon the farm. Workers get sick, demand concessions, and strike. Managing a meatpacking operation involves maximizing the value extracted from growers and workers without driving them into debt, the picket line, or the grave.[6] To achieve this balancing act, the meat industry has reorganized the factory floor, busted unions, moved its operations, recruited vulnerable workers, skirted the law, and pushed for legislation that silences whistleblowers.[7] It hides the violence inherent in its production behind factory walls, nondisclosure agreements, detention centers, and national borders. The geography of chicken work—and meatpacking in general—reflects not only the voracious demands of accumulation but the ways humans have resisted that accumulation.

Far from an inevitable outcome, the industrialized nature of modern meat production is born of conflict: violent battles between capitalists, government regulators, workers, growers, consumers, and the animals themselves.[8] It is a conflict over geography, and one that continues today.

• • •

A chicken growing farm is a small city that goes through boom and bust every seven weeks. Nearly all US chickens grown for meat (known as "broilers") are raised by farms under contract with "integrators," agribusinesses like Tyson Foods, Perdue, and Sanderson.[9] Large contract farms may have up to six broiler "houses," each holding up to 27,000 birds at a time. The integrator's job is to supply day-old chicks and feed, the grower's job is to raise the integrator's birds, and the chicken's job is to eat and remain calm. Each house's automated feeders provide the birds with food, overhead pipes supply water, and heat lamps keep the house warm on cold days. Enormous exhaust fans suck out toxic air saturated with dust, bacteria, and ammonia from bird droppings. Think of a broiler house as an artificial guano island but with more birds per square foot. If living downwind of a broiler house is unpleasant, being inside one can be deadly.

But that is precisely where the catchers must go. At seven weeks old and weighing roughly 5.6 pounds, the chicken's metabolic work is done and a "catching crew" comes to take them away. Catchers do their work at night with the lights off, grabbing birds by the handful and putting them into cages. The cages are loaded by forklift onto a waiting truck and driven to the integrator's nearby processing plant. It takes three hours for a crew to finish "catching" a house, and skilled crews can catch three houses a night. A US Department of Labor survey found that 60 percent of integrators failed to pay overtime to their workers despite the twelve-hour shifts.[10] As you might expect, catchers are at increased risk of respiratory illness.[11]

On arrival at the plant, dead birds are removed and live chickens are counted and weighed. The count and weigh-in determine how much the grower will be paid using a formula that pits growers against each other. Meanwhile, the chicken's work is done. Each live bird is hung on a hook attached to an overhead chain that winds around the factory. A human worker cuts the chicken's throat and submerges it in boiling water. The serpentine chain then whisks the bird through various stages of disembowelment, cleaning, inspection, and disassembly. As pieces of its flesh move in different directions through a maze of machines and workers, the chicken's seven-week work product is swiftly refined into shapes and arrangements more recognizable as food. Within hours of leaving the crowded broiler house, the chicken is once again packed together with its compatriots, this time in a refrigerated truck.

As smooth as this operation may seem, the machine-human-animal interface is far from perfect. Machines break down, workers get tired, and animals resist giving up their life's work. The extraordinary pace of disassembly leaves little room for error. Sometimes workers "fail to make sure [chickens'] throats are cut before their bodies are thrown in scalding water." When this happens, still-conscious chickens "flop, scream, kick, and their eyeballs pop out of their head in pain." In beef slaughterhouses, one cow in twenty fails to be stunned on the first try, sometimes resulting in animals that get "skinned and dismembered while still conscious."[12]

Ironically, such "human" errors happen most when humans are forced to behave like machines. Factory workers spend long shifts standing shoulder to shoulder, in sweltering heat and numbing cold, repeating the same motions under the relentless demands of the chain. Breaks are few and short. Evisceration lines run at speeds of up to 140 birds per minute. Deboning and cut-up lines run at 35 birds per minute. Like the twenty-hour light cycle for broilers, high line speeds have a downside: Poultry workers experience twice as many serious workplace injuries and eight times as many work-related illnesses as the industrial average. At one South Carolina plant in 2014, 42 percent of workers suffered carpal tunnel syndrome (CTS) and 57 percent reported at least one musculoskeletal symptom.[13] Why run the lines so fast? Because forcing humans to process a bird every two seconds may increase injuries, but it also increases throughput. The average meatpacking plant generates over $700 million in annual revenue. The average company CEO takes home $5,000 an hour. Workers make 1/350 of that.[14]

This is the story of how meat animals and meatpackers came to lead similar working lives. It is the story of how a reformist president and an upstart socialist helped bring an end to the neighborhood butcher in 1906, how a geographic revolution in the 1960s altered the demographics of exploited lives, and how meatpacking tumbled from one of the best paid industrial jobs to one of the worst. At the center of this story is how the industry depends upon both the vitality of living beings and the subjugation of those living beings—a paradox it has historically resolved through geography.

The US food system has consistently moved its inherent violences into distinct elsewheres—Indian reservations, guano islands, labor camps, sharecroppers' homes, and Central American farms. The meat industry is no different, externalizing its harmful animal, human, and environmental costs within the United States and beyond. But the industry exemplifies two other spatial strategies: *aggregation*, an increase in the concentration of production and consumption, and *segregation*, the division of production into separate functional categories. Together, these complementary processes create what are euphemistically known as "economies of scale." Aggregation and segregation redirect violence into hyper-managed parts of the supply chain, between animals in the same feedlot, or workers on the same killing floor. In meatpacking, elsewhere is a place not just on the periphery, but in the dense, enervated center of the supply chain.

This chapter begins and ends with chickens, but the processes it describes are not limited to poultry. In fact, chicken is a latecomer to industrial meat production, benefiting from the best practices of beef and pork, a finely tuned distribution infrastructure, well-developed retail markets, and a compliant policy environment. As such, the story of sleepy chickens and exhausted humans is a history of meatpacking more broadly: the abuse it perpetuated, the resistance it encountered, and the way it responded. Many of the best examples come from beef and pork, commodities that set the table not only for chicken, but for the industrialization of corn, soy, fruits, vegetables, and even automobiles.

## THE NATURE OF LABOR AND THE LABOR OF NATURE

Meatpacking operations, like all capitalist enterprises, exist to accumulate wealth through a never-ending cycle of production, exchange, and reinvestment. For this cycle to continue, companies must make a profit, and to do this, they must add value through the production process. For Marx, the source of that value is labor: "[Man] develops the potentialities slumbering within nature" so as to "appropriate the materials of nature in a form adapted to his own needs."[15] Breeders add value by selecting chickens that will accumulate flesh as quickly as possible, growers add value by creating the conditions for chicken growth, factory workers add value by disassembling the animals into edible parts, and transporters add value by moving those edible parts to market. The value generated by all that labor is owned by the company, embedded in the product, and realized when that product is sold.[16]

In exchange for their labor, workers receive a wage, and with it, attempt to purchase the means of survival: food, water, clothing, and shelter. A portion of the value generated by the worker each day is needed to cover the wage. But for the company to profit, workers must generate more value each day than what is necessary to maintain their survival—they must produce a *surplus*. To maximize the extraction of surplus value, companies have forced laborers to work faster, for longer, with machines that magnify their effort. Workers come to be surrounded by devices designed to increase their throughput and minimize distractions. Companies reduce their labor costs by suppressing wages, hiring workers whose survival costs are lower, and externalizing these costs onto the state.[17] Over time, meatpacking labor has become increasingly integrated with efficiency-improving capital, even as the worker becomes increasingly isolated and desperate.

Where is the animal in all this? Does the chicken labor? Precisely what the animal *does* for capital is the subject of lively debate in academic circles.[18] Following the anthropocentrism of the period, Marx distinguished between the use values inherent in nature (rocks, trees, and animals) and the realization of those values through human labor (coal, timber, hides, and food). For Marx, labor is a uniquely human characteristic: "A spider conducts operations which resemble those of a weaver, and a bee would put many a human architect to shame by the construction of its honeycomb cells. But what distinguishes the worst architect from the best of bees is that the architect builds the cell in his mind before he constructs it in wax."[19] Others have challenged this anthropocentrism, arguing that "working" animals like oxen, horses, and dogs—creatures bred to simplify and speed up human tasks—do, in fact, labor.[20] Some have suggested that animals are not laborers because they are not paid, others offer that food and shelter serve as wages, and still others argue that the proper metaphor for animal work is not wage labor but slavery.[21]

The issue gets messier for food animals in which the creature's productive activity is essentially to eat and sleep (or not sleep, as the broiler hen will attest).[22] If the criterion for labor in the animal world is that it must resemble human work, it is difficult to account for what sedentary cows, pigs, and chickens do for capital. Likewise, if an organism must understand the consequences of its activity to be considered a laborer, as Marx's bee metaphor suggests, food animals seem more like raw materials than workers. It is tough to imagine a cow building a burger "in its mind" before it walks into a slaughterhouse.

But what if these are the wrong criteria? Rather than asking whether the activity of a chicken looks like work, we might ask, *what relationship does the chicken have to capital?* What value does it provide the company? And who (or what) is the source of this value?[23] In fact, the value of a food animal to capital is its flesh—flesh accumulated through that animal's biological work. Cultural anthropologist Les Beldo refers to this work as "metabolic labor," the process of building a body that will become an input for production.[24]

Like the human worker surrounded by machines designed to increase her throughput, the food animal is surrounded by capital designed to reduce its labor time—automated feeders, protein-rich food, and twenty-hour lighting. Like the human worker who endures long, uninterrupted hours with few breaks, the food animal is isolated from things that could distract it from production—pathogens, physical activity, and access to the enriching outdoors. Like the human worker who produces greater value than is necessary for her survival, the food animal accumulates more flesh than is necessary to sustain itself. And in both cases, capital owns the surplus; in fact, it is the demand for surplus that subjects humans to the violence of faster chain speeds and animals to the violence of faster metabolism. Neither one receives adequate compensation.

When we compare the nature of their labor, food animals and the human workers who disassemble them appear to do vastly different things. But when we compare the *relationship* that each living being has to capital, when we inquire about the role each plays in generating surplus value, the food animal and the human worker stand together. Down on the killing floor, sharp knifes intercede between their respective working lives.

But the food animal has a leg (or hoof or claw) up when it comes to job security. Whereas the human meatpacker has been steadily replaced by more efficient and less argumentative machines, the food animal remains essential. The chicken's metabolic labor does one thing that no human or machine can accomplish: It produces flesh. This ability to turn soy and corn feed into fat and muscle is at the center of capitalist meat production, but also beyond its reach.[25] As the magician responsible for this trick, the broiler hen's continued existence is secure. At least for now.

## LIMITATIONS AND RESISTANCES

In the pursuit of profit, companies try to maximize throughput; that is, greater surplus value extracted per unit of time and space. In row agriculture, this means getting more plants to grow per acre and per season. In meatpacking, it means getting more animal parts distributed and sold, more animals processed per factory and per worker, more pounds of flesh per animal, and more animals per year, cubic foot, and pound of feed. Because moving animal flesh around takes time and kilowatts, greater throughput often entails more workers per unit of space on the factory floor.

To achieve this, the industry relies on packing animals into concentrated animal feeding operations (CAFOs) where they can be fed and managed for lower cost. Companies pack as many workers as possible into common structures and place them around long, serpentine disassembly lines to enable more hands and flesh to share machinery. Slaughterhouses are centrally located to reduce the costs associated with moving live animals and butchered meat. As much as possible, meat is transported in bulk along publicly funded infrastructure. The shared capital that supports this aggregation is said to produce "economies of scale," the special sauce that turns run-of-the-mill accumulation into runaway accumulation.

But strategies to increase throughput often meet resistance. Nature is notoriously uncooperative with capital. Animal growth rates reduce the risk of early death but lengthen the time before slaughter. Confined together, animals mate, give birth, and fight, complicating their management. They become anxious at the sights and sounds of the killing floor and fight back. Their bodies are not all the same size and shape and can foul intolerant machinery. Microbial undertakers begin breaking down flesh the moment it dies, leaving traces that can sicken consumers.

Capitalist agriculture has attempted to address each of these "resistances." Animals are bred to accommodate the machines that will process them. Selective genetic traits speed up metabolic labor time for food animals while refrigeration slows it down for microbes. Cow testicles and chicken beaks are removed, lights are dimmed, and—thanks to animal welfare efforts—killing floors now conceal the sights and sounds of slaughter behind curved walls.[26]

Strategies to increase throughput also encounter resistance from human laborers. The human body can perform only so much work before needing to rest and refuel. A wage can be pushed only so low before the worker's survival is threatened. Humans will accept only so much demeaning and punishing labor before they lose motivation, get frustrated, fall physically sick, and die. Some have demanded changes in the industry and have organized with other humans to obstruct production. Workers are also consumers who make food choices, vote, and write letters to influential people. Like chickens, humans have a *joie de vivre* that leads them to seek more life and less suffering.

To overcome these human resistances, companies have threatened layoffs, attacked picketers, blacklisted union leaders, and silenced whistleblowers. Companies have moved factories to places where policies are more welcoming, workers are more desperate, and unions have less sway—a spatial fix for violence. Within the factory, the industry has worked hard to divide its workforce along racial, ethnic, gender, and functional lines, then pitted these groups against each other. If aggregation creates economies of scale, segregation protects it. As thousands of nearly identical animals wait to be slaughtered, thousands of human workers, systematically divided by language and country of origin, wait to dismantle them.

### HUB OF THE WHEEL

Maximizing throughput is easier in some parts of the supply chain than in others, thanks to the kinds of limits each productive activity faces and the dangers involved in exceeding those limits. Raising animals is risky business. Coyotes, foxes, and racoons feast on chickens. Cold snaps freeze cattle, and dry summers desiccate them. Livestock diseases tear through CAFOs. Raising animals also involves substantial fixed capital in the form of land, a commodity vulnerable to changing markets and interest rates. Since the mid-1800s, periodic boom and bust cycles have sent the livestock sector from feeding frenzy to bankruptcy. Growers have a great deal to lose from things beyond their control. At the other end of the supply chain, retailers face similar challenges: considerable fixed capital in the form of warehouses and storefronts vulnerable to changing land values. Being at the end means the wholesale prices they pay reflect all the bottlenecks and surpluses in the rest of the chain. Like growers, retailers have much to lose from things outside their control.

The middle of the supply chain is where risks are lowest and where technologies for overcoming resistance can be best applied. Refrigeration and publicly funded highways and railroads allow production to be centrally located. Government subsidies like tax breaks and welfare and immigration programs allow companies to reduce overhead and target cheap labor. Over time, meatpackers have colonized the supply chain from the middle out, pushing growers and retailers toward the precarious ends in a process called "vertical integration." Today's contracted chicken growers own nothing but the land and the broiler house. The integrator owns the chickens, the feed, the catching and transport labor, and the meat. In some cases, the integrator also controls the supply chains that produce feed and fertilizer.[27]

The middle of the supply chain also benefits from an increasingly influential force—financialization. Political economist Jennifer Clapp notes how financial markets have substantially increased the number of actors in the food system: institutional and private investors, banks, and financial services firms.[28] These globe-spanning actors, often with little knowledge of—or material connection to—the system, direct financial resources into these "middle spaces" where it benefits large agribusinesses at the expense of producers and consumers.

As the hub of the wheel, meatpacking fortunes stay relatively stable even as fluctuations in prices send growers and retailers into the red. The COVID-19 pandemic was a crisis for growers and retailers, but a boon for large meatpackers. As packing plant bottlenecks slowed the supply of meat, wholesale prices rose 101 percent. As demand for animals fell and CAFOs swelled, cattle prices dropped 5 percent. The beneficiaries of higher wholesale prices and cheaper cattle were the big meatpacking companies. Being at the hub of the wheel allowed companies like JBS, the largest meatpacker in the world, to benefit from a crisis of their own making.[29]

Vertical integration is often presented as a defining characteristic of industrial agriculture, the inevitable result of unceasing demand for greater throughput. But the evolution of a vertically integrated supply chain was far from inevitable. In fact, corporate control over the middle of the supply chain ebbed and flowed in response to key moments of resistance from workers, consumers, politicians, competitors, and unruly animals.[30] This is where the story of meat and migrants begins.

## THE FARMER TAKES THE LAND

Like so much else in US agricultural history, the story of modern meatpacking begins with genocide. The metabolic machinery of billions of animals requires a steady source of food. Depending on the species, eventual use, and stage of its life, a food animal may graze on living fodder or eat meals made from a variety of harvested crops. But in the end, it is *land* and its vegetable bounty that sustains meat. Today, 80 percent of US agricultural production goes to feeding food animals—half a million square miles of corn, soy, alfalfa, and hay.[31]

Before this land could support meatpacking capital, the United States had to overcome the resistance of its traditional caretakers. The settlers that established Western ranches in the mid-1800s provided provisions for US Army expeditions against Arapaho, Cheyenne, Comanche, Kiowa, Lakota, and others. Some cowboys and ranchers fought alongside military units, while others launched their own campaigns. Armed gangs targeted people and settlements, but also food sources, especially plains bison, a creature at the center of Indigenous dietary, spiritual, and economic life. "The quickest way to compel the Indians to settle down to civilized life," General William T. Sherman explained in 1869, "was to send ten regiments of soldiers to the plains, with orders to shoot buffaloes until they became too scarce to support the redskins."[32] By the 1870s, starvation and disease forced many groups into devastating land concessions, ultimately "transforming the Great Plains ecosystem from a grass-bison-nomad system to a grass-cattle-rancher one."[33]

With ranches came railroads. In 1830, there were 100 miles of railroad in the United States. By 1860, the country boasted 30,000. Early legislation supporting rail construction—like the Railroad Act of 1862—fueled the violent conversion of bison lands into cattle lands. Railroads carried feed to hungry animals, animals to slaughterhouses, and meat to hungry urbanites. By the end of the Civil War, the expanding rail network linked ranches in the Western plains with a growing urban population in the East.

Live animals carry more than the flesh that sustains the meatpacking business. A thousand-pound steer contains 450 pounds of flesh-making infrastructure: bones, organs, and fluids that net far less money. To reduce shipping costs, companies established slaughterhouses in Chicago, Kansas City, and Cincinnati—centralized locations between the pasture and the dinner plate.[34] To prevent spoilage, slaughtered animals were transported in railcars packed with ice. But ice is also heavy, and it wasn't until the late 1870s that innovations in refrigerated transport allowed meat processed in the Midwest to reach Eastern cities in large quantities. For the next two decades, railroads were so essential for the movement of live animals and processed meat that railroad companies could name their own freight prices (and they did).

Prior to the 1870s, butchering was a skilled trade developed over years of practice and passed down to successive generations. To increase throughput and avoid the cost of skilled laborers, meatpacking companies pioneered the disassembly line, a long chain that carried carcasses from the killing floor around the factory through an array of low-skilled, repetitive operations. Industrial histories often attribute the modern assembly line to Henry Ford, but Ford credited Chicago's meatpacking houses with the innovation: "The first step forward in assembly came when we began taking the work to the men instead of the men to the work. . . . The idea came in a general way from the overhead trolley that the Chicago packers use in dressing beef."[35]

Cities offered packing companies a bottomless supply of cheap labor. On Chicago's South Side, livestock fresh off the rails awaited slaughter by workers fresh off the boat from northern Europe. Animal waste overflowed into enormous, fetid lagoons around and between the overcrowded dormitories of immigrant workers. And each morning, hundreds of desperate men and women lined up outside the recruiting offices of Chicago's big meatpackers hoping for a few hours of paid work.[36]

The work they got was deadly. On the killing floor, where the steamy air often exceeded 100°F, workers moved and carried heavy loads across floors slick with blood and effluent. In other areas, temps below 40°F congealed fat on the walls and floors. Wastewater leaked from overhead pipes. Slippery conditions in dark rooms led to puncture wounds from knives as tired workers moved swiftly about the floor. The rank air bred respiratory illnesses like tuberculosis and pneumonia, repetitive motions led to rheumatism, and heavy carcasses made for debilitating back injuries. Those working in curing departments lived with open sores from constant contact with salt.[37] A Chicago packinghouse worker in the late 1800s might outlive the animals he slaughtered, but not by much.

From the plant, processed meat rode the rails in ice-cooled comfort toward Eastern cities. Before the Civil War, most meat animals were reared within a few hundred miles of where they were consumed. Urbanites ordered cuts from neighborhood butchers who received their animals from nearby farms. But as technological innovations in transport overcame microbial challenges, Chicago beef started to undercut the urban market. Meatpacking companies contracted directly with retailers to cut out small-scale, mostly immigrant butchers. By the 1880s, a pound of beef raised on Texas grass could be sold in Boston for less than what a local butcher could offer.[38] In 1860, over two thousand butchers supplied the million residents of New York City with meat; by the end of the century, the city had grown to 3.4 million, but only a few hundred butchers remained.[39]

On December 28, 1890, three hundred starving Lakota men, women, and children were gunned down by US soldiers at Wounded Knee, marking the culmination of US imperialism in the Great Plains. That year, a record 2.2 million cows passed through Chicago's disassembly lines, and four meatpacking companies—Armour, Swift, Cudahy, and Wilson—succeeded in capturing nearly 90 percent of the country's beef market.[40] But as railroads, rifles, and refrigeration overcame continent-sized challenges and propelled industrial meat production to new heights, the industry also made new enemies.

## THE MONOPOLY PICKS A FIGHT

By 1900, Chicago's "Big Four" meatpackers had established the blueprint for what would become the centralized, vertically integrated, high-throughput production system we see today. To accomplish this feat, the largest packing companies formed a collaborative known as the Beef Trust (though "cabal" might be more accurate). The trust coordinated attacks on competing companies and industries, illegally fixing wholesale prices, designating retail territories, and blacklisting butchers who purchased from other packinghouses. To control the cost of animal feed—thereby suppressing the price of cattle—the trust purchased grain elevators and invested in wheat futures.

By the 1890s, the Beef Trust was powerful enough to take on the railroad companies that had controlled shipping prices. Refrigerated transport was expensive: Specially designed, insulated rail cars needed access to ice, and meat had to be carefully loaded and unloaded to ensure it did not spoil. At first, railroad companies burdened with the cost of building new lines were content to let meatpackers supply the specialized cars. Under private car contracts, the Big Four received a discount on freight rates for meat shipped in these "trust cars." Railroads began using trust cars to carry fruits, vegetables, and other perishables from California and Florida, and with each load the trust received a cut. Later contracts obligated railroads to transport all perishables in trust-owned cars and required them to pay a surcharge whenever a car was moved, even if it was empty. In 1905, US railroads moved 390,000 tons of refrigerated fruit and paid $25 million in car surcharges to the Big Four.[41] With the proceeds, the trust invested in political influence.

In the 1905 exposé *The Greatest Trust in the World*, investigative journalist Charles Edward Russell described the Beef Trust as

> a power greater than the government, greater than the courts or judges, greater than legislatures, superior to and independent of

> all authority of state or nation. . . . It owns steam and electric railroads, it owns the entire trolley-car service in several cities, and is acquiring the like property elsewhere. It owns factories, shops, stock-yards, mills, land and land companies, plants, warehouses, politicians, legislators, and Congressmen. . . . It terrorizes great railroad corporations long used to terrorizing others. . . . It fixes for its own profit, the prices the farmer of the West shall receive for his cattle and hogs, and the prices the butcher of the East shall charge for his meat. It fixes the price that the grower of California shall receive for his fruit, and the price the laborer of New York shall pay for his breakfast.[42]

Half a century before the United Fruit Company controlled the fortunes of Central American economies, the Big Four—Chicago's own El Pulpo—controlled the fortunes of growers, ranchers, factory workers, butchers, and consumers from California's Central Valley to Boston's Quincy Market. And like the Guatemalans who struggled under United Fruit Company control, the public did not take this lying down.

In fact, Russell's exposé tapped into deep and long-standing public frustrations over the dominance of monopolies in general: The Grange, an early farmers' advocacy group, was formed in the 1870s in part to protest railroad price-fixing; Texas cowboys struck in 1883 to protest wage suppression by powerful ranchers; and meatpacking workers struck in 1904.[43] By then, populist anger was increasingly directed at the Beef Trust. Chicago's Big Four defended their image, portraying centralized packing operations as modern, efficient, and essential for lowering food prices. Small-scale packers countered that their operations were "family centered, nonindustrial, and authentically American."[44] In cities across the country, local butchers launched campaigns to paint Big Four meat as inedible and unsafe, potentially contaminated by inattentive workers, and vulnerable to spoilage on the long train ride from Chicago.

But fears of moldy, tainted meat sickening millions of unsuspecting consumers did little to slow the growth of the Big Four. The Beef Trust may not have been popular, but it could call its own shots. US demand for meat was at an all-time high and growing. Politicians adopted—or resigned themselves to—"a consumerist mentality that prioritized cheap beef over price fixing, worker exploitation, and butcher displacement."[45] As the price of calories decreased, the surplus value that could be extracted from each urban laborer grew. Cheap meat was good for economic growth, and not just in the United States. Millions of pounds of canned beef left Chicago each year to satisfy soldiers in Europe's imperial battlegrounds, particularly in the tropics where refrigeration was a challenge. Canned beef fed US soldiers suppressing independence movements from the Philippines to the Caribbean following the Spanish-American War. As ships carried Western-raised beef to Europe, European bankers invested in Western ranches.[46] British-controlled tea and sugar may have defined globalization in the eighteenth and nineteenth centuries, but American red meat defined it in the early twentieth.

Public outrage alone could not deter Chicago's Big Four. It did, however, attract the attention of two people who would pose the next obstacle to unfettered capital: an upstart socialist named Upton Sinclair and a fanboy of Western ranching named Theodore Roosevelt. Their unlikely partnership would substantially expand the role of government in meatpacking and inadvertently seal the fate of those they sought to protect.

## THE PRESIDENT READS A BOOK

Package a few million cans of beef under sketchy conditions and at some point one or more of those cans will catch up to you. That is exactly what happened in 1898 when Army Major Nelson A. Miles—dispatched to Cuba to fight the Spanish—became convinced that canned beef was sickening his soldiers. Nelson claimed that nearly 200,000 pounds of beef sent to Cuba and Puerto Rico to feed fighters was tainted with waste. Newspapers like the *New York Journal* wrote exculpatory stories "designed to humiliate not only the meat packers but also the official charged with provisioning US troops."[47] To get ahead of the firestorm, the US Army launched an investigation. One of the witnesses who gave testimony in support of Miles was then-Colonel Theodore Roosevelt, and even though the final report cleared the beef industry of wrongdoing (and punished Miles for going public), the process left Roosevelt "harboring deep suspicions about the safety of industrial beef production."[48] Two years later, he was elected president and quickly set his sights on the Beef Trust.

But like the US Army investigation, the USDA agents Roosevelt sent to Chicago did not find anything incriminating. It was, after all, not in the USDA's interest to draw scrutiny to its most productive sector. The president would have to wait until his second term before taking action. By then, public outrage was once again directed at the Big Four, this time because of a book. Upton Sinclair's *The Jungle* took the public on an unflinching tour of Chicago's meatpacking plants through the eyes of immigrant workers. Sinclair, a socialist, wanted to expose industrial violence to public scrutiny, "a crime," he explained, "committed in a dungeon, all unseen and unheeded, buried out of sight."[49] With memorable prose, Sinclair described the brutality of packing labor under the unrelenting demands of throughput: "The speeding-up seemed to be growing more savage all the time; they were continually inventing new devices to crowd the work on—it was for all the world like the thumb-screw of the medieval torture-chamber."[50] Sinclair noted how visitors to the plant were awed by the speed and efficiency of production, but unsettled by its unsanitary conditions and horrific treatment of animals: "It was porkmaking by machinery, porkmaking by applied mathematics. And yet somehow the most matter-of-fact person could not help thinking of the hogs; they were so innocent, they came so very trustingly; and they were so very human in their protests."[51]

It would have been hard for Roosevelt to avoid reading *The Jungle*: The book was wildly popular, and for his part, Sinclair made sure to send Roosevelt a copy. Though the president strongly disagreed with Sinclair on socialism, he was appalled by the description of conditions in Chicago and saw an opportunity to protect ranchers and small businesses from the monopolistic excesses of the Beef Trust.[52] In March 1906, Roosevelt invited Sinclair to the White House and asked him to join a new team of investigators. This time, the final report corroborated Sinclair's evidence and gave the administration enough leverage to crank start the legislative machine.

Sinclair and Roosevelt were a political odd couple: The author hoped to end capitalism through a workers' revolution; the president hoped to *preserve* capitalism by regulating it. The Beef Trust provided a common enemy. But in the end, neither got their wish. The Pure Food and Drug Act and the Meat Inspection Act both passed in 1906. Together, these laws prohibited the adulteration and false labeling of meat transported across state lines and authorized federal officials to ensure sanitary slaughtering and processing conditions. But the acts imposed no such restrictions on meat raised, slaughtered, and consumed *within a state*. The effect was to bolster public trust in centralized meat production but erode trust in local butchers. After all, Chicago's factories were now regulated, but no one was watching the neighborhood market.[53] Over time, the same regulatory oversight actively opposed by the Big Four helped eliminate their competition. As urbanites enjoyed government-inspected meat, the animals and workers whom Sinclair hoped to protect got essentially nothing. The frustrated revolutionary would later lament, "I aimed at the public's heart, and by accident I hit it in the stomach."[54]

The new laws did little to limit the reach of the Beef Trust. By 1919, Big Four companies owned shipping infrastructure in 25,000 towns and cities, including 90 percent of railroad branch houses and car routes.[55] Trust companies slaughtered over 94 percent of all cattle and 81 percent of all pigs in the United States. By leaving the trust's expansive network in place, Roosevelt inadvertently created barriers to competition. And not just in meatpacking: Big Four companies used their networks to market lard, hides, milk, margarine, coffee, oil, and

**Figure 6:** Union Stockyards, Chicago, circa 1947. Photograph by John Vachon. Courtesy National Archives. NAID 541823.

even soft drinks, pushing out smaller competitors.[56] The Packers and Stockyards Act of 1921 forced meatpackers to abandon these non-meat products, but as before, it left the networks intact. By 1937, the Big Four controlled 78 percent of the total value of meat products in the United States, not because it won the battle against regulation, but because it lost.

With the collapse of local butchering, urban neighborhoods lost a highly skilled and distinctly immigrant trade. As shops shuttered, industrial meatpacking opened its doors: The place for the country's newest arrivals would not be at the helm of their own businesses, but on the killing floor of someone else's.

## THE WORKER GOES ON STRIKE

As the Big Four consolidated control over the middle of the supply chain, company recruiters combed Chicago's immigrant neighborhoods in search of cheap labor. The late 1800s saw meteoric growth in US cities driven by immigration from southern and eastern Europe. In 1840, Chicago was a small town of 4,400 people. By 1890, it was the second largest city in the country with over a million, and by 1920, it boasted 2.7 million, more people than live there today.

Ethnic bases in the Midwest drew families escaping violence and unemployment in Europe toward Big Four recruitment offices. Capital aggregated around labor power, and labor aggregated around capital. Growth in the number of workers did not translate into better working conditions. A 1940 government study found twice as many disabling injuries among meatpackers as the industrial average, with one in five killing floor workers experiencing a disabling injury each year.[57] So long as the line outside the factory remained long, conditions inside remained dangerous.

Aggregation allowed workers to talk, overcome ethnic and religious differences, and commiserate. Big cities may have offered packing companies attractive labor surpluses, but they also offered workers access to unions and political groups like the Communist Party.[58] In the 1910s and 1920s, working class frustration over factory conditions led to frequent strikes. During the 1918 flu pandemic, packed conditions on the factory floor facilitated disease transmission. Heavy mortality among the working class may have fueled subsequent labor actions: There were 3,600 strikes across the United States in 1919, with over 4 million participants (one out of every twenty-five people in the country).[59]

To discourage labor mobilization, companies divided workers along ethnic, racial, and gender lines. In Chicago, older immigrant stock (German, Irish, and Scandinavian) were given higher-paid nonproduction jobs, while eastern and southern Europeans (Slovaks, Polish, Croatians, Serbs, Italians, and Greeks) received lower-paid skilled positions on the killing floor. Black men were limited to killing floor jobs, rendering, and work in the hide cellar for little pay.[60]

US-born white women could be found in sliced bacon, the cleanest department and the one most frequently exhibited to visitors. But the worst jobs and lowest pay went to Black and immigrant women, who "trimmed the meat off animal heads, cut out the tongues . . . washed out the feces by pulling the intestine over a water pipe," stood all day in 40°F water, and performed other jobs that men refused to do.[61] To endure the cold, some created makeshift leggings out of gunny sacks. To endure the odor of offal and excrement, women in some departments "learned to breathe exclusively through their mouths."[62] And to keep their jobs, women invented ways to navigate pervasive sexual harassment. For all this, they earned 30 percent less than men, and the threat of replacement by a cheaper female worker was frequently used to keep restive male workers in line.[63]

These processes of *segregation* allowed meatpacking companies to pit workers against each other in competition for better wages and working conditions.[64] Even when workers succeeded in organizing, management found help beyond the factory. When skilled white workers struck in 1894 and 1904, companies recruited thousands of Black men as strikebreakers. To the resentment of its white workforce, the Big Four opened offices in Chicago's segregated neighborhoods to hire newly arrived Southern Blacks. It is not surprising that 1919—a year that witnessed widespread strikes—also saw race riots in Chicago.[65] When the strikes ended, few Blacks were allowed to keep their jobs. Hoping to escape racial terror lynchings and the Klan in the South, Black migrants suddenly found themselves enlisted as pawns in a class war between whites.

The Big Four's power over labor would start to wane in the 1930s, thanks to two structural developments. The first was the Wagner Act of 1935, which established the National Labor Relations Board (NLRB) and guaranteed workers the right to engage in collective bargaining. As the Great Depression brought industrial working conditions under greater public scrutiny, unions took advantage of a more welcoming policy environment and pushed back.

The second development was the realization among union leaders that the only way to overcome labor segregation was to bridge racial and ethnic barriers. Unions took advantage of aggregation on the factory floor to build camaraderie between German and Polish, Irish and Italian, white and Black workers. Meanwhile, unions like the International Union of All Workers (IUAW), founded by Minnesota meatpackers, took advantage of urban aggregation to forge alliances across industries, a model pioneered by the International Workers of the World (IWW).[66] These efforts were far from perfect—it took years for unions to bridge the gender gap—but the "inclusive unionism" of the 1930s through 1950s ushered in an era of better conditions for many packinghouse workers.

Inclusive unionism allowed leaders to win "model contracts" that governed wages, benefits, and chain speeds not just in individual factories but across the industry. In the 1960s, model contracts ensured that meatpackers were making 15 percent more than the average for industrial workers. Unionization gave workers a sense of not only security, but pride. Labor historian Roger Horowitz writes, "At Chicago's Armour plant, 'You felt like you were a slave,' recalled union founder Sophie Kosciowlowski. 'I wanted to feel like I was a human being, that I had some dignity.' In Austin, a white Catholic recalled that 'the morning when I put on the union button, I felt like a man when I walked through the gate; before I felt like a dog.'"[67] But as it had with resistant railroads in the 1890s, meatpacking companies responded to union resistance with a new, geographical innovation. Unions won their battles in the 1930s through 1960s thanks to strong public support; in response, capital increasingly turned to people who lacked it.

## THE COMPANY MOVES AWAY

During agriculture's golden years (1900–1918) rural, midsized farms benefitted from high demand and an expanding railroad network.

But over the next three decades, many were thrown off the agricultural treadmill as large agribusinesses profited from capital-intensive inputs. By midcentury, small towns from Iowa to the Texas panhandle were losing people and livelihoods.

In the early 1960s, Iowa Beef Packers (IBP) became the first major meatpacking operation to move production out of the city and into the countryside. What became known as the "IBP revolution" saw companies take advantage of a labor force that had never been unionized: poor, rural whites. For local governments, the arrival of meatpacking plants offered a tantalizing source of income; for packing companies, it offered access to public funds earmarked for "community development."[68]

Over time, companies set up shop in "right-to-work" states where laws discouraged unionization.[69] Factory floors became increasingly automated and workers' roles even more specialized. Process standardization allowed companies to temporarily shift production from one plant to another in the event of a strike, without retooling or retraining workers. As historian Joshua Specht notes, this made meatpacking "bigger than any particular geographic place," undercutting one of the primary tools by which unions had secured improved conditions.[70]

Meanwhile, the carousel of vulnerable labor kept turning. When the poultry industry shifted from the Mid-Atlantic to the rural Southeast after World War II, it hired a predominantly Black and female workforce. But when Black women began unionizing, companies recruited labor from other sources.[71] In the 1970s, IBP targeted refugee camps holding Vietnamese and Laotians who had escaped the Vietnam War. In the 1980s and 1990s, companies targeted Hispanics. Recruiters set up offices in Mexico, broadcast ads on radio stations heard across the border, and paid one-way bus fares to factory towns. Later, when Mexican workers began unionizing, packing plants pivoted once more, taking advantage of thousands of Central Americans fleeing civil war and poverty.[72] Guatemalans who had lost their land to a US fruit company in the 1930s and their children to

US-backed death squads in the 1960s, lost their grandchildren to US meatpackers in the 1990s.

By the early 2000s, half of the workers employed in meatpacking were undocumented, a vulnerability that played into company hands.[73] Managers used the threat of deportation as a weapon against unionization and as "[retaliation] against women who sought to organize to fight sexual harassment and assault."[74] But by 2006, political discourse and public sentiment had turned against undocumented workers, presenting another challenge for capital. Between 2008 and 2011, Immigration and Customs Enforcement (ICE) agents deported 1.5 million people. In the largest immigration raid in US history, agents rounded up 12,000 meatpacking workers at six Swift and Company plants. When ICE raided slaughterhouses in Texas, Iowa, Nebraska, Utah, Colorado, and Minnesota, agents separated parents from children and forced many into hiding.

In *The Mayans Among Us*, Ann Sittig and Martha González provide a wide-ranging ethnography of Guatemalan workers at a Nebraska packing plant: Mayan women who had escaped the bodily violence of the country's civil war and the economic violence of DR-CAFTA. Through their interviews, Sittig and González observed the deep psychological wounds exposed by immigration enforcement: "Martha sadly narrated that during the [ICE] raids many Mayans fled to the cornfields to hide in fear, just as they had done when their communities were burned and massacred in Guatemala. The ICE raids in the meatpacking plants have devastating effects on Mayan immigrants in rural Nebraska, creating panic and massive fleeing, a Mayan exodus that recalls the genocidal civil war for them."[75]

Crackdowns on undocumented workers and public antagonism toward Latin American immigration pushed companies to look for other labor sources. In the late 2000s, JBS began recruiting newly arrived refugees from East and Central Africa. Resettlement agencies "struggling to place newly arriving refugees in jobs that did not require English language skills . . . were eager to send refugees to JBS and Cargill."[76] Since then, people escaping violence in Somalia, Ethiopia, South Sudan, and Myanmar have been at the top of the meat industry's hit list.

As they were in 1900 when German, Lithuanian, Irish, and Serbian immigrants crowded their blood-slick floors, meatpacking plants today are some of the most diverse workplaces in the country. In 2024, 51.5 percent of all meatpacking workers were foreign-born, three times the average for the general workforce. In some plants, 90 percent are foreign-born and as many as thirty languages can be heard in the breakroom.[77] As it did in the early 1900s, the industry still segregates, using "race, ethnicity, gender, and immigration status as ways of dividing up tasks, sowing discord among workers, breaking unions, and ensuring a continuous supply of replaceable labor."[78] The strategies are often subtle. At one plant, "managers fostered distrust . . . by enclosing the paychecks of African refugee workers in envelopes, while leaving those of Latino workers in the open. . . . Many Latino workers assumed that the African workers were receiving a higher wage for the same work, although this was not actually true."[79]

These nefarious strategies work. At a Colorado JBS plant in 2008, Somali workers requested a change in break times to observe prayers during Ramadan. When the company refused, the workers struck. JBS relented and rescheduled one of the shift breaks by fifteen minutes. But when Hispanic workers complained, JBS reneged on its agreement. Eighty Somali workers who left the line to pray were promptly fired. Following the incident, JBS began recruiting Southeast Asian refugees of Karen, Chin, and Rohingya backgrounds. Like the Somali workers they replaced, Rohingya are Muslim. But plant managers figured that Rohingya (victims of genocide in Myanmar) would be less likely to complain, for fear of losing "their green cards, TANF benefits, and Section 8 housing vouchers."[80] Stress makes chickens fat, trauma makes workers docile.

Meanwhile, meatpacking remains one of the most unpleasant jobs in the country. Outside the plant, workers take whatever cheap housing they can get, including "dilapidated mobile homes rented from the slaughterhouse companies . . . [and] abandoned railroad cars and shanties."[81] Inside the plant, workers are subjected to high line speeds under conditions eerily similar to the ones that precipitated the great strikes of the 1910s and 1920s.

One thing did change in the intervening years: wages. Under the union contracts of the 1930s through 1960s, meatpacking was one of the best paid industrial jobs in the country. But IBP's escape to the countryside signaled the long decline of union power. Before rural relocation, workers in urban plants averaged $30,000 a year; after relocation to "right-to-work" states, some workers saw less than half of that. In 2024, the average manufacturing job paid $29.51 an hour; the average meatpacking job paid $18.[82]

For many migrants and refugees, $18 an hour does more than feed their households; it supports an extended network of people in their home countries through remittances. These transfers of money are now a significant contribution to foreign economies. Between 2008 and 2011, $40 billion in remittances were sent from the United States to Guatemala—more than the total of Guatemala's annual exports and tourist revenue.[83] Sittig and González note how Mayan meatpackers in Nebraska live in fear of temporary layoffs caused by outbreaks of *E. coli* and mad cow, and public concerns over "pink slime." Without paychecks, the women must put wire transfers on hold and rely on social services to feed their households. Thanks to the meat industry and Western Union, the welfare of distant *campesinas* is now hitched to Midwestern livestock diseases and consumer fears.

The IBP revolution did more than disarm labor unions. The shift to rural meatpacking coincided with unprecedented grain surpluses, the result of advances in chemical fertilizers and pesticides. As the country looked for ways to dispose of its stockpiled corn, a meat industry anchored in expansive rural feedlots offered a solution.[84] With more space, lower population density, and fewer environmental regulations, rural production encouraged enormous aggregations of grain-consuming animals. On the upside, Chicagoans today can eat steak without smelling the stockyard. On the downside, rural CAFOs are ecological disasters, producing 130 times more excrement than the human population (Victor Hugo would have been impressed). Those living near CAFOs suffer higher rates of gastrointestinal and respiratory ailments.[85]

Today's packing plant may be in flyover country, but it occupies the gravitational center of the meat industry. Pulled into its orbit are animals and humans—both migrants of a sort. The former are drawn in under the compulsion of chutes and prods, the latter under the compulsion of poverty and desperation.

## THE CHICKEN PECKS BACK

The industrial model that beef and pork pioneered in the late nineteenth century would be perfected by poultry in the mid-twentieth. Large-scale chicken farming officially began on Maryland's eastern shore in 1927. By the early 1930s, the region was producing seven million broilers a year. In 2017, large poultry operations—those that raised over 100,000 chickens per year—accounted for 99.6 percent of all broilers in the United States, a feat accomplished by an unprecedented aggregation of metabolic laborers.[86] A large beef CAFO may contain 1,000 cows; a large hog CAFO may hold 5,000 pigs. A single chicken house can top 30,000 birds.

Aggregation increases throughput but introduces problems of its own. Packing chickens together indoors keeps predators away and prevents the birds from wasting metabolic labor on exercise, but it also keeps them out of the sun. Chickens need sunlight to make vitamin D for bone growth. By the 1930s, farmers were giving chickens cod liver oil and later vitamin-enriched feed. Experimentation in feed was big business: Companies like Pillsbury and Ralston Purina captured the market by funding research into recipes that accelerated flesh accumulation. In 1927, an average broiler weighed 2.5 pounds after sixteen weeks; by 1941, broilers weighed 2.9 pounds after twelve weeks. Over that period, the industry halved the amount of feed necessary to grow a pound of flesh.[87]

Keeping chickens indoors also means inventing novel ways to control airflow and waste. Cram enough warm-blooded animals into a small space and the result is an abundance of heat. Today's chicken and pig CAFOs require large ventilation systems to prevent the animals from dying of heatstroke. In some places, nearby row agriculture provides a market for the nitrogen-rich guano that chicken houses generate. Sanderson Farms, the third largest poultry producer in the United States, effectively owns a share of the nitrogen cycle by making chicken feed from grains fertilized by the company's own chicken houses. But this symbiosis cannot be reproduced everywhere. In North Carolina's Lumbee River Basin, aggregated waste from hundreds of Sanderson CAFOs is sprayed onto a three-hundred-acre plot of land bordering an ecologically important wetland. At an EPA-approved rate of 1.4 million gallons per day, the operation effectively doubles the region's annual rainfall. But unlike rain, this CAFO "irrigation" contains "pathogens . . . and residual feed additives, including pharmaceuticals and heavy metals."[88]

Packing chickens together leads to fighting. Fighting leads to wasted calories and dead birds. To protect profits, it is common for growers to remove the beaks of layer hens, a painful process that can lead to sensory loss. A $100 machine can slice through 1,800 beaks an hour. In cattle operations, it is common for growers to cut or burn off the animals' horns, sometimes without anesthesia. On some dairy farms, cows' tails are amputated to prevent the spread of disease to handlers during milking, a process that involves "acute pain . . . the possibility of chronic (long-term) pain in the tail stump . . . and altered ability of the cow to avoid flies."[89]

To enable more consistent machine processing at the plant, chicken bodies are bred to be genetically similar. But genetic similarity, crowded CAFOs, and continuous air circulation conspire to spread communicable diseases. Thanks to aggregation, the meat industry has given the microbial world an extraordinary gift: a vast, warm petri dish filled with identical media in which to perfect its DNA and mingle with other microbes. Today, CAFOs are prime suspects in the emergence of novel, hyper-transmissible pathogens.[90] In response, the meat industry is prodigious in its use of prophylactic antibiotics, but this has the paradoxical effect of making pathogens more resistant.[91]

When human and animal disease outbreaks do occur, the industry has two very different plans. On the human side of the knife, meatpacking plants embrace a "work if you're sick" culture.[92] During the COVID-19 pandemic, repeated denials by company officials, failure to implement Centers for Disease Control and Prevention (CDC) guidelines, and a presidential order to stay at work created ideal conditions for transmission. Inside the plants, "close proximity on the line, cold and humidity, infrequent access to bathroom breaks," lack of sick leave, and inadequate health care conspired to sicken workers. By May 2020, rural counties with meatpacking facilities had infection rates five times higher than the national average.[93] Nearly 60,000 packing plant workers fell ill in the first year—people like Miska Jean Baptiste, father of three and a poultry worker in Maryland:

> [Miska] didn't realize he'd contracted Covid-19 in April [2020], when he began to feel feverish. The plant checked his temperature, but his widow . . . says they didn't inform him about his fever; instead, they gave him ice cream to reduce the reading. . . . This went on for four days; the plant kept it secret, she says, to prevent other workers from getting scared and leaving work. . . . When Baptiste finally went to the doctor, his temperature was 105 degrees. . . . In the hospital, he went into a coma, and was put onto a ventilator; he passed away alone. . . . After his widow told her story to the media, the company reached out. "They tell me they're going to send an envelope." She pauses. "It was a card, and $100 cash . . . I was so mad."[94]

On the animal side of the operation, the industry's plan for disease outbreaks involves far more than ice cream and hush money. When a pathogen is detected in a CAFO, animals are executed en masse to prevent the disease from spreading to other locations. The culling of animals as a disease firewall is now a global strategy. In 1996, four million cows were killed to prevent the spread of Bovine spongiform encephalopathy (BSE). Millions of birds were killed in Southeast Asia and Canada to prevent the spread of avian flu (H5N1) in 2004, and again in 2015 when so many chickens were killed in the Midwest that "disposal of them became an environmental hazard."[95] Every pig in Egypt was killed in 2009 to prevent the spread of swine flu, and in 2018 African swine fever (ASF) killed or led to the culling of a quarter of the world's hogs.[96]

In an age of fast animal growth and even faster pandemics, new technologies have emerged to kill quickly, sterilize pens, and refill CAFOs with a fresh cohort of animal laborers. Through a devilishly clever repurposing of life-support systems, one culling strategy involves turning off the CAFO's ventilation systems and letting the animals cook themselves with their own body heat. In experiments perfecting this technique, known as "ventilation shutdown" (VSD), hens were observed "struggling as they died over the course of 1.5 to 3 hours."[97] VSD is sometimes aided by the addition of $CO_2$ or steam to speed up death, presumably reducing animal suffering, but conveniently reducing lost moneymaking time as well. In fact, technologies like VSD are not restricted to disease outbreaks, but often used on perfectly healthy animals that are no longer profitable to sell. Ten million pigs in the United States were culled during the COVID-19 pandemic, in some cases through VSD—not because they were unhealthy, but because selling them would have been unhealthy for the bottom line.[98]

Rapid contagion, antibiotic resistance, fighting—these obstacles to capital arise not from a pristine or primordial Nature (capital N), but from the nature of accumulation. Beak removal, tail amputation, and VSD are responses to the industry's self-inflicted wounds. Rather than challenge the CAFO, the industry invents new technologies to maintain its violent animal geographies: new genes, new antibiotics, and new ways to render the animal docile. Domestication is not a fixed quality of the cow, pig, or chicken, but an *ongoing process to extract more surplus value.* Domestication is also imperfect. As Dinesh Wadiwel writes, "Animals press against, disrupt, and leak value from even the most apparently complete and relentless models of authoritarian subordination that we can devise."[99] It is this act of pushing back that has forged the meat industry into its current form.

Unlike domestication, there is already something perfect about the chicken: It can grow flesh. The meat industry may make the labor of its human workers increasingly obsolete through automation, the labor of local butchers obsolete through centralization, and the labor of growers obsolete through contract farming. But so long as the food system continues to consume meat, the industry will continue to rely on the vital, irreplaceable metabolic labor of food animals. The chicken has the most secure job in the business.

Or does it? It turns out, food animals may soon get laid off from the job of producing their own flesh.

## THE CELL STANDS ALONE

If you thought the chicken's metabolic magic trick was a good one, wait till you see what Uma Valetti, CEO of Upside Foods, can do. In 2023, Upside Foods became the first company to receive government approval to sell "cultivated meat," animal flesh grown outside of an actual animal, variously known as "lab-grown," "cultured," or "vat-grown" meat. The work of Upside Foods is part of the emerging field of cellular agriculture, a synthesis of advances in chemical, genetic, and biological engineering. In contrast to plant-based and fermented meat alternatives, cultivated meat involves growing flesh from animal stem cells in a bioreactor. The cells are bathed in a liquid medium that provides the necessary nutrients for metabolism. Once grown, flesh cells are structured around a gelatin or soy-based scaffold that provides texture and recognizable shape.[100]

According to Valetti, the big upside of cultivated meat (hence the inspiration for his company's name) is the "opportunity to preserve the choice of eating meat . . . at the same time we preserve life."[101] Indeed, cellular agriculture is a response to growing public concerns over traditional meat production, in particular issues of animal welfare, environmental degradation, and water use. Rural CAFOs are at the center of these concerns. With flesh being grown in "clean" laboratories, cellular agriculture promises a recentralization of production to cities (Upside Foods' facility is fifteen minutes from San Francisco). Valetti insists, "You should be able to have [cultivated meat production] in your neighborhood without having a feedlot or a slaughterhouse," an echo of Chicago's concerns a century ago.[102]

Of course, cultivating meat in the suburbs still relies on expensive and environmentally harmful inputs—enormous feedstocks of glucose, sucrose, nitrogen, and phosphorus that must be sourced elsewhere.[103] It also requires plenty of energy, far more than what keeps broiler hens awake. In a familiar refrain, Valetti insists that input costs will decrease as production scales up.

Meanwhile, public attention has focused on taste and nutritional value. The new industry goes to great lengths to assure the public that there are only upsides to this revolution, promising that you can have your meat and your ethics, too. To be sure, there appear to be sizeable benefits in replacing the known suffering of CAFO animals with the ethical gray area of a bioreactor. But is this, as cellular agriculture claims, a *revolution* in the meat industry?

As unique as cellular agriculture seems, the fundamental relationship between the cell and capital has not changed. To extract greater surplus value from metabolic labor, cultivated meat outsources the chicken's productive infrastructure—circulation, respiration, digestion, and waste removal—to a sophisticated mechanical-molecular partnership. Like the contract farmer whose job is increasingly surrounded by the aggregator's property, the cell is increasingly surrounded by the company's scientific capital. Like the Big Four that took distributed production and centralized it under one roof, cultivated meat takes flesh once distributed across multiple animals and aggregates it into vats. And like the IBP revolution that moved production out of cities to reduce union labor costs, cultivated meat has moved the cell out of the chicken to reduce *metabolic* labor costs. Cultivated meat may have accomplished a twenty-first-century revolution in tissue engineering, but it did so to perfect a nineteenth-century dream of vertical integration.

Cellular agriculture offers a chance to overcome the resistance of unruly animals, concerned environmentalists, and restive workers all in one fell swoop. Without legs to run, beaks to peck, and claws to scratch, disembodied flesh is rendered docile to the machine and worker (who now wears a lab coat). Without a stressed-out chicken, the vociferous indignation of welfare activists fades. With neither a body to dismantle nor a degree in bioengineering, the meatpacker (whose birth was not governed by supply and demand) must look elsewhere for survival: Upside Foods, like its Chicagoland forebears, does not specify how or where these workers should look. Presumably they will be absorbed into other elsewheres.

What seems a fresh idea with revolutionary potential is in fact grown from the stem cells of Big Agriculture. It is no surprise that Upside Food's investors include Tyson Foods, the largest poultry company in the United States; Cargill, the largest agribusiness in the world; and Bill Gates, the largest landowner in the country.[104]

## AGGREGATION, SEGREGATION, AND THE VIOLENCE OF SCALE

The history of meatpacking is replete with spatial fixes, from the conquest of the plains to the conquest of railway branch houses, neighborhood butcher shops, and rural towns. But it also exemplifies *aggregation* and *segregation*, two other spatial strategies that have allowed the industry to pursue greater throughput and manage resistance. Aggregation is an increase in the density or concentration of production and consumption. Aggregating systems like the CAFO or disassembly line pull more animals and workers toward the same fixed capital, increasing economies of scale. These systems also introduce new *violences of scale*: exposure to injury and disease, chronic illness, and isolation from loved ones; for the CAFO-bound, the loss of beaks, horns, tails, and sleep.

Aggregation is accompanied by segregation, the division of production and consumption into distinct parts. By dividing the workforce into separate tasks, companies reduce the need for skilled labor and limit worker mobility. By segregating workers along linguistic, ethnic, racial, and gender lines, companies undermine collective action. Segregation and aggregation are complementary: Precarious labor is pulled from elsewhere, confined to a controlled space, then categorized in ways that magnify differences and breed animosity.[105]

## FRONTIERS OF ACCUMULATION

Before they filled out CAFOs and killing floors, animals and humans occupied other spaces—spaces that are now drastically changed. Aggregation decoupled animals and their metabolic circulations from grazing land, and workers from their home communities. In each case, new industries emerged to profit from the rift: Mills turned Midwestern corn and soy into animal feed, banks sent money home, "coyotes" trafficked in vulnerable laborers.

To keep the cycle of accumulation going, the meat industry must continue to bring new territories into its orbit, like the plains cleared of Natives and bison, or the labor surpluses of Latin America recruited for the line. Pushing back these geographic *frontiers* is seen as a defining characteristic of capitalism, which tries to appropriate and transform whatever it encounters into a commodity.

One such frontier of accumulation is the body.[106] Chicago companies tried to wring more value from each working body before it broke down and pitted racialized ones against each other to break down picket lines. Geneticists redesigned the animal body to give it more flesh at the expense of life, and biochemists designed hormone supplements to accelerate the body's metabolic engine. Today, tissue engineers are domesticating the chicken out of its own existence. And surrounded by diets high in fat and sugar, consumers' gut biomes are increasingly tuned to the frequencies of the industrial food system, craving its endorphin-releasing commodities.[107] Even our colons—metabolic extensions of the disassembly line—are busy working for capital.

## MISSED OPPORTUNITIES

The history of meatpacking is sometimes presented as an inevitable consequence of modernization, a logical evolution toward greater efficiency. Such a story seems filled with geographic coincidences: CAFOs and grain surpluses, the IBP revolution and the collapse of rural economies, falling wages and incoming refugees. Of course, these are not coincidences but *connections*, the result of conscious decisions to fix capital and labor in new arrangements. Joshua Specht notes that these outcomes were not preordained, but the result of myriad "individual choices and hard-fought policies."[108] And that means there were alternatives.

Teddy Roosevelt could have dismantled the Big Four's network rather than give them a weapon to use against their competition. Meaningful decentralization during this period could have improved working conditions by forcing more employers to compete for labor, improved animal welfare by discouraging the shift to CAFOs, and prevented future supply bottlenecks.[109] The government could have used regulation to protect small processors, rather than expose them to monopoly power as it did in 1906. Instead, the trend toward centralization in meatpacking continues as smaller operations fall through the grate: In 1980 there were 1,400 beef processing plants in the United States; in 2024, there were fewer than half that number. Today, four companies slaughter 85 percent of grain-raised cows and control 70 percent of total beef production, numbers that have barely changed since Teddy Roosevelt read *The Jungle*.[110]

The United States could have supported its working classes, rather than allow agribusinesses to undermine unions in the 1970s and 1980s. Labor and immigration policies could have aligned to provide meaningful security for the country's foreign-born meatpackers, rather than trapping them between exploitative employers and ICE. Political leaders could have fostered constructive discourses on immigration rather than weaponizing it for political gain. Today's meatpackers are trapped in an abusive relationship with a public that alternately ignores them, despises them, and insists they are critical infrastructure.[111]

These alternatives would have involved baking more humane relations into the system—between cells, animals, workers, consumers, and the capital that sits at the center. This would have meant countering the spatial strategies of capital, perhaps slowing down the chain, making and eating less meat. To do this, policymakers would have had to envision alternatives to industrial mechanization, rather than assume its inevitability.[112]

## SLEEPY CHICKENS AND THE PURPOSE OF THE FOOD SYSTEM

So long as we presume the food system's purpose is to produce *food*—and the meat industry's purpose is to produce meat—we cannot explain why chickens are systematically deprived of sleep. Does exhaustion make a chicken taste better or cook more readily? Does it cut better? Absorb flavors better? Pair with wine better? So long as we see the chicken only as food, we cannot explain why its beak is removed, why—if it is male—it is strangled at birth, or why it and millions of its perfectly edible compatriots are intentionally culled and composted before they reach our stomachs.

So long as we assume the food system's purpose is to nourish *life*, we cannot explain why packing companies keep pushing up chain speeds as their workers' bodies break down. Neither can we explain why some refugees receive their paychecks in envelopes and others do not, or why—as meat industry profits have grown—those paychecks have shrunk.

To the meat industry, the chicken and the worker provide surplus value, and what the industry does at any given point is designed to maximize that value. Life (a prized moral category) and food (a substance necessary for life) are bound into these transactions, but neither life nor food is the *reason* behind the meat industry's choices. We know this because of how the industry responds when forced to choose between profit, food, and life. Facing supply chain bottlenecks, the industry dumps edible milk into rivers and buries healthy pigs in the ground.[113] Facing workers who wish to pray, the company hands out pink slips and recruits others to take their place. Facing agitated birds, the industry deploys the debeaking machine.

The implications for building a nonviolent food system are profound. You cannot make meat production less violent with a quieter killing floor or a better-smelling CAFO. You cannot make it less violent by giving workers more breaks or simply paying them more. Resistance is always met with innovations—technologies, policies, and practices—that embrace *new* violences and relocate old ones. Whatever you do, new strategies to increase throughput will emerge. Like different skins on the same old snake, new workers will be recruited from new elsewheres, new factories will open in places more friendly to exploitation, and new broiler houses will pack a bit more flesh onto feeble birds.

• • •

Back at the packing plant in Nebraska, Sittig and González offer a glimpse into the inner lives of the Mayan women they work with: "Their dreams are basic, their characters profound, and their manner is dignified . . . [they] would live and work anywhere, in any job, to ensure their children have a better life than they did struggling in poverty in Guatemala." To keep this dream alive, the women work together and look out for each other. "Some of the Guatemalans have started growing food and sharing it with others, taking up the practices from back home, and thus creating a new model."[114] The spirit of sharing that inspired the 1976 reconstruction committees in Guatemala lives on in Nebraska. Precious seeds smuggled across the border and planted in the fertile soil of Section 8 housing have sent up new shoots.

To protect these dreams, the women resist talking about problems at the packing plant or with the meat industry in general. When the lives of relatives back home depend on the next paycheck, arguing over animal ethics seems a pastime of privilege. "They appreciate having the meat and don't harbor negative sentiments about the conditions at the meatpacking plant," write Sittig and González.[115] They fear plant closures, slowdowns, and layoffs. Like precarious workers across the sector, they worry about the relocation of the plant to another town or country, the arrival of new populations able to work for less, and the potential for wage increases to attract workers whose language and skin color make the public more comfortable. Like other agricultural workers, migrant meatpackers fear that their ethnicity may be the next one in the crosshairs of the country's freewheeling xenophobia. They know their dreams rely on elsewheres far beyond Nebraska.

This is the paradox of a capitalist food system: Ending one violence inevitably creates another. A century since the rise of labor unions, the promise of structural reform in the meat industry never truly changed the violence experienced by workers, it merely changed which people and places experienced it. Before we know if a less violent food system is possible, we need to know if making the food system less violent is a zero-sum game. And if it is, does it have to be?

# 7

# HUNGER IS NOT A POLYGON

## *Making Place for Malnutrition on the National Map*

Through the summer and fall of 1906, patients at the Hospital for the Colored Insane in Mount Vernon, Alabama, developed unusual rashes on the backs of their hands, feet, and necks. Physicians assumed a bad case of sunburn, but when the insides of their mouths broke out in painful sores, they knew something was awry. Over several weeks, profuse salivation and persistent diarrhea sapped victims of strength. They became listless, disoriented, obstinate, and depressed, unwilling to eat and unable to get out of bed. By the end of the year, eighty-eight patients had been sickened by this mystery ailment and fifty-seven had died. From its distinctive skin rashes, mouth sores, and cognitive changes, a medical team led by Doctor George Searcy diagnosed the condition as acute pellagra, a disease of unknown origin.

The name pellagra comes from the Italian *pelle agra* meaning "rough skin." First identified in Spain in 1739, pellagra spread throughout southern Europe during the eighteenth and nineteenth centuries. By the 1880s, hundreds of thousands of peasants around the Mediterranean were affected. One of the earliest and most persistent theories for pellagra's origin blamed spoiled corn. High in carbohydrates and yielding more calories per acre than wheat, corn transformed the diet of Europe's peasantry when it was first introduced from the New World. But as hotspots of pellagra spread across northern Italy, doctors began to suspect that a toxin, bacteria, or fungus in spoiled corn was sickening people. In the 1890s, the Italian physician Cesare Lombroso claimed to have successfully extracted the toxin and used it to induce pellagra in healthy subjects. But consensus in the European medical community remained elusive.[1]

Before the Mount Vernon outbreak, pellagra was essentially unknown in North America. Only a handful of cases had ever been reported: a sailor in the 1880s, a Georgia farmer in 1902, and isolated cases in psychiatric wards. But the sudden appearance of eighty-eight cases in one place caught the attention of health officials. As Searcy studied the disease, he adopted Lombroso's theory of toxic corn and warned that many more deaths were to come. After all, cornbread and grits were centerpieces of nearly every Southern dinner table.[2]

Searcy was wrong about the mechanism, but right to suspect diet. He was also right about the death toll: Over the next two years, mental hospitals, prisons, and orphanages from Texas to Florida reported an alarming number of cases. Before long, pellagra appeared in the general population. It ravaged the crown jewels of the industrializing "New South"—cotton milling and mining towns where poor families lived in company dormitories and bought food from the company store. Between 1907 and 1911, eight Southern states reported 15,870 cases. Nearly four in ten died. By 1920, there were 100,000 new cases per year and pellagra became the fourth leading cause of death from disease, after tuberculosis, pneumonia, and typhoid fever.[3] As researchers argued over its cause and cure, "pellagraphobia" gripped the nation. Healthy people flooded doctors' offices convinced they had contracted the disease. Some chose suicide over fears of a pellagra diagnosis.[4]

Pellagra would affect a disproportionate number of Black people, women, and poor laborers. It would infect patients but not their nurses, prison inmates but not guards, some family members but not others. It plagued certain company towns, farming districts, and upper-class households, but not all. The hardest hit towns were located near railroads and surrounded by cotton fields, while out-of-the-way farming districts stayed pellagra-free.[5] This uneven geography would eventually lead researchers to the cause, but only after years of confusion and acrimonious debate. Meanwhile, Southern leaders, pundits, and eugenicists warned against "pellagrous districts," unhygienic neighborhoods, and the "constitutionally deficient" people who occupied them. Some Northern social commentators, "noting the debility of its slow-witted victims," referred to pellagra as the "Southern Scourge" and the "lazy" disease.[6]

Long after the cause had been identified and treatments perfected, pellagra continued to claim lives, eventually affecting over three million people and killing one hundred thousand. In the late 1930s, public health campaigns and improved treatments brought the death rate down, but it was a change in the food system that ultimately ended its epidemic status. This is the story of how pellagra's origins were uncovered: a tale of tortured prisoners, starved dogs, and injured pride. It is the story of extraordinary geographic insights and entrenched geographic prejudices. Long before the advent of GIS, medical researchers were drawing lifesaving connections between cases and places. They were also forging powerful ideas about people and the communities they live in, ideas that continue to influence the where, how, and why of public health.

• • •

In the century since the mysterious disease appeared at Mount Vernon, other diet-related epidemics have taken place. Today, heart disease is responsible for 20 percent of deaths in the United States and type 2 diabetes affects over 13 percent of adults.[7] Since the 1980s, the increasing prevalence of these diseases has been linked to a so-called obesity epidemic. Between 1980 and 2023, the country's obesity rate rose from 15 to 42 percent, and with it calls for bans on soft drinks and fast food, more bike and walking paths, and increased nutritional education.[8]

Like pellagra, heart disease and diabetes are not distributed evenly, with higher rates in inner cities, rural towns, and mountain hollers. Like pellagra, these diseases persist among marginalized populations: predominantly Black, Indigenous, and Hispanic, but also poor whites. And like the pellagrous districts of the early twentieth century, today's policymakers trace obesity to distinct places—so-called food deserts where healthy foods are hard to find. According to the USDA, approximately 10 percent of US census tracts qualify as food deserts, home to 23.5 million people.[9] Here, too, geography tells a story, one that influences how we understand the causes of malnutrition and where we look for a solution. Like the case of pellagra, powerful geographic ideas may be doing more harm than good.

This is the story of how the food system continues to *make place* for malnutrition: how dietary policy, industrialized agriculture, and the pursuit of greater throughput can introduce novel diseases; how those diseases come to be associated with certain people and places; and how those people and places are stigmatized with the qualities of the disease. It is also the story of how capital benefits from these pathologies—in the "cure" for pellagra and, a century later, in the supposed cure for food deserts.

So far, most of the violence described in this book has been directed at food producers: farmers and farmworkers, sharecroppers and meatpackers, Navajo shepherds and Mayan smallholders. Much of the discussion has examined how violence emerges from structures: policies, laws, markets, and treaties. This chapter explores how geographic ideas, metaphors, and ways of seeing are deployed to *justify* those structures by directing attention away from causes, inviting harmful interventions, and foreclosing on beneficial alternatives.

## THE "SCOURGE OF THE SOUTH"

When Searcy first reported on the outbreak at the 1907 meeting of the Alabama Medical Society, he reiterated Lombroso's theory of a toxin or biological pathogen in damaged corn. Daily meals at Mount Vernon consisted of the "three Ms": meat, meal, and molasses. Cornmeal mush, syrup, salt pork, and dried beans were staples in institutions throughout the South and among the poor. The nurses, however, ate much less cornbread and supplemented their meals with outside food. Searcy noted that none of the nurses contracted pellagra, despite having "handled the patients [and] slept in the halls near them."[10] On this basis, Searcy ordered cornbread and grits removed from daily meals and replaced with wheat bread and potatoes. No new cases emerged. As a control, Searcy kept eight patients on the old diet: All eight ended up in "poor general health" and one contracted pellagra.[11]

For the next three years, the spoiled corn theory dominated. Among Alabama's doctors, there was growing consensus that pellagra was caused by "eating damaged maize . . . which has not matured, not been properly dried or meal which has not been kept dry after grinding."[12] The lowest grade cornmeal was also the cheapest cornmeal, a relationship that explained higher incidences of pellagra among the poor.

But the spoiled corn theory had problems. Not everyone who ate the same cornmeal got sick. Some who ate very little cornmeal still got sick. The disease cropped up consistently in spring when diets were improving, and not in winter when diets were at their worst. Several observers noted that pellagra suddenly appeared in places where diets had recently improved, while other places with poor diets remained pellagra-free.[13] If the disease was caused by bad corn, there had to be more to the story.

Perhaps the biggest problem with the spoiled corn theory was a political one. Postbellum champions of the New South promoted a region ripe for industrial investment. By 1900, civic boosters were touting the success of industrialization with the rapid growth of textile milling and mining. But by 1910, pellagra was running free in the South's mill and mining towns—hardly a ringing endorsement. Furthermore, any public backlash against cheap corn could have serious economic consequences. In the Alabama Medical Association meeting of 1910, Dr. Stephen Hale connected the dots for his peers: "We use corn because it contains about six times more fat than the average cereal we use, and fat, especially in the poorer classes, is needed a great deal, and with it we are able to maintain the proper amount of calories necessary to keep the economy in proper working condition."[14]

This period marked the height of the Lost Cause narrative, a retelling of Southern history that downplayed slavery and elevated the Confederacy as a heroic attempt to defend the Southern way of life. As health officials struggled to understand pellagra, Southern politicians commissioned monuments to Robert E. Lee and Nathan Bedford Forrest, founder of the Ku Klux Klan. To blame pellagra on poverty—as spoiled corn theorists did—was to indict the socioeconomic policies of the post-Reconstruction South. By nearly every social, economic, and educational measure, the South lagged far behind: Pellagra would now be added to slavery and racial terror lynching as subjects of Northern scorn. As embarrassment turned to anger, Southern boosters went after the messengers: Through the 1910s, researchers suggesting a dietary origin for pellagra were publicly humiliated. Many lost their funding. Some were fired.[15]

The stigma of a socioeconomic cause led some states to withhold reporting on pellagra cases. In fact, despite its statistical association with the South, the disease was widespread in the Northeast, Midwest, and West. Stigma also motivated the search for new theories. For some, the presence of rashes on the backs of the hands, neck, and feet suggested exposure to the sun. The correlation between cases of pellagra and hookworm led to theories that the two were related.[16] Entomologists were called in to investigate whether biting flies and mosquitoes could be vectors, along with bedbugs and fleas. Even mangy dogs did not escape suspicion.[17]

One dietary theory blamed sugar. "I claim that certain products of sugarcane contain an active poison," wrote Dr. Ray Blosser, "and that all that is necessary for the development of pellagra is the consumption of a sufficient quantity of these poisons."[18] To be sure, it was difficult to tease apart the influence of sugar and cornmeal on pellagra, given that molasses and cornbread were frequently consumed together. Ultimately, Blosser's sugar theory did not pass scrutiny, but his explanation for the disease's sudden appearance was astute: "The comparative modernity of pellagra . . . is accounted for by the coincident change in the method of refining sugar whereby large quantities of cheap sirups . . . have been thrown on the market."[19] Blosser may have gotten the key ingredient wrong, but he got part of the recipe right.

As progress on dietary theories languished, another explanation grew in popularity. By 1910, health officials were turning their attention to *contagion*—the possibility that pellagra resulted from an "infectious epidemic germ" passed from one person to another, much like the familiar diseases of tuberculosis, typhoid, and cholera.[20] The contagion theory helped explain clusters of pellagra in households, towns, and institutions. Infectious diseases like malaria and yellow fever were associated with particular climates and proponents of the contagion theory saw a connection between the warm weather of the Mediterranean (where pellagra was endemic) and the balmy South. Debates raged between the two camps until a well-funded medical commission attempted to set things straight.

## THE THOMPSON-MCFADDEN COMMISSION

In 1912, with the financial backing of Robert Thompson (a mining magnate) and Henry McFadden (a wealthy cotton trader), a team led by Drs. Joseph Siler and Philip Garrison of the Armed Services Medical Corps embarked on a systematic study of pellagra in Spartanburg County, South Carolina. Over the course of two years, the Thompson-McFadden Commission examined 847 cases in thirty

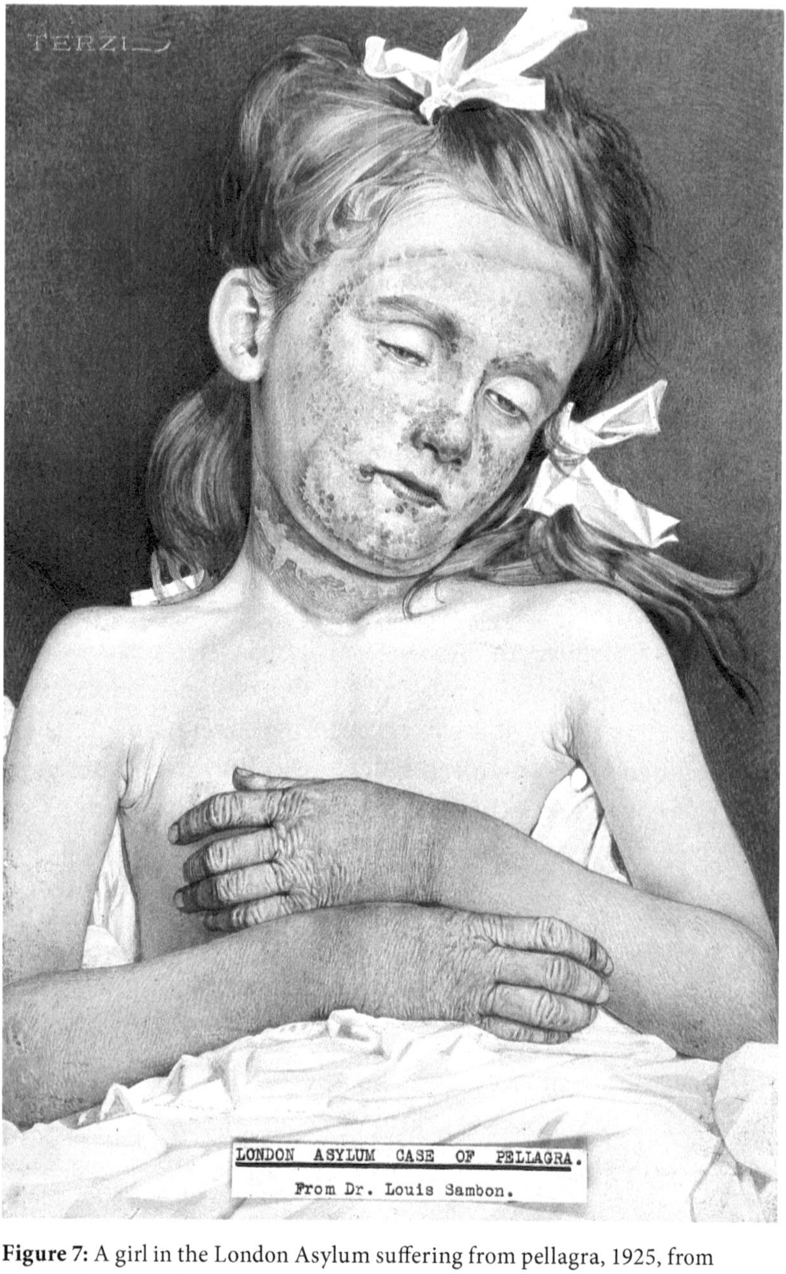


**Figure 7:** A girl in the London Asylum suffering from pellagra, 1925, from Dr. Luis Sambon. Watercolor by A. J. E. Terzi. 41705i. Courtesy Wellcome Collection.

cotton mill towns. Among its findings, rates of pellagra were higher in urban areas than in rural areas despite diets being roughly the same. Overall, Black people and white people showed similar case rates. In the mill towns, the highest rates were among younger women, followed by men over fifty. And in contrast with earlier studies, the commission found households with multiple cases to be quite common. But the commission's most controversial finding concerned the connection between victims: 80 percent of cases had a "close association" with a previous case of pellagra.[21]

Siler's team concluded that contagion was the likely cause: "Pellagra is in all probability a specific infectious disease communicable from person to person by means at present unknown."[22] The commission suspected that unhygienic conditions played a role, tracing higher incidences to neighborhoods where residents used "surface" or "pail" privies rather than sewers. As for the higher rates among young women, Siler speculated that the infectious agent could live "in the soil surrounding the immediate household, perhaps the garden or flower yard."[23] "We can absolutely exclude corn meal as the essential cause of pellagra," Siler declared at a meeting of the Medical Association of Alabama, "there is no question that pellagra most frequently attacks people who are below par . . . [and] does not confine itself to the lower classes of the population."[24] Politicians and civic boosters could breathe a sigh of relief: A nefarious germ was to blame—not a nefarious society. Southern pride had been restored.

Siler's insistence that pellagra "attacks people who are below par" was less of a clinical conclusion than an assumption necessary to explain the data. Differences in vulnerability to infection were crucial for explaining how patients got sick but their caretakers didn't, how young women fell ill but their husbands stayed healthy. How, despite lower rates of infection, Black people had higher rates of mortality. Assumed differences in group vulnerability played into discourses of race and gender. The infamous eugenicist C. B. Davenport even weighed in on the report:

> That there should be a correlation between mental insufficiency and pellagra is not strange, since the mentally insufficient are, on the whole, less likely to appreciate the importance of sanitary surroundings and less able to avail themselves of them. . . . Persons who are mentally well developed are, on the whole, more likely to care for their bodies and keep themselves in good condition than are the mentally deficient or unstable. Other things being equal, pellagra is more liable to make headway in "Nam Hollow" than in the cottages on the cliffs at Newport.[25]

For Davenport and other proponents of the contagion theory, it was not just the people who were sick, but the places they occupied. Deficient genes made for deficient people, deficient people made for deficient places, and deficient places—the unsanitary, blighted, and impoverished neighborhoods of Spartanburg County—made people sick. Proponents of the contagion theory rested their case on the commission report, wielding it "like a well-worn catechism of faith."[26] With corn no longer in the frame, antagonism turned toward the presumed source of contagion: immigrants from Italy. Perhaps ironically, Italian immigrants to the South did not have pellagra. Nor did they eat Southern cornbread.[27]

## DINNER AT THE GOLDBERGERS'

Broad support for the contagion theory did not solve its fundamental problems. Other than a correlation with open privies, no vector had been identified (though Siler was open to several possibilities). The theory's reliance on gender or racial vulnerability was both unscientific and dubious. And contagion theorists could not explain why pellagra was often cured by substituting cornbread in the diet with meat and dairy—a fact that rural doctors had noted from the outset.

In 1914, the surgeon general appointed Dr. Joseph Goldberger to lead a US Public Health Service (PHS) investigation into pellagra. From the start, Goldberger focused on Southern asylums, orphanages, and prisons where he became convinced that poor diets were to blame. Goldberger's team tested the nutritional theory by reducing the amount of cornmeal and increasing the amount of milk for a group of orphans. Existing cases of pellagra cleared up and no new cases emerged. Wherever the "Goldberger diet" was implemented, case rates began to decline.

But Goldberger's success did little to convince his powerful detractors. Doctors like J. F. Yarbrough warned that adopting Goldberger's prescriptions would be "crucifying their patients upon a cross of error."[28] Contagion theorists accused Goldberger of trying to undermine the economic development of the South. Chris Leslie, a historian of science and technology, writes, "Those who advanced the infectious paradigm held so much power at the time that they were able to impose their beliefs over existing knowledge."[29] Despite Goldberger's insistence that they were on the right track, his team struggled to secure continuous funding.

In 1915, an irritated (and perhaps desperate) Goldberger changed tactics. If curing pellagra through diet did not convince his peers, maybe *inducing* pellagra would. In one horrific experiment, Goldberger subjected twelve healthy male convicts at a Mississippi prison farm to a strict diet of biscuits, gravy, cornbread, grits, and syrup. Within weeks, half of the test subjects showed signs of pellagra. In excruciating pain, prisoners begged to be released from the study; some even asked to be killed. Goldberger had proven his point, but the validity of his claims was drowned out by public outcry and accusations of torture. Even Goldberger's longtime supporters began to abandon him.[30]

Undeterred, Goldberger doubled down and sought to discredit the contagion theory directly. In another audacious experiment, he injected himself, his wife, and his colleagues with blood drawn from pellagrins, then hosted "filth parties" in which he and his colleagues "wrapped skin scales, feces, dried urine, and other dirt from pellagrins in dough and swallowed repeated doses."[31] Goldberger's special dumplings gave his guests nausea and diarrhea, but no one came down with pellagra. The doctor saw this as proof that pellagra was noncontagious, but in the end, those who "partied for science" at the Goldberger house probably suffered in vain. Beyond turning the stomachs of his medical contemporaries, Goldberger's brazen experiments did little to sway public opinion.

In 1916, Goldberger teamed up with statistician Edgar Sydenstricker to conduct a new study of Spartanburg County's mill towns, this time focusing on what residents ate and where they acquired their food. Mill towns were an ideal subject of study: Company stores maintained meticulous records of who bought what, giving Sydenstricker a clear picture of household diets. In urban mills where workers relied on the company store, Goldberger and Sydenstricker found a strong positive correlation between poverty, poor diet, and pellagra. The poor, with no other option, ate the only food they could afford at the store: cheap Midwestern cornmeal brought in by rail. But in rural mills where food from the store was supplemented with locally grown vegetables and dairy, pellagra was less prevalent.[32]

Not only did the data suggest a nutritional deficiency, Sydenstricker suspected that pellagra was connected to a broader change in Southern agriculture. Between 1900 and 1914, cotton acreage increased 30 percent as plantations took over non-cotton areas. As the rural poor abandoned their farms for jobs in mill towns, they inadvertently traded a traditional (and more diverse) diet for imported grain. If Sydenstricker was right, pellagra was not an unfortunate coincidence of a modernizing "New South," it was *caused* by it.

Goldberger's and Sydenstricker's findings were met with outrage. "Zealots of food deficiency as the sole cause of pellagra," wrote Dr. Seale Harris, "have . . . left the impression that the South is backward and poverty-stricken," thereby hindering the "agricultural development of the most fertile lands in the United States [and] preventing many industries from locating in the favored South."[33] By blaming pellagra on poverty, and poverty on agricultural change, Goldberger and Sydenstricker had implicated King Cotton. For some Southern leaders, this was an unforgiveable heresy.

Southern pride would come at a cost. In the early 1920s, as agricultural prices fell and rural poverty set in, Goldberger and Sydenstricker warned of increasing pellagra rates. In 1927, when flooding hit the Mississippi Delta, the PHS advised that food aid was needed to prevent a pellagra outbreak. In both instances, Southern health departments ignored the warnings, and between 1925 and 1930, deaths from pellagra quadrupled.

## THE CURIOUS INCIDENT OF THE DOGS THAT ATE YEAST

Goldberger would eventually turn his attention toward the search for a "pellagra preventive factor" (PPF), a dietary additive that could prevent the disease. His work took an unexpected turn in 1916 when nutritional researchers at Yale University reported that some of their canine test subjects developed hemorrhages and mouth sores, a condition known as "black tongue" disease. Around that time, Goldberger learned of a curious geographical connection long observed by Southerners: Wherever people had pellagra, dogs developed black tongue. Goldberg theorized that dogs, well-suited to cleaning table scraps, were inadvertently picking up the same pathogen that caused pellagra or being deprived of the same vital nutrients as their human companions. When he learned that hunting dogs belonging to an employee at the Georgia State Asylum had developed black tongue, Goldberger took note.[34]

Per a long-standing (and mistaken) belief that dogs are keener hunters when hungry, the dogs at the asylum were being deliberately starved, their rations cut in half and supplemented with cornbread from the asylum's kitchen. Over time, the dogs started refusing their meals: Black tongue makes it painful to eat, much like the mouth sores characteristic of pellagra. To stimulate their appetites, some of the dogs were given brewer's yeast, an old breeder's trick. Remarkably, those that received brewer's yeast were also cured of black tongue. Wondering if the cure might translate to pellagra, Goldberger dosed his patients with yeast, and like the dogs, they recovered.[35]

It turns out, pellagra is not born of a pathogen, nor a toxin lurking in "damaged" or "low quality" corn. The disease is caused not by the *presence* of something pernicious but by the *absence* of something necessary: vitamin B3, or niacin. Grains like wheat and corn contain small quantities of niacin, but yeast and milk have it in abundance. Meat and dairy also contain tryptophan, an amino acid from which the liver can synthesize niacin. When doctors noticed pellagra in institutional residents but not the staff, it was due to the staff getting niacin from other sources. When Siler observed higher incidences of pellagra in women than men, he was probably seeing differences in the allocation of niacin-rich food within the home—with men eating a greater share of the family's meat and dairy. Siler failed to distinguish women who were pregnant or breastfeeding, biological processes with high niacin demand. And when he saw a connection between pellagra and unhygienic neighborhoods, he was overlooking the intervening variable of social class and its connection to diet.

Using a treatment regimen of brewer's yeast, a high-protein diet, and (after 1937) synthesized niacin, the nutritional clinic at Birmingham's Hillman Hospital began treating a steady flow of patients. Sensitive to their financial circumstances, director Dr. Tom Spies offered free transportation to and from the clinic and established an outreach program to rural households. Decades after Goldberger observed hunting dogs with black tongue, Spies's rural outreach program still used the condition of the family dog as a "clinical test" to determine if pellagra was present.[36] By 1945, the annual death toll had fallen to 865 from a peak of nearly 7,000 in 1928.

But one mystery remained: pellagra's timing. "The disease did not occur in the South before the Civil War," observed Dr. Sydney Leach in 1909, "notwithstanding the fact that nearly all of the negroes lived on corn bread and bacon."[37] Only a handful of cases had been identified before eighty-eight patients fell sick at Mount Vernon in 1906.

So, why did pellagra explode onto the scene in 1906 and not earlier? The answer lies a long way from the asylums, prisons, and company towns of the Deep South.

## IT'S THE GERM, NOT GERMS

Goldberger is credited with proving the nutritional origins of pellagra and inspiring the search for PPF. His mill town study with Sydenstricker in 1920 is considered a classic in social epidemiology (his filth parties are notable for other reasons). But Goldberger was not the first to connect the dots between pellagra and social changes taking place in the South. That credit goes to a lesser-known health official in rural Alabama.

The year before Goldberger and Sydenstricker began their work in Spartanburg (and four years before it was published), Dr. Carl Grote conducted a study of pellagra in the coal mining regions of Walker County, Alabama, where Grote was county health officer. At one mining operation, workers were split between two camps a quarter of a mile apart: Camp A, which had no cases of pellagra, and Camp B, in which pellagra was widespread and had been "endemic for several years."[38] Based on these facts, company officials were convinced that pellagra was contagious. Grote noted that the residents of both camps shared the same open privy and purchased food in the same company store.

Camp A was home to "company officials, the bookkeepers, the clerks, the skilled miners who had been imported, and the negro miners [who were] good miners, made more money than the average white miner."[39] With their higher wages, Camp A's residents bought the best quality food from the store. Meanwhile, Camp B had "one class of people—the poor white natives who had moved from the farms, sold their cows and chickens and . . . brought their large families to the mining camps."[40] Camp B's residents bought the lowest quality food from the store, mostly canned goods and cornmeal. Grote noted, "The only vital difference that we could discover in these two camps was in the people, their economic conditions, and the difference in the foods they ate."[41] Grote uncovered a similar connection between poverty, diet, and pellagra at another mining camp and among the county's struggling farmers.

There were, of course, outliers: a preacher's wife with two children who lived "under good surroundings"; a forty-year-old unmarried woman who lived in a home "spotlessly clean and perfectly screened against flies." Like a good detective, Grote chased them down. The preacher's wife reported developing "an abnormal craving for corn flakes" around the birth of her last child and ate nothing else for eight months. The fastidious woman complained of a digestive issue that had made it impossible to eat meat, eggs, dairy, or vegetables for the past twenty years. Grote's outliers were in fact exceptions that proved the rule.

Though Grote did not find the smoking gun, he found a piece of the puzzle that others had missed. Walker County was once a rural, agricultural district where farmers grew and consumed their own crops and livestock. But when coal deposits were discovered in 1890, mining companies and railroads arrived. Grote writes, "Then a large percentage of our natives [farmers], especially our unprogressive farmers, began to leave the farms, sold their cows and chickens, and moved to the mining camps, only to make poor miners and a poorer living for their families. And while the corporations and many individuals have grown richer, a larger percentage of our people have grown poorer."[42]

Profits from the coal boom were not reinvested. By 1915, nearly 70 percent of the taxed property in the county was owned by outside entities. Grote described how the influx of wealthy families into the county drove up the price of healthy foods (meat, dairy, and vegetables), while the price of cornmeal stayed low. Over time, the lower class lost the means to eat well. It was not poverty per se that led to pellagra—as Sydenstricker would later claim—but the combination of income inequality and a free market for food. What Grote uncovered was a battle of buying power waged between the shelves of the company store.

As inequality grew, lower-class diets became less diverse, reducing the intake of essential nutrients. But the cornbread itself changed, too. Before 1900, most of the corn eaten in the South was grown locally and turned into meal at nearby grist mills.[43] But as sharecroppers and smallholders abandoned the farm for wage labor and their acres were replaced by cotton, local corn production disappeared. No longer able to feed its population of landless laborers, the Southern food system switched to imported grains. The timing was favorable: Production in the Midwest was on the rise. Despite the loss of corn acres in the South, between 1886 and 1906, US corn production grew 50 percent. This imported "Western" corn was a finely ground meal produced by large milling operations and transported to the South by rail.

Most of the niacin in a corn kernel is contained in the germ. But the germ is also the first part of the kernel to decompose. To accommodate lengthy storage and long-distance transport, milling companies removed the germ. Technological innovations like the Beall degerminator, patented in 1900, vastly increased the throughput of milled corn. This was the cornmeal that flooded the South and found its way into institutional kitchens and company stores. "Within a few years, the corn meal and other staple cereal foods of millions of people changed from what was essentially a vitamin-rich whole grain product to a highly refined carbohydrate that contained practically no essential vitamins or other nutrients."[44] Mill towns with the highest rates of pellagra were those located near railroads and surrounded by cotton fields. Meanwhile, rural areas less suitable for cotton held onto their local produce and had the lowest rates of pellagra.[45]

In fact, pellagra's connection to food processing could have been uncovered earlier. Niacin's cousin is thiamine (vitamin B1), and thiamine deficiency leads to beriberi, a similarly fatal disease encountered decades earlier in East and Southeast Asia. In the 1870s, centers of rice production in French Indochina and the Dutch East Indies began adopting new, high-throughput rice mills that polished away the thiamine-rich bran from the grain. Populations that relied on imported rice started dying from beriberi.[46] By the time pellagra emerged in the United States, beriberi had already killed tens of thousands in East Asia and suspicion had settled squarely on the milling process. Military physicians in the Philippines were some of the first US doctors to examine beriberi: The disease plagued Filipino scouts who fought alongside US troops and received rations from the US Army. In 1909, the Army Medical Board in the Philippines recommended "that Filipino number 2 rice (an undermilled article) [replace] choice Saigon rice (highly milled)" in scout rations.[47] A year later, cases of beriberi were down 90 percent.

Pellagra and beriberi were the first epidemics of an industrializing food system. Propelled by economies of scale, these "diseases of scale" sickened agricultural populations transitioning from subsistence to wage labor. The history of pellagra is often taught in schools of public health and epidemiology, and rightly so. But it is also a parable of violent modernization in the food system.

In the 1940s, the poor nutritional condition of men enlisting in World War II led to federal laws requiring processors to add niacin and other vitamins back into highly milled grains, a move that many companies resisted.[48] Today, fortification is not required, though the Food and Drug Administration (FDA) encourages it and regulates products labeled as "enriched." This balancing act was designed to protect both consumers and a system built around centralized processing, long supply chains, and vertical integration. Over time, the poor would increasingly receive their nutrients through industrial processes, while their once-humble vegetables became the icons of a health-conscious middle class.

## WANDERING IN THE (FOOD) DESERT

The work of Searcy, Grote, Goldberger, and others helped foster state interest in nutrition, both as a science and social intervention. It also contributed to a growing "place-consciousness" in public health. Today, researchers examine how our social and environmental surroundings affect our vulnerability to disease by mediating exposure to pathogens and influencing behavior. Altering the landscape has become a key policy tool in the fight against chronic ailments and disease epidemics.

In the intervening decades, nutritional diseases and their associated places have changed. Pellagra has given way to heart disease and diabetes, comorbidities that together claim over 800,000 lives each year.[49] These diseases are associated with growing rates of obesity, defined as having a body mass index (BMI) of over 30. Today, obesity rates range from 25 to 60 percent depending on location. In an echo of Southern political discourse, economists warn that this "scourge of obesity" is "a tremendous drain on the U.S. economy . . . [a] drag on attendance and productivity at work" that costs the country "$1.4 trillion annually."[50] Since the 1980s, the so-called obesity epidemic has drawn attention to the proliferation of unhealthy foods in low-income areas. The company towns, mental hospitals, and unhygienic neighborhoods of the pellagra days have given way to food deserts where—it is claimed—lack of access to healthy foods is condemning residents to lower educational achievement, higher health care costs, and an earlier grave.[51]

A resident of Scottish public housing in the 1990s was the first to coin the phrase "food desert" to describe the dearth of healthy options in her neighborhood.[52] Since then, the metaphor has come to occupy a privileged place in policy circles, justifying government investment in health education, bans on fast foods and soft drinks, "improvements to school lunches, safer and more walkable environments, and greater regulation of food advertising and nutritional claims."[53]

Food deserts are defined as places where household incomes are low and foods are expensive, difficult to acquire, or of poor quality.[54] Mapping household income is straightforward: Tools like the USDA's Food Access Research Atlas (FARA) use census data to identify low-income areas. But mapping barriers to food access is more challenging. For simplicity, supermarkets are often used as a proxy for affordable, healthy food. Thus, if a significant share of residents in an urban census tract live more than a mile from a supermarket (or more than ten miles in the case of rural tracts) the USDA classifies the tract as a food desert.

Today's food deserts can be found in every state and proliferate in rural areas with widespread poverty: Appalachia, the Mississippi Delta, the border with Mexico, and Indian reservations, among others. But the most sustained—and controversial—use of the concept is in cities where food deserts consistently overlap with minority populations. Like the emergence of pellagra in mill and mining towns, the absence of supermarkets in urban Black and Hispanic neighborhoods is the product of social and economic history.

## MAKING THE DESERT

Back when Midwestern grain companies and Chicago's Big Four were consolidating control over the middle of the supply chain, the country's urban population was fed by small, full-service grocers. But by the 1930s, these independent businesses were being squeezed out by large-footprint self-service stores. With greater variety and lower labor costs, these proto-supermarkets began to undercut and buy out the competition. The meteoric growth of suburbs after World War II saw middle-class whites abandon the urban core for the racial homogeneity of new suburban developments, a transition known as "white flight." Urban grocery stores followed—some even led the way, anchoring suburban shopping malls. Through the 1950s, supermarkets expanded their share of retail food sales from 35 to 70 percent, almost entirely on the urban fringe.[55]

"Retail flight" was coupled with residential redlining, a formal process that limited access to home financing and insurance based on race. The 1933 redlining map for Birmingham divided the city's neighborhoods into color-coded zones of "best" (green), "still desirable" (blue), "definitely declining" (yellow), "hazardous" (red), and "Negro concentrations" (gray).[56] As white residents enjoyed privileged access to the green and blue polygons on the urban fringe, nonwhite families were pushed toward the red and gray polygons near the center, areas more likely to experience industrial pollution, flooding, and inadequate infrastructure.[57] Minority-owned businesses were labelled high-risk and struggled to get loans. For many Black entrepreneurs, the redlining era was worse than Jim Crow. Former Birmingham city councilman Darrell O'Quinn explains, "Under Jim Crow, people could still at least amass wealth, and you had Black-owned businesses, including local grocers, in many of these communities. Redlining undermined those businesses and almost ensured that desegregation would harm African American communities."[58]

Today, Birmingham has the poorest neighborhood in the state, Fountain Heights, a community that is 83 percent Black and was designated "Negro concentrations" on the redlining map of 1933. Birmingham also has the second wealthiest neighborhood in the state, Mountain Brook, which is 96 percent white. Five miles separate these places on the map, and $140,000 separate their median household incomes.[59]

As wealthier consumers moved out and disparities increased, profit margins for city grocers shrank. Greater purchasing power in the suburbs translated into higher wholesale prices for healthy foods. Inner-city residents increasingly struggled to afford—or even find—fresh produce as their grocers stopped carrying it. Meanwhile, prices for highly processed nonperishables remained low. In the same way that the wealth of Camp A limited Camp B to cheap cornmeal, suburban purchasing power effectively limited inner cities to cheap, processed foods. Between 1978 and 1984, Safeway closed over six hundred inner-city stores, many of which were the "only source of reasonably priced (and minimally processed) meat and produce in their neighborhoods."[60] As the Business Enterprise Trust explained in 1993, "It makes no sense to serve distressed areas when profits in the serene suburbs come so easily."[61]

## DISMANTLING THE DESERT

Food deserts describe a landscape devoid of supermarkets, but can they explain inequities in food access? For that matter, does the desert metaphor support polices that reduce those inequities? Despite its popularity, the food desert concept has come under sustained criticism over the past twenty years, with commentators finding four broad categories of fault. First, maps like USDA's FARA create the impression that food deserts are *homogenous* and that inequities in access are uniform. Color-coded polygons make it seem like everyone in the census tract acquires food the same way and experiences the same challenges—convenient fictions that belie important variations both within and between food deserts.[62]

It would have been a mistake for Grote to treat Walker County's mining operations as homogenous units when each camp was different, or to treat each company store as a homogenous unit when what could be bought there differed for each customer. When Siler treated each Spartanburg household as a unit, he missed the fact that men and women ate different foods at home. In fact, such hyper-local variations were key to uncovering the cause of pellagra. What local variations are we overlooking today because access inequities have been milled down to uniform polygons?

Second, by focusing on *absences* (supermarkets), the food desert metaphor ignores local resources, practices, and resistances that influence food access. As geographer Ashanté Reese notes, at the root of the metaphor is an assumption of *deficit* that obscures all the possible ways a person may choose to eat.[63] Food might come from mini-marts, dollar stores, and pharmacies, but also farms, community and school gardens, cooperatives, and food sharing networks.[64]

The desert metaphor "makes us think of an empty, absolutely desolate place," says activist Karen Washington, when in fact "there is so much life and vibrancy and potential. Using that word runs the risk of preventing us from seeing all of those things."[65]

Third, the metaphor fails to capture the complex and unique *structural* realities that shaped each neighborhood.[66] Fountain Heights and Mountain Brook are vastly different places because of residential redlining and retail flight, but you would not know it from the USDA's map. The food desert framing condenses the complex racial, gender, and class origins of food inequity down to a shortage of Walmarts, Krogers, and Albertsons.

The metaphor overlooks the racist structures that maintain inequity today, even in places that are not food deserts. A 2022 study in Ohio and North Carolina found that 88 percent of middle-class whites living in mixed-race neighborhoods did not shop at their local grocery stores, opting instead to buy from supermarkets in wealthier, whiter suburbs.[67] Respondents cited a litany of reasons, including greater variety, higher quality produce, friendlier employees, and better aesthetics. One person explained that at his preferred supermarket, "they have chefs there, a sushi station, and even a jewelry store . . . they have special foods you can't get at the neighborhood Safeway [because the Safeway] *caters to a background of people* who might need cheaper prices."[68] One white homeowner in Ohio was more direct: "Some of the people in [the Greenfield Safeway] make me lose my appetite. And that sounds awful but I just can't, I can't [go there]. It honestly makes me feel like the food is gross."[69]

Like the white flight that redirected investment from cities to suburbs, white *retail* flight redirects investment from local grocers to suburban supermarket chains. This is partly why Karen Washington and others prefer to describe the processes that create inequity as food "apartheid," a term that emphasizes the racial foundations of food access. In South Africa, apartheid divided the country into a patchwork of Black townships and white cities. Undoing apartheid entailed dismantling legal, economic, and social structures *at the scale of the country*, not tweaking race policy here and there. As a metaphor, "food apartheid" invigorates systemic change, inviting us to identify and undo the historical and contemporary structures that shape each place. The addition of a few supermarkets will not cut it.[70]

Finally, deficit framing makes it seem like food deserts are *the site of the problem*. Geographer Jerry Shannon notes how projects designed to address obesity "pathologize low-income communities . . . by locating the cause of obesity within their geographic boundaries."[71] Places and their residents come to be stigmatized and shamed for their conditions. The poor are blamed for not knowing what to eat, or for continuing to eat an "unhealthy" diet. Echoing the cornbread stigma of the pellagra days, malnourishment is blamed on cultural traditions like "soul food."[72] Obesity becomes the "lazy disease" of Northern critique and the food desert becomes the "Nam Hollow" of Davenport's racialized map. With the sickness neatly quarantined within the clean boundaries of the polygon, we need not concern ourselves with other places, especially those that *benefit* from inequity.

## THE DESERT KILLS (JUST NOT THE WAY YOU THINK)

So long as poor neighborhoods without grocery stores are the problem, wealthier neighborhoods must contain the solution. The food desert metaphor idealizes both the food landscape of the middle class and the bodies that occupy it, justifying policies that facilitate gentrification. Interventions seek to terraform food deserts with amenities found in upscale, predominantly white neighborhoods: bike lanes, pedestrian streets, parks, boutique food stores, and trendy restaurants.[73] The unhealthy neighborhood is not changed so much as it is erased and replaced.

Today, "gastro-development" is big business in the South, where poor inner cities are being renovated into foodie destinations. From Charleston, South Carolina, to Ocean Springs, Mississippi, trendy restaurants have dominated the podium at recent James Beard

Awards. In 2018, celebrity chef Andrew Zimmern declared Birmingham the "hottest small food city in America."[74] Even as middle-class foodies headed into town to dine at award-winning gastropubs, other neighborhoods were getting prepped for renovation: Between 2015 and 2019, the number of census tracts in greater Birmingham that qualified as food deserts *increased* 25 percent.[75] With nearly half of the city now a USDA-defined food desert, Birmingham is a gentrifier's market. "Hottest small food city" indeed.

Even when gentrification is not the goal, the food desert metaphor invites other forms of intervention. Big-box retailers like Walmart and Target routinely use the concept to encourage government support for the construction of new stores. Programs like the Healthy Food Financing Initiative (HFFI) provide federal funds to build stores in urban food deserts.[76] When First Lady Michelle Obama announced the initiative in 2011, Walmart, SuperValu, and Walgreens promised to open fifteen hundred stores to help solve the "food desert problem."[77] After all, the most direct way to remove a food desert from the map is to build a supermarket, and who better to do that than a supermarket chain.

Building it, however, does not mean they will come.[78] Research shows that the presence of supermarkets does not automatically translate into the consumption of healthy food.[79] Once again, the case of pellagra is insightful: The miners in Camp B lived the same distance from the store as those in Camp A, and yet they still got sick. Being able to reach the shelf doesn't mean you can pay for what's on it. Even if you *can* pay for it, that doesn't guarantee you can keep it fresh, have time to prepare it, or find it appealing. Of the fifteen hundred new stores promised in 2011, fewer than half had been built or renovated by 2018, and many that had opened were already closed.[80]

Finally, maintaining the geographic status quo has its advantages. In 2014, Walmart opened a free bus service to transport desert-bound residents in Grand Rapids, Michigan, to one of its suburban supercenters. Twice a week, the bus would stop at low-income housing complexes, take residents to the store, and return them home with their shopping. "Tuesdays are my dialysis days," explained one resident, "Thursdays I would ride the bus and get what I needed." For the elderly, disabled, and those who lacked transport, the service was a lifeline—until Walmart ended it without warning in 2019. "The shuttle service became cost prohibitive," a company spokesperson explained, "[but] we are aggressively rolling out Grocery Delivery across the country and hope to offer the service in Grand Rapids soon."[81] Of course, delivery costs more money, requires internet access, and does not accept WIC—new obstacles for those without access to a local store.[82]

For some, bussing and grocery delivery may offer immediate access to healthy food, but for others, these solutions perpetuate existing inequities. Supermarkets rely on enormous tracts of underserved customers to support their footprints. In Detroit—a city that is 85 percent Black and has zero Black-owned grocery stores—inner-city residents spend $200 million a year buying groceries at *suburban* stores, a boon for chains like Walmart and an enormous opportunity cost for local food enterprises.[83] By bussing customers and dropping off groceries, the supermarket has perfected its ability to extract wealth and undermine alternatives in a predominantly Black community.

The pellagra epidemic of the early 1900s was a symptom not of places but of *relationships*—between members of the same household, between milling companies and cornmeal wholesalers, between Southern leaders desperate to industrialize and Northern grain companies eager to profit. For the residents of Grote's mining camps, whether you contracted pellagra had less to do with your proximity to a store than with the store's relationship to wealthy customers. The same may be true of obesity and its associated diseases in today's supermarket-challenged neighborhoods, and if so, the problem is far more complex than dropping a supermarket on a troublesome polygon.

## THE VIOLENCE OF GEOGRAPHIC METAPHOR

In June 2022, the USDA repeated its commitment to the HFFI with a substantial increase in funding to address the problem of food deserts. This commitment is somewhat surprising given sustained criticism of the food desert concept and the program's lack of success.[84] Why does the metaphor of a desert continue to dominate?

Beyond the material advantages that accrue to renovation-happy developers and big retailers, the concept offers several rhetorical benefits. Unlike the apartheid metaphor, food deserts do not implicate racist, classist, and sexist structures. By centering the problem around supermarkets, the definition both depoliticizes and corporatizes hunger. As a census tract, the food desert is invented at a scale that encourages policy intervention. Each hungry polygon neatly inscribes a neighborhood: a humanizing backdrop for groundbreaking ceremonies, photo-ops, and quotable interviews.

### PROBLEM ENCLOSURE

The metaphor also obscures other ways of seeing the problem. Sociologist Maarten Hajer calls this "problem closure," when an issue is defined in a way that rules out alternative understandings and limits possible solutions.[85] In 1912, the Thompson-McFadden Commission framed pellagra as a contagious disease. Siler expected to find cases clustered together in place, in pellagrous households, neighborhoods, and mill towns. He expected these places to have conditions conducive to the spread of infection: namely, fewer sanitary sewers, an expectation based on experience with cholera, dysentery, and typhoid fever. When he found what he was looking for, Siler recommended better sanitation—not a bad suggestion, but not the solution to pellagra.

Likewise, the desert metaphor creates the impression that malnutrition resides in place, inviting researchers to make maps. As "deserts," such places are assumed to be "deficient" the way a desert is seemingly devoid of life. The opposite of deficiency is abundance, and the paragon of food abundance is the supermarket. Not only is the desert devoid of life, it is seemingly antagonistic to it, requiring outside intervention to "green" it. What begins as an attempt to discover inequities in access ends up as a map of missing grocery stores and a plan to intercede. Long before any inquiries are made about the place or its people, the metaphor has blinded the decision-maker to diverse ways of growing, acquiring, sharing, and selling food.

Problem closure is accompanied by a corresponding process of *en*closure, when a problem's definition limits its spatial extent in ways that *exonerate* other places from responsibility. By defining where the problem *is*, the map also defines where it *isn't*. So long as the scourge was "Southern," the rest of the country was absolved. So long as the inner-city food desert is the problem, suburban food biomes must be in the clear.

In fact, the headwaters of inequitable food access are vast, encompassing the boardrooms of grocery store chains, the suburbs to which middle-class shoppers escape, and the rural prison farms where a disproportionate number of inner-city Black and Hispanic people grow food for white-owned companies.[86] Places of nutritional violence are intimately connected to places of abundance and privilege. The poor of Camp B would not have contracted pellagra but for the purchasing power of Camp A. Degerminated corn would not have flooded into Southern poorhouses had money not been flowing into Northern banks. Urban grocery stores would not have gone under had capital not moved to the suburbs with white disinvestment. Problem enclosure exonerates these participants, isolating food deserts as sickly anomalies in an otherwise healthy landscape.[87]

Today, activists are making concerted efforts to "unmap" the food desert. Community asset mapping focuses on the resources that are present, not the middle-class amenities that are absent. Participatory mapping helps residents convey their lived experiences without the constraints of conventional geographic categories. These and other strategies counter the stigma associated with food deserts and undermine the assumption that such places are identical, homogenous, and sick.[88] But they do not capture the *sources* of inequitable access. A map of pellagra sufficient to explain its cause would have had to include burgeoning cotton plantations and Midwestern grain companies. A map of food deserts sufficient to explain their existence would have to include Wall Street.

## MISSED OPPORTUNITIES

In the case of pellagra, Southern politicians and civic boosters could have set aside injured pride long enough to take seriously the nutritional theories forwarded by their own health officials. For their part, the Northern press could have kept its hungry dogs in check. Researchers could have identified the similarities between pellagra and beriberi sooner, implicating degerminated corn and speeding up the search for PPF. Southern policymakers could have slowed the pace of agricultural transition and import dependence. Instead, they doubled down on cotton and accepted that imports of degerminated corn were the way of the future.

In contemporary policy discussions, there remains a misconception that the loss of urban grocery stores was inevitable, when in fact it relied on sustained government support for white, middle-class flight. Today's supermarkets retain this privileged status, a "sheen of inevitability" that prejudices policymakers against viable alternatives.[89] The closure and enclosure of the problem ensures that wealth extraction from urban neighborhoods continues much as it did during redlining, this time under the guise of fixing malnutrition. Meanwhile, the places that facilitate this extraction go unmapped.

The uneven emergence of pellagra could have been understood as a localized symptom of systemic causes. Instead, attempts to associate the disease with discrete spaces perpetuated errors and slowed progress. Southern scourge, pellagrous districts, and unhygienic neighborhoods—the "Nam Hollows" of pellagra—helped conceal the disease's complex origins. So, too, with food deserts, which emerge in the public imagination as ugly tumors lurking in healthy tissue. Isolated in space and severed from history, the sickly neighborhood awaits government and corporate rescue. Meanwhile, those who benefit from its infirmity reinvest in their own "best" and "still desirable" polygons. It is this ongoing failure to link places of hunger with places of abundance—for that matter, places of violence with places of prosperity—that is the biggest missed opportunity of all.

• • •

Pellagra is rare today, but its associates are making a comeback within industrialized food systems. Since 2007, cases of malnutrition in England have quadrupled, with alarming increases in scurvy (vitamin C deficiency) and rickets (vitamin D deficiency)—diseases more familiar to Queen Victoria than today's doctors.[90] Rickets is also on the rise in the United States. Meanwhile, heavy consumption of ultra-processed foods (UPFs) has been associated with increased risk of "overall cardiovascular diseases, coronary heart diseases, cerebrovascular diseases, hypertension, metabolic syndrome, overweight and obesity, depression, irritable bowel syndrome, overall cancer, postmenopausal breast cancer, gestational obesity, adolescent asthma and wheezing, and frailty."[91] From degerminated corn to high-fructose corn syrup, the actors have changed, but the story remains the same.

On the one hand, we might ask what places and people experience this disease *burden*, a line of inquiry that leads us to map deserts and supermarket shadows—the pellagra mill towns of today. On the other hand, we could ask what people and places experience the disease *benefit*, a line of inquiry that directs us to the neighborhoods where purchasing power drives up the price of healthy produce, the gentrifying projects where corporate profits are reinvested, and the policy circles where a narrative of deficient deserts is used to protect the status quo.

# CONCLUSION

*If It Ain't Broke, Break It*

"Yes, whatever we choose to do will harm someone. But given the choice between people in your own community and those on the other side of the world, wouldn't you rather support people more like yourself?"

This was the panelist's answer to my question at the Kent locavore meeting on that snowy night in 2012. There was something reassuringly simple about it, that it was okay to privilege "us" and "here" over "them" and "there." That it was okay for empathy (like sound and light) to fade with distance. That community and country were natural barriers to care. It was reassuring, but also unsettling.

"Wouldn't you rather support people more like yourself" could have come from a white planter in the Delta justifying the starvation of Black sharecroppers in the 1950s. It could have come from Austin Anson justifying the incarceration of Japanese Americans in 1942. It might have come from any number of founding "fathers" justifying the removal of Indigenous people west of the Mississippi, or from a State Department official picking which countries should receive food aid under PL-480. It could have been said by a meatpacking foreman hoping to prevent workers from reaching across ethnic lines.

Instead, it came from a well-educated activist passionate about changing the food system to *repair* past wrongs, not perpetuate them.

Of course, neither the locavore movement nor its representative in this instance is out to harm anyone: Violence is the opposite of what the movement stands for. But that is precisely the point: There was an implicit geographical violence in the panelist's response that he could not escape—a geographical violence that *none* of us who participate in the food system can escape. If we are to build a nonviolent system, it is this reality we must change.

## EATING IS AN ACT OF VIOLENCE

To eat is to nurture life, sustain culture, and show love for others. It is also to concentrate wealth and power in the hands of a few, let die, and sometimes kill. It is easy to think that this paradox is new: a characteristic of industrialized agriculture, post-Depression food policies, or so-called late stage capitalism. The system's capacity to destroy environments and malnourish populations may have ebbed and flowed, but the fact of violence has stayed stubbornly constant.

The land that anchors the US food system was acquired through genocide. The nitrogen and phosphorus that fed it enslaved thousands on distant islands, starved Southern sharecroppers, and dispossessed Guatemalan farmers. Today, it shapes war and peace in the world's oil-producing regions. The value of its vegetables justified the incarceration of Japanese Americans in the interior as its demand for tropical fruits provoked foreign wars. Demand for meatpacking labor exploits migrants from distant countries as suburban supermarkets take advantage of poverty next door.

The food system produces violence because it is built on *expropriation*—the taking of something without equivalent return, whether soil nutrients, wages, cheap commodities, or years of life. Expropriation drives other processes: accumulation through dispossession, the agricultural treadmill, and the development treadmill. These processes magnify the disparities that capitalism depends on, making expropriation all the more profitable. Over time, chronic crises set in: metabolic rift, overproduction, farm loss, and malnutrition. The food system does not produce violence coincidentally, accidentally, or maliciously: Built on a capitalist logic, it produces violence because it *has to.*

The system mitigates this unpalatable outcome by pushing its violence into myriad elsewheres. Carceral fixes, extraterritorial fixes, domestic rearrangements, aggregation, and segregation redirect harm away from privileged places into manageable, forgettable, and expendable elsewheres. When the system fails to do this, its privileged places suffer: Workers take over factories, protestors take over airbases, and countries take back their land. Meanwhile, these strategies are justified through the tactical use of geographic metaphor and problem enclosure.

These fixes are not confined to history. In 2025, the Trump administration tried to suspend federal funding for the Supplemental Nutrition Assistance Program (SNAP)—a program that helps feed 41 million low-income people in the United States. It raided farms, grocery stores, restaurants, and meatpacking plants to round up alleged undocumented migrants, separated parents from children, and confined them to "detention centers."[1] In a reprise of Japanese American incarceration, there is growing concern that detainees may be used as leased or captive labor in agriculture.[2] In mid-2025, as over a million farmworkers fled the fields, the administration *lowered* wages for migrant labor—precisely the opposite of how a free market would have responded. And in a reversal of decades of liberalization, tariffs were slapped on agricultural products, punishing domestic and foreign producers and consumers alike.[3] These policies (and threats) are creating new elsewheres and repurposing old ones. Through it all, violence remains, even if the zip code, congressional district, or country code changes.

Today, we are unwitting conspirators in this violence whenever we grow, sell, buy, and eat food. We aid and abet expropriation even in our efforts to challenge it. We are asked to bound our empathy in ways that keep the game going—through old standbys of nationalism, racism, and sexism. We are asked to privilege groups who are "more like ourselves." Like meatpacking companies, expropriation benefits from a patchwork of segregated consumers, alienated by antagonism, but all linked together through the market. This geographic shell game is not a symptom of a broken food system, but an essential feature of a *working* one. That is why its most egregious violences—like some in this book—are often hailed as victories.

If the food system isn't broken, then insisting it *is* broken is a red herring. It draws attention to transient policies and away from deep and persistent structures. It points us to the places that seem dysfunctional when the real work involves places that seem to function perfectly. If we are to change the relationships that allow violence to fester, then the parts that aren't broken will need breaking. There are two ways to approach this, and one is a dead end.

## THE TIMID PATH

The timid path tweaks the market, adjusts food policy, and awaits the next technological messiah. It insists we can consume our way into a better system by shopping in different aisles, buying smart refrigerators, or investing in humanely raised meat. The timid path seeks to invent seemingly new actors within existing, inequitable structures: a trendy urban farm that sells produce beyond what its neighbors can afford; a farm-to-table movement in a gentrifying neighborhood; a harvesting machine that only giant agribusinesses can afford. Or it updates existing models for new times: a slightly better Farm Bill; an office of food loss and waste at the USDA; a Walmart that buys some of its produce from local growers. Under the timid path, the old food system reemerges each election cycle in a new skin. It eeks out small wins by moving its losses elsewhere. It still plays the geographic shell game, just better.

**EAT YOUR WAY TO A BETTER SYSTEM**

One popular prescription for healing the food system claims that the market can do it for us. We can get the food system we want by choosing to buy and eat different things—to pick one notable example, the foods that our great-grandmothers ate.[4] Through the magic of supply and demand, the invisible hand will mold the food system to our liking. Of course, for consumer choice to work, buyers need to know more about where their food comes from. This is why authors like Michael Pollan emphasize building awareness: "Our food system depends on consumers' not knowing much about it beyond the price disclosed by the checkout scanner," writes Pollan, "it's a short way from not knowing who's at the other end of your food chain to not caring."[5]

Knowledge may be power, but knowing more—even caring more—will not cause healthy alternatives to materialize on our plates. Nor will it reduce the proliferation of violence. There are two reasons for this. First, when consumers manage to change the food system, the result is not always what they expect. In response to growing demand for organic food and humanely raised meat, the food system has created parallel, "certified" product lines priced for elite consumption: It has *not* cut back on pesticides nor shut down CAFOs. Privileged retail spaces enjoy a multitude of certified options, while ultra-processed foods and cheap meat fill the shelves in other neighborhoods.[6] These forms of market *segregation* undermine broad-based food movements—precisely the kinds of divide-and-conquer strategies that inspired labor leaders in the 1930s to insist on inclusive unionism and model contracts.[7]

Second, consumer influence is proportional to purchasing power. Wealthier consumers can afford higher value items and support larger margins. This is why wealthy neighborhoods attract the healthiest grocery stores, the trendiest farmers markets, and the best-stocked meal delivery boxes. Meanwhile, these benefits are out of reach for the single mother who delivered that box and who feeds her family from the drive-through window, the refugee on the disassembly line who keeps silent for fear of losing TANF benefits, or the farmer who grew the food but cannot afford to eat it. No amount of food shaming over ancestral eating habits will change the mealtime reality for tens of millions of Americans, even if they live next door to a supermarket. So long as free markets are allowed to govern the production and consumption of food, disparities in quality and availability will reflect disparities in the power of consumers themselves.[8]

## VOTE FOR A BETTER SYSTEM

Acknowledging the inequalities built into markets, others hope to level the playing field through improved food governance. Democratizing approaches aim to give each person a greater say in what food is available to them and how it is made.[9] "Food democracy is about citizens having the power to determine agro-food policies and practices locally, regionally, nationally, and globally," writes activist Neva Hassanein, resulting in a system that responds more readily "to demands from below."[10] Democratizing approaches include decentralizing food policy, improving the dissemination of information, increasing resource sovereignty, and increasing citizen involvement in policy development.

In the abstract, democratization sounds good, but the devil is in the details. There are many ways to make the food system more democratic; some may reduce spatial inequalities, while others will create new ones. Would a new "food democracy" be representative and majoritarian? Would it be organized territorially? If so, this will privilege local participants over nonlocal ones. Would participation be organized by sector or function? Everyone is an eater, but a shrinking proportion are growers. If this trend continues, the influence of those who consume food will increase relative to those who produce it. How does a food democracy avoid exacerbating the inequalities between spaces of consumption (cities and suburbs) and spaces of production (primarily rural areas)? One of the attractions of localism is that smaller territories of production and consumption seem easier to translate into territories of food governance. But this is not always the case, nor does it eliminate harmful power gradients between locales.

Unless accompanied by substantial gains in equity, the spatial structure of a food democracy could exacerbate violence. After all, the political democracy examined in this book starved its own sharecroppers, stole its citizens' land, and imprisoned them to labor in the name of *supporting the public interest*—a quintessentially democratic ideal. How would a "food democracy" avoid these pitfalls?

Ultimately, access to nutritious food builds the foundation for everything else that supports a robust democracy: education, health, social mobility, and civic engagement. Today, the places that struggle with access also struggle with political visibility and often lose the contest for public resources. It is easy to foresee a "treadmill of influence" in which the places that struggle to have their voices heard never catch up. As political scientist Kenneth Dahlberg notes, "The reforms needed to democratize society and to democratize food systems are parallel and reinforcing."[11] For now, these reforms are also languishing.

## BUILD A BETTER FARM(ER)

From billionaire Bill Gates to influential journalist George Monbiot, there are increasing calls to "green" agriculture through new technologies.[12] Artificial intelligence, cultivated meats, perennial grains, a new generation of chemical inputs, and climate- and drought-tolerant GMO seeds are just some of many proposals to address a cluster of problems caused by the food system, including biodiversity loss, deforestation, and greenhouse gas emissions. Not simply "technological," these solutions are "technocratic" in that they involve deep partnerships between financial backers, technology companies, scientists, government actors, and nonprofits.

In a retread of past revolutions, proponents hope new technologies will reduce future harm, even if the transition is painful. But there are unanswered questions. What solutions are we overlooking because we have framed agriculture's problems as technical ones?[13] In the 1960s, poverty was framed in a way that justified more fertilizer and benefited oil companies. Today, food deserts are framed in a way that justifies more supermarkets and benefits big retailers. Some places will reap a greater share of the rewards from new technologies, but what places will suffer? The United States lost most of its farms to the agricultural treadmill because farmers could not afford the latest innovations. Those who could were bound to the business plans of powerful companies and the designs of powerful governments. Every technological innovation from guano to glyphosate has made something obsolete. What will be made obsolete next?

Agriculture contributes up to 33 percent of global greenhouse gas emissions, a considerable fraction that must be reduced, especially in the industrialized places that contribute the most. But tech evangelists like Gates look well beyond the wasteful CAFOs and monocultures of the North to the smallholders of South Asia and sub-Saharan Africa. Here, multinational agribusinesses advocate for a new green revolution that will once again transform smallholder agriculture.[14] Last time it was to save national economies, this time to save the world.

Such proposals contain an insidious spatial fix. Yes, the food system contributes heavily to global carbon emissions, but it also produces *food*. Food production *should* be a major portion of our carbon footprint because eating is a major part of *being alive*. What productive activities are *not* a major part of being alive? Building megayachts, mining cryptocurrency, and fabricating the miscellany of petroleum-based extravagances that delight the jet set—that is, a good chunk of the remaining 67 percent of our carbon footprint. A megayacht emits tens of thousands of tons of $CO_2$ each year, making it "the most polluting activity a single person can possibly engage in."[15] In 2024, there were 10,000 of these leviathans and growing. A UN study in 2023 found that Bitcoin mining consumed 173 terawatt hours of energy annually, more than Pakistan, a country of 234 million people.[16] The wealthiest 1 percent (who generally do not farm) contribute as much greenhouse gas as the poorest 65 percent (who mostly do).[17] But by enclosing the problem around agriculture, technocratic solutions inadvertently put elite consumption in the clear.

It seems perverse that farming, the economic activity that supports the poorest and most vulnerable people, is the one that must be upended to save the world. This follows an old colonial script—of Indigenous North Americans who needed to be replaced by European monocultures (when in fact Indigenous knowledge saved the early colonists), and of poor Guatemalans who needed to be modernized by Green Revolution technologies (when those farmers supplied the genes for Green Revolution seeds). Whenever powerful societies need a spatial fix, other people and places cover the cost. This was true in 1492 as Columbus sailed into his New World; it is true today as billionaire technocrats envision theirs.

If we can't eat, vote, or invent our way into a less violent food system, what then? Is violence in the food system a zero-sum game? Is it inevitable? Was the panelist at the locavore meeting right, that no matter what we do, we will have to harm somebody?

The timid path has ensured better food access and nutrition for some places around the country and for some countries around the world, but because it does so within a system built on expropriation, it does so at a geographic cost. The timid path will keep improving the food system for some at the expense of others. We know this because it already has—in the Meat Inspection Act of 1906 that cleared the way for the Big Four, in the minimum wage increase of 1967 that put sharecroppers out of work, and in PL-480 food aid that bankrupted Guatemalan smallholders. Good intentions, violent outcomes. Same old snake, different skin.

## THE BOLD PATH

It may be possible to build a system that provides everyone with access to healthy food, does so in an environmentally friendly way, and is humane to animals and humans. But for it to work, it can't *also* be expected to make the rich richer and powerful countries more powerful. It must become a *food* system first and foremost, not a wealth-generating system that happens to make food. Whereas the timid path seeks to tweak the underlying sources of expropriation, the bold path relentlessly targets and dismantles them. Only then can we break and rebuild the relationships that perpetuate violence. The bold path is exactly that—a path, not a destination—because its work is never done and its form will change from one moment and place to the next.

In 1993, a diverse group of Latin American organizations representing farmers, rural women, landless workers, fishers, migrant laborers, and Indigenous communities formed a coalition called La Via Campesina (LVC), "the Peasant's Way." Since then, LVC has become "one of the largest and most vibrant social movements in the world," with chapters in eighty-one countries and over 200 million participants.[18] In 1996, LVC members coined the term "food sovereignty" to describe "the right of peoples to healthy and culturally appropriate food produced through ecologically sound and sustainable methods and their right to define their food and agriculture systems."[19]

Today, food sovereignty emphasizes seven principles: protecting food as a human right, reforming agrarian policies to support smallholders, preserving natural resources, modifying trade to support local producers and consumers, reducing global concentrations of power, nurturing peace, and promoting greater democracy in the food system.[20] As an organizing concept, food sovereignty differs from dominant discourses of "food security" in that it "takes power and inequality seriously."[21]

Since its inception, LVC (and others organizing under the banner of food sovereignty) has inspired an extraordinary number of initiatives around the world, from reformative to revolutionary. Urban farms in cities across the United States are addressing alienation and metabolic rift by "bridging the distance between production and consumption."[22] From Hawai'i to New Zealand, alternative food networks are re-centering women's and Indigenous voices in food provisioning.[23] In the United States and around the world, governments are incorporating the human right to food in their constitutions.[24] And in 2018, the United Nations passed the Declaration on Rights of Peasants and Other People Working in Rural Areas, a landmark resolution pushed through in large part by LVC and growing interest in food sovereignty discourses.

LVC's emphasis on power and structural inequality seem more applicable to the pursuit of nonviolence than timid-path discourses that assume tweaking capitalism will make it humane. But what about the food system's addiction to elsewheres? How can we ensure that agrarian reforms, food federalism, alternative food networks, and locally determined trade policies don't simply push violence onto others' farms and dinner plates? To address violence is to tackle expropriation, but it also demands a keen sensitivity to how structural violence behaves; how it is justified, concealed, and celebrated; how it emerges, moves, and (critically) where it goes.

How might we hone our sensitivity to spatial fixes, and what lessons does a history of food violence offer?

## STOP LOOKING FOR PANACEAS

The bold path toward nonviolence may entail more local food consumption, more farmers markets and community-supported agriculture, but it doesn't have to. In places with few barriers to access, markets may work well; in others, they won't. In places where land and labor are limited, long supply chains will make sense; in others, decentralized mutual aid regimes will make more sense. Some places may need to adopt a different idea of private property; others won't. Environmental challenges get easier as more people give up meat, but this does not mean that everyone needs to be vegan. *There is no panacea*. In fact, the best way to maintain the status quo is to insist that a solution must work everywhere, all the time. Policy monocultures can be as destructive as agricultural ones.

Flexibility will be important. Supply and demand never stay the same for long; a warming planet will continue to impact food production into the foreseeable future. The food system needs to be organized more like an ecosystem, embracing both complexity and diversity—qualities that build resilience.[25] A dynamic federation of producers and consumers that differs depending on location and condition. Of course, this is not how the system is organized today, where its complexity is used to conceal violence, and its rigidity is used to justify profitable interventions.

Once we stop looking for panaceas, a universe of possible solutions opens. To evaluate them for each place and circumstance requires us to change how we define the problem, and that means asking different questions.

## ASK DIFFERENT QUESTIONS

Following the bold path means no longer asking "What new program or policy or widget would allow this person to get what they need?," and instead asking "What stands in the way of them getting what is already here?"[26] When we find someone who cannot access healthy food or has to denigrate themselves to starvation wages, we must ask who *benefits* from their hunger. When we find places that suffer under the social, economic, and environmental burden of the food system, we must look for the places that are enriched. What actors receive the spoils, and if we are a part of these networks, how can we alter them?

Capital asks different questions: What is the best way to exploit these disparities? Can expropriation be managed in a way that prevents financial loss? If a crisis occurs, what is the best way to profit from it? In fact, taking the bold path often means asking "What would capital do?" and then doing the opposite.

## DISENTANGLE FOOD FROM CAPITAL

The bold path to nonviolence will involve the steady untangling of capital from food. In some places this will mean a redistribution of land, wealth, and power. In others, a substantial reinforcement of private and communal land tenure. Agriculture and finance capital must end their codependency. So should commodity markets and speculators. Production and consumption regimes must become antagonistic to agribusiness monopolies and monopsonies. There is no single way to accomplish this but a diversity of options depending on geography, culture, history, social structure, politics, and human need. It will take time, and progress will be erratic.

Centralization under a command economy is *not* the solution. In fact, the bold path is suspicious of state power, not only for its tendency to collude with capital, but for its habit of pursuing elite interests. Today, state and federal laws prevent food from being given away to the poor, prevent the sharing of seeds, and prevent investigation into agricultural violence.[27] Since the land-speculating days of George Washington and Thomas Jefferson, national leaders have worked hard to protect concentrations of capital (much of it their own). To be sure, environmental, consumer, and labor laws have protected people from capital's more violent excesses, but these violences emerged *because capital was involved in the first place.*

If the best solutions do not rely on billionaire financiers, giant agribusinesses, or all-powerful governments, where are we going to find them?

## START WITH WHAT'S THERE

The bold path looks at each resource that already exists in a place (or in its larger network) and finds a better way to get it where it needs to go. A school that feeds and educates its students by partnering with a struggling farm next door. A public transport system that stocks basic groceries on its buses. A library that partners with local chefs and gardeners to offer cooking classes and free food. A farmers' network that shares equipment, inputs, knowledge, and even land. A union (or any group) that stands in solidarity with migrant meatpackers and farmworkers. Two people who realize they have a common culinary interest and agree to share recipes. Landowners who give acres, gardeners who give seeds, and others who can provide labor and know-how.

These new alliances emerge not because the people involved are more "like" each other, live in the same census tract, or share a common language, ethnicity, or creed. They emerge because the thing each one offers satisfies the other's need.

"Here" is where it starts, but not everything that needs to be done can be found here. Now what?

## LOOK BEYOND THE EDGES OF THE PROBLEM

It is tempting to think that places with sufficient healthy food are exonerated: Sure, there are problems in inner cities, on Indian reservations, and in the Mississippi Delta, but things are fine in my zip code, where everyone can afford to eat exotic items from around the world. In fact, these places conspire to move violence elsewhere. Domestic demand for sugar helped crush Aguinaldo's hope for an independent Philippines. Demand for bananas helped crush Árbenz's dream of a democratic Guatemala. The purchasing power of Camp A kept Camp B from getting niacin-rich meat and dairy. Today, white flight to suburban supermarkets deprives racially diverse neighborhoods of investment. You cannot address scarcity without *changing its relationship* to abundance, and that means taking action elsewhere.

The point is not to renounce geographical fixes (if such a thing were even possible) but to use fixes to counter expropriation. In fact, the resources of a new food system will almost certainly come from dismantling something *outside* the system. Food is necessary for bodily survival and social reproduction; it is the material basis for human life. But there is a great deal that isn't, and the production and consumption of this excess drives violence in the food system. The bold path involves redirecting time, thought, and resources from the systems that benefit a few, to the one that feeds us all.

Since the Industrial Revolution, the narrative of modernization has involved freeing people from the drudgery of farming so that they can do more lofty things (notably, accumulating wealth for others). Perhaps the drudgery from which we need to free ourselves is one that alienates us from the product of our labor and alienates our labor from the processes that give us life.[28]

## DEVELOP NONVIOLENT TECHNOLOGIES

This does not mean that farmers should go back to pulling weeds by hand. Innovation—the process of experimenting, learning, and developing new tools and practices—has always been essential to the food system. But when the benefits of that process have been captured by a few, its inventions have harmed the many. It is easy to get starry-eyed with the productive power of new technologies. One economic advisor I spoke with boasted that capitalism and innovation had given the United States "the most productive agricultural system in the history of the world." When I asked if agriculture should value something other than productivity, he chided me: "Well, *now* you're talking about ethics." I had neglected to silence my conscience while the economist was talking.

Following the bold path means developing technologies that are ethical *first*. This means improving access, nutritional value, and quality of life for all people by design. Such innovations must make food workers' jobs better, not obsolete. Following the bold path means building harvesters sized for smallholders (as in Japan), inventing fertilizers by combining what is already available on farms (as in India) and not locking up these innovations behind a patent paywall.[29] Those who use new technologies must have a say in how they are developed and receive a greater share of the benefits. And because every machine is built on a long history of small improvements—the unacknowledged labor of generations—it should be the property of all.[30] From novel molecules and genes to the next robotic harvester, engineers should assume that the final product is a gift to humanity: After all, everything that went into it was a gift *from* humanity. Innovations deserve to be shared, the way generations of farmers have shared seeds and best practices. The way generations of parents have shared family recipes with their children.

But caution is due here. Technologies are culturally specific, and sharing technology may risk cultural appropriation. The bold path must include a full, honest, and continuous accounting of what has been stolen and how to make amends. After all, it opposes a system that normalizes such theft.

Capital will ask how such an enterprise can be funded. The bold path involves supporting a scientific culture that is also disentangled from capital, one that serves the social and biological interests of its beneficiaries, not the political and fiduciary interests of its funders; a science that is *always* talking about ethics, not just when the economist is out of earshot. This is where research in new food system technologies aligns with the decolonization of science more broadly—a bold path of its own.[31]

## FIND NEW WAYS TO MEASURE SUCCESS

In a widely publicized report, the Rockefeller Foundation found that the "true cost" of the US food system, after factoring its impact on "human health, the environment, and social and economic inequity," was $3.2 trillion, almost three times the amount spent annually on food.[32] In 2023, the FAO announced that the "hidden costs" of the global food system topped $10 trillion annually.[33] Once again, it seems perverse that institutions choose to measure the problem in *dollars* when lost lives and livelihoods should be enough to justify change. Counting money makes sense when the purpose of the system is to generate profit. But different measures of success and failure are necessary if the food system is to nurture health and well-being.

If transnational institutions will be useful on the bold path, the way they understand the relationship between capital and people must change. Development goals should be rewritten to promote disentanglement from international finance capital. National and regional food systems should be evaluated not on their productivity, but on their *expropriating potential*: the degree of monopoly control, dependence on international markets, concentration of power, and influence over other food systems. This means the places that need the most work are in the industrialized North. Places like the United States. As the cases in this book demonstrate, this has been true for a while.

## BE PATIENT AND PERSISTENT

Following the bold path will take a very long time. Each place will go through an evolution, try on different strategies, and attempt to balance its needs and resources with other places. In the meantime, climate change will upend the food system. The tendency will be to fall back on abusive relationships, to cry out for rescue from agribusinesses and powerful governments. If we do this right, however, an upended food system will make the bold path easier. If we don't take advantage of the opportunity, capital surely will. In fact, it is already doing so.[34]

We have been conditioned to see "the system" as something massive and beyond our control, operating outside of our awareness and above our paygrade. It is hard to imagine that such a colossal edifice could ever be different. But if we asked Indigenous farmers in the 1700s, yeoman farmers in the 1800s, or even (at Pollan's bidding) our great-grandmothers in the 1900s to picture a country fed by end-to-end corporations, cold supply chains, robotically slaughtered animals, and vat-grown flesh, their imaginations might be challenged too. And yet, here we are.

To increase throughput, meatpacking companies imagined the disassembly line. To capture land, Anson and his colleagues imagined a national enemy. To overcome an uncooperative government, United Fruit's leaders imagined a coup. Capital has never had a problem imagining the next great thing—*for capital.* Why should our imaginations be limited? In fact, there is plenty to inspire, we just need to know where to look.

## FIND INSPIRATION ELSEWHERE

Saul Alinsky, a Chicago activist who worked in the meatpacking neighborhoods of *The Jungle*, writes, "Change means movement. Movement means friction. Only in the frictionless vacuum of a nonexistent abstract world can movement or change occur without that abrasive friction of conflict."[35] The bold path assumes that change will be obstructed and pushes forward anyway. Greenville protestors, Navassa workers, Guatemalan farmers, and unionized meatpackers followed the bold path. The Indigenous communities that are returning to collective ownership despite legal obstacles are following the bold path, as are those rediscovering a material and spiritual connection to food nearly erased by settler colonialism.[36]

Inspiration can be found in the Japanese farmers' cooperatives that resisted racial discrimination by building new networks; the Guatemalan collectives (and their Nebraska offshoots) that fed each other when their leaders wouldn't; the Southern counties that organized to distribute food when cotton pushed out local farms; and the meatpackers who set aside racial and ethnic differences to find a common humanity.[37] In these and other elsewheres we saw the bold path being taken. Our task is to keep that inspiration alive.

• • •

A final note on the use of "we." By now, I know you a little. You have endured burned villages, breathless guano workers, marasmic children, beakless chickens, and village wells filled with bodies. You have worked backward from corpses to culprits along blood-slick disassembly lines and trails of tears. Making it to the end of a book like this involves some self-selection. This is why the words at the end of this book are said the way one adventurer might speak to another: with recognition for where we have been and hope for where we might go.

In *your places* there are ways to make and obtain food. Some nurture life, others do not. Within your networks there are possibilities for preparing, sharing, and enjoying food that have yet to be discovered: ways that are less dependent on expropriation, less vulnerable to co-option by political and social ideologies, and less infused with metaphors that poison life. Some of these possibilities have been there all along, others are about to be imagined, possibly by you. There is hope, in whatever elsewheres you find yourself.

# ACKNOWLEDGMENTS


> Every machine has had the same history—a long record of sleepless nights and of poverty, of disillusions and joys, of partial improvements discovered by several generations of nameless workers, who have added to the original invention these little nothings, without which the most fertile idea would remain fruitless. . . . Every new invention is a synthesis, the resultant of innumerable inventions which have preceded it . . . Each discovery, each advance, each increase in the sum of human riches, owes its being to the physical and mental travail of the past and the present. By what right then can anyone whatever appropriate the least morsel of this immense whole and say—This is mine, not yours?
>
> **PETER KROPOTKIN,** ***The Conquest of Bread***

As with Kropotkin's machines, so too with books. The idea for this book took shape during conversations with James Tyner, V. Kelly Turner, Babacar M'Baye, Josh Stacher, and Mandy Munro-Stasiuk—people whose innumerable inventions and fertile ideas have been the building blocks for so many rewarding projects. Since then, others have added new discoveries: Zsuzsa Gille, Jesse Ribot, Joshua Sbicca, Ashanté Reese, Don and Alex Ufford, Heather Zenone, Annia

Ciezadlo, José Diaz-Garayua, Michelle Green, Chella Bluth, the volunteers at the University of Illinois organic farm, and of course a locavore activist in Kent, Ohio. This book is the least morsel of their immense contributions.

This book is dedicated to those whose tireless efforts are steering the food system toward nonviolent ends, and to those long passed whose physical and mental travail showed us the path ahead. I am grateful to all the storytellers who have kept these songs alive. In particular, I'd like to thank the librarians and archivists at the National Archives and Records Administration, the Library of Congress, the University of Alabama at Birmingham Medical Library, Harvard Business School's Baker Library, the High Museum, the University of Maryland Library System, and Anne Arundel Community College. A project like this relies on information-sharing networks that are easy to take for granted. I am grateful to all who keep these systems running under challenging circumstances: webmasters, programmers, interns, and scores of helpful strangers who process interlibrary loans.

This book is also dedicated to friends and family who are no longer with us: David Giltrow, who helped build agricultural extensions in East Africa; Diana Lavik, who fought for the economic rights of fruit pickers in Southern California; Betty Baker, who worked as a Peace Corps volunteer in Ghana; Arnie Gibson, who grew up picking almonds as a migrant worker in California; and Wally and Ellie, two retrievers whose happy wags cannot be forgotten nor replaced.

This book would not exist without Than Saffel, Kristen Bettcher, Marguerite Avery, Sarah Munroe, and Derek Krissoff at West Virginia University Press. Nor would it exist without the reviewers and copyeditors who helped shape its final form. One reviewer in particular—Jamey Essex—offered invaluable feedback at every phase in its development.

Books like this are written in liminal spaces, tiny elsewheres tucked into busy schedules where small moments are translated into priceless reflection. I am grateful to all who have created these breathing spaces over so many years: Alex Peimer, Alex Colucci, Claire Welty, Crista Johnson, Emily Midgley, Stewart Pharis, Sokvisal Kimsroy, Peggy Giltrow, Greg and Donna Smith, Kevin and Susan Dodd, Roy Lavik, and Albina David, among others.

Most of all, I am grateful for family: for my mother who gave me a love for food, my father who gave me a love for storytelling, and my sister who gave me a love for words and the kaleidoscope of ideas they imperfectly conjure; for Erin, who endured those sleepless nights and disillusions and responded with joys, reassurances, and unwavering optimism; for William, my partner in reading and reflection, and the inspiration for bolder solutions; and for Zippy and Misty, two food-driven mutts whose obsession with early breakfast meant that much of this book was written before dawn.

Finally, I am grateful for the many who entered my life as students only to become collaborators, advocates, devil's advocates, and friends: Lauren Neville, Ella Westerfeld, Madison Jenack, Deepi Ramesh, Zoe Lemke, Maura Campbell, Bronwyn Burgess, Kate Callin, Shawn McMahon, Jon Boles, John Tevis, Gabriella Jarboe, Sarah Maller, and Frank Lascala. Each in their own way ensured this book took the bold path and not the timid one.

# NOTES

## PREFACE

1 See Duell, "Is 'Local Food' Sustainable?"

2 For coffee, see Teixeira, "Picked by Slaves." For spinach, see Environmental Protection Agency, "Symptoms of Exposure to Pesticide Products." For grapes, see O'Neill, "Inmates Will Soon Be Growing Wine Grapes." For chicken and meatpackers, see chapter 6.

3 For example, see Pollan, *In Defense of Food*; Pollan, *Omnivore's Dilemma*; Schlosser, *Fast Food Nation*; Lorr, *Secret Life of Groceries*; Egan, *How to Be a Conscious Eater*; Kenner, *Food, Inc.*; Monbiot, *Regenesis.*

4 Galeano, *Open Veins of Latin America*, 8.

## INTRODUCTION

1 United Nations FAO, *Agricultural production statistics 2010–2023*

2 Lemkin, *Axis Rule in Occupied Europe*, 45; See also Madley, "Reexamining the American Genocide Debate."

3 See Oshinsky, *Worse Than Slavery.*

4 Davidson, *Broken Heartland.*

5 Sinclair, *The Jungle.*

6 McWilliams, *Factories in the Field.*

7 Steinbeck, *Grapes of Wrath*; Steinbeck, *In Dubious Battle.*

8 Villarejo, "Health of US Hired Farm Workers"; Svensson et al., "Migrant Agricultural Workers."

9 Cochrane, *Development of American Agriculture.*

10 See Chen et al., "Consumption of Ultra-Processed Foods and Health Outcomes." Nine million is the number of adults and children living in households classified as having very low food security. US Department of Agriculture (USDA) Economic Research Service (ERS) data for 2022.

11 See Muir, *Broken Promise of Agricultural Progress.*

12 Buller and Roe, *Food and Animal Welfare.*

13 Speech delivered on May 10, 1865, at the American Anti-Slavery Society.

14 One could argue that the majority of slaves in the South should not be considered part of the food system because they labored under cotton. However, slaves working in cotton were also involved in the operation of the food system, in fruit and vegetable cultivation, animal husbandry, food preparation, and consumption. Furthermore, the decision to grow cotton precluded other cultivation options (like corn). As such, cotton production impacted the food system through the limits it placed on land available for food crops.

15 On the temporality of violence, see Nixon, *Slow Violence and the Environmentalism of the Poor.*

16 McIlwaine, "Geography and Development."

17 See Galtung, "Violence, Peace, and Peace Research," 168.

18 For a detailed look at this line of reasoning, see Tyner and Inwood, "Violence as Fetish."

19 Jigsaw puzzle: Wageningen University, "New Course Shows Bigger Picture of the Food System." Web: Bloom et al., "Integrating Food Systems and Local Food." Battlefield: Patel, *Stuffed and Starved.* Rubber band: Lawrence, "Pigs' Revenge." Vast machine: Leonard, "Whole Food System in Britain." Complex adaptive system: Nesheim et al., *Overview of the U.S. Food System.* Runaway train: Aubrun et al., "Conceptualizing US Food Systems."

20 Rice, *Famine in the Remaking.*

21 The volatility and decentralization of food provisioning has led some to call this a food "network" rather than a "system." See McDonald, *Food Power.*

22 Albritton, *Let Them Eat Junk.*

23 Other strategies include changes in the commodity portfolio and the emergence of new uses for agricultural products. But extensification and intensification have historically driven growth in US agriculture.

24 Cochrane, *Development of American Agriculture.*

25 Cochrane.

26 Cochrane, *Curse of American Agricultural Abundance.*

27 Steinbeck, *Grapes of Wrath*, 449.

28 Laudan, "Words Matter." As Laudan reminds us, calling the food system "broken" has repercussions that may help or hinder productive action.

29 Garth and Reese, *Black Food Matters.*

30 USDA ERS data as of December 22, 2021. This figure does not include industries that supply inputs to agriculture.

31 See Harvey, "Spatial Fix"; Harvey, "Globalization and the 'Spatial Fix.'"

32 Following Dunbar-Ortiz, *Indigenous Peoples' History of the United States.*

33 Following the terminological guidelines established by Densho.org: Densho, "Terminology."

34 See section 5.7 of American Psychological Association, *Publication Manual of the American Psychological Association.*

35 Younging, *Elements of Indigenous Style.*

## CHAPTER ONE

1 See Diné Policy Institute, "Diné Food Sovereignty"; Stern, "Navajo Nation."

2 For a discussion of the applicability of genocide in the Native American context, see Madley, "Reexamining the American Genocide Debate."

3 Today, the US military still uses "Indian country"—or its shortened form, "In country"—to denote enemy territory. See Dunbar-Ortiz, *Indigenous Peoples' History of the United States.*

4 National Congress of American Indians, "NCAI Response to Usage of the Term, 'Indian Country.'"

5 Over one hundred books with "American agriculture" in the title have been published since 2000. Exactly whom those books are about goes broadly unquestioned.

6 Wolfe, "Settler Colonialism and the Elimination of the Native."

7 Williams, *Savage Anxieties.*

8 Younging, *Elements of Indigenous Style.* On the refusal to fetishize suffering, see also Tuck and Yang, "R-Words: Refusing Research." As a non-Indigenous person educated in a Western tradition, it is all too easy for me to repeat a long tradition of knowledge expropriation. To avoid this, I focus primarily on the structural relationships between settler and Indigenous societies, rather than the lived experiences of individuals and communities. Such memories are integral to Indigenous oral traditions and traditional knowledge. Where Indigenous authors have published these experiences, I reference them, but I try not to retell them. After all, these stories belong to the communities that have kept them alive, often against extraordinary odds.

9 For example, Midtrød, *Memory of All Ancient Customs*; McDonnell, *Dispossession of the American Indian*; Ouden, *Beyond Conquest*; Edwards and Petrikowski, *Relocation of Native Peoples of North America*; Immerwahr, *How to Hide an Empire*; Kluger, *Seizing Destiny.*

10 Dunbar-Ortiz, *Indigenous Peoples' History of the United States,* 105–6.

11 For other critiques of the "hunger" metaphor, see Blaakman, *Speculation Nation.*

12 See Calo, "Yeoman Myth."

13 See Phillips, "American Agrarianism's Answers to the Nation's (In) Securities."

14 McEwan, *Thomas Jefferson, Farmer,* viii.

15 Wetherington, *American Agriculture.*

16 Historian and economist Willard Cochrane describes land speculation in this period as a "national disease." See Cochrane, *Development of American Agriculture*; Blaakman, *Speculation Nation.*

17 For example, see Pelzer, *Planters Against Peasants.*

18 The Indigenous origin of some US state names is not always clear. See Bureau of Indian Affairs, "Origin of Names of US States"; National Park Service History, Electronic Library and Archive, "History of the Daniel Boone National Forest."

19 For this sequence of events, see Wetherington, *American Agriculture*; Hurt, *American Agriculture*.

20 Pauline Wakeham refers to these types of explanations as settler colonialism's "calculated alibis." See Wakeham, "Slow Violence of Settler Colonialism."

21 Blaakman, *Speculation Nation*.

22 Cession 11, cession 75, and cession 203, respectively. See Kappler, *Indian Affairs*.

23 See, for example, article V, Treaty with the Cherokee (1785), in Kappler, 8.

24 See Dunbar-Ortiz, *Indigenous Peoples' History of the United States*. For a contrasting view, see Prucha, *American Indian Treaties*.

25 See, for example, article I, Treaty with the Choctaw (1802), in Kappler, *Indian Affairs*, 63.

26 See, for example, article I, Treaty with the Delawares (1804), in Kappler, 70.

27 Dunbar-Ortiz, *Indigenous Peoples' History of the United States*, 101.

28 Report of an Onondaga chief who witnessed the massacres, quoted in Kiernan, *Blood and Soil*, 325.

29 Kiernan, 327.

30 Thomas Jefferson memorandum to Henry Dearborn on Indian policy, December 29, 1802, Founders Online.

31 Thomas Jefferson letter to Henry Dearborn, August 13, 1802, Founders Online.

32 Thomas Jefferson letter to Henry Dearborn, August 28, 1807, quoted in Madley, "Reexamining the American Genocide Debate," 109.

33 Article I, Treaty with the Chickasaw (1805), in Kappler, *Indian Affairs*, 79.

34 Cession 46, ceded in Treaty with the Choctaw (1802).

35 This principle, known as the Doctrine of Discovery, dates back to the Spanish conquest of the New World. See Williams, *Savage Anxieties*.

36 Smith and Cole, *Fluctuations in American Business*.

37 Dunbar-Ortiz, *Indigenous Peoples' History of the United States*.

38 Immerwahr, *How to Hide an Empire*, 39.

39 The territory surrendered in the Treaty of Guadalupe Hidalgo included what is today California, Nevada, Utah, New Mexico, and parts of Arizona, Colorado, Oklahoma, Kansas, and Wyoming.

40 Dunbar-Ortiz, *Indigenous Peoples' History of the United States*, 37.

41 Quoted in Eichler, "Ecocide Is Genocide," 116.

42 Madley, "Reexamining the American Genocide Debate"; Madley, *An American Genocide.*

43 Cochrane, *Development of American Agriculture*; Hurt, *American Agriculture.*

44 Hurt, *American Agriculture.*

45 See section IV, "An Act Donating Public Lands to the Several States and Territories Which May Provide Colleges for the Benefit of Agriculture and Mechanic Arts," US Congress, July 2, 1862.

46 Hurt, *American Agriculture.*

47 Cochrane, *Development of American Agriculture*, 59.

48 Vantrease, "Commod Bods and Frybread Power," 57.

49 See chapter XIV, "An Act Making Appropriations for the Current and Continent Expenses of the Indian Department," US Congress, February 27, 1851.

50 Cochrane, *Development of American Agriculture*, 77.

51 Statistics aggregated from US Congress, Board of Indian Commissioners, 1891, *Annual Report*, 52nd Cong., 1st session, pgs. 843–882. Values: 461,000 pounds of corn; 7,908,500 pounds of wheat flour; and 35,697,100 pounds of beef.

52 Act of Congress, March 3, 1871, 16 Statute 566, in Kappler, *Indian Affairs.*

53 Indian Land Tenure Foundation, "Land Tenure History."

54 Native Governance Center, "Blood Quantum and Sovereignty."

55 See Kappler, *Indian Affairs*; Indian Land Tenure Foundation, "Land Tenure History."

56 Dunbar-Ortiz, *Indigenous Peoples' History of the United States.*

57 Wolfe, "Settler Colonialism and the Elimination of the Native," 400.

58 Speech by Captain Henry Richard Pratt at the National Conference of Charities and Corrections, Nineteenth Annual Session, June 23–29, 1892, Denver. Pratt was a military veteran of the Indian Wars and

founder of the Carlisle Indian Industrial School in Carlisle, Pennsylvania.

59 For more on the definition of cultural genocide and its applicability in the settler-colonial context, see Madley, "Reexamining the American Genocide Debate"; Rice, "Genocide."

60 McKinney, "History of the Albuquerque Indian School."

61 US Congress, Board of Indian Commissioners, 1891, *Annual Report*, 52nd Cong., 1st sess.

62 City of Albuquerque, "Albuquerque Indian School Cemetery at 4H Park."

63 Sims, "Indian School Graves Rediscovered Under City Park."

64 Cochrane, *Development of American Agriculture*, 100.

65 From Franklin D. Roosevelt, Fireside Chat 8, "On Drought Conditions," September 6, 1936.

66 See Institute for Government Research, "Problem of Indian Administration."

67 Wolfe, "Settler Colonialism and the Elimination of the Native," 400.

68 Diné Policy Institute, "Diné Food Sovereignty."

69 Kelsey Timler and Dancing Water Sandy refer to these practices as "culinary imperialism." See Timler and Sandy, "Gardening in Ashes," n.p.

70 Diné Policy Institute, "Diné Food Sovereignty," 48.

71 See Diné Policy Institute.

72 Hormel and Norgaard, "Bring the Salmon Home!," 356.

73 Data are from 2011. See Diné Policy Institute, "Diné Food Sovereignty"; Norgaard et al., "Continuing Legacy."

74 Nesterak, "1950s Plan to Erase Indian Country."

75 Norgaard et al., "Continuing Legacy."

76 Nesterak, "1950s Plan to Erase Indian Country."

77 From Ruth Underhill 1957 radio series *Indian Country*, quoted in Nesterak.

78 Walls and Whitbeck, "Intergenerational Effects of Relocation Policies on Indigenous Families."

79 Quoted in Nesterak, "1950s Plan to Erase Indian Country."

80 For a more complete definition of the carceral fix, see Rice, "Divide and Cultivate."

81 For more on surplus life and labor, see Tyner, *Violence in Capitalism.*
82 Grey and Patel, "Food Sovereignty as Decolonization."
83 See Polanyi, *Great Transformation.*
84 National Congress of American Indians, "NCAI Response to Usage of the Term, 'Indian Country.'"
85 Vizenor, "Aesthetics of Survivance"; Dunbar-Ortiz, *Indigenous Peoples' History of the United States.*
86 See Coté, "Hishuk'ish Tsawalk—Everything Is One."
87 Renwick, "Healing the Land and Spirits Through a Community Garden," n.p.
88 Bock, "Planting the Seeds of Change."
89 Coté, "Hishuk'ish Tsawalk—Everything Is One," 47.

## CHAPTER TWO

1 For example, see the different accounts in James, "'Buried in Guano'"; Lesher, "Load of Guano"; Skaggs, *Great Guano Rush.*
2 Cushman, *Guano and the Opening of the Pacific World.*
3 President Millard Fillmore, State of the Union Address, December 2, 1850.
4 From "Chinese Coolie Trade," *Christian Review* (1862), quoted in Foster and Clark, *Robbery of Nature*, 60.
5 *Jones v. United States*, 137 US 202 (1890).
6 In addition to nitrogen and phosphorus, potassium (K) is often added to agricultural soils.
7 Cushman, *Guano and the Opening of the Pacific World*; Foster and Clark, *Robbery of Nature.*
8 Crook and Short, "Marx, Lemkin and the Genocide–Ecocide Nexus," 300.
9 Crook and Short, 300.
10 This breakdown of strategies is derived from Cushman, *Guano and the Opening of the Pacific World.*
11 See Kapayou et al., "Reuniting the Three Sisters."
12 Melillo, "First Green Revolution."
13 Tajima, "Marketing of Urban Human Waste"; Melillo, "First Green Revolution."

14 Foster and Clark, *Robbery of Nature.*

15 Hugo, *Les Miserables*, bk. 2, i.

16 Marx, *Grundrisse*, 527.

17 Quoted in Mann, *1493*, 214.

18 Justus von Liebig, quoted in Bear, *Theory and Practice in the Use of Fertilizers*, 190.

19 Marx, *Capital*, 860.

20 Jordan, "Peruvian Guano Gospel in the Old South."

21 Foster and Clark, *Robbery of Nature.*

22 Cochrane, *Development of American Agriculture.*

23 Crawford, "Cotton, Land, and Sustenance," 228.

24 Crawford.

25 Foster and Clark, *Robbery of Nature*; Cochrane, *Development of American Agriculture*; Hurt, *American Agriculture.*

26 Quoted in Crawford, "Cotton, Land, and Sustenance," 230.

27 Quoted in Crawford, 245.

28 Crawford, 216.

29 Hurt, *American Agriculture*; Genovese, *The Political Economy of Slavery*, 85.

30 Munita, "Guano."

31 Vigne, *Travels in Mexico*, 2:153.

32 Cushman, *Guano and the Opening of the Pacific World*; Foster and Clark, *Robbery of Nature.*

33 Foster and Clark, *Robbery of Nature.*

34 Quoted in Jordan, "Peruvian Guano Gospel in the Old South," 219.

35 Foster and Clark, *Robbery of Nature.*

36 Melillo, "First Green Revolution."

37 Quoted in Higgins, *Lima*, 113.

38 Foster and Clark, *Robbery of Nature*; Mann, *1493.*

39 *Vermont Watchman and State Journal*, December 27, 1844, quoted in Immerwahr, *How to Hide an Empire.*

40 Immerwahr, *How to Hide an Empire*, 49.

41 Vigne, *Travels in Mexico*, 2:154.

42 Melillo, "First Green Revolution."

43 In Peru, the coolie trade was known as *la trata amarilla* ("the yellow trade"). The peak of the coolie trade coincided with the Taiping Rebellion in China (1850–64), a civil war that claimed the lives of 20 to 70 million people.

44 Mann, *1493*.

45 Quoted in Melillo, "First Green Revolution," 1039.

46 Stewart, *Chinese Bondage in Peru*, 96.

47 Nash and Parish, *Plough, the Loom and the Anvil*, 72.

48 Dispatch from D. J. Williamson to Secretary of State, Callao, September 20, 1870, quoted in Stewart, *Chinese Bondage in Peru*, 98.

49 Cushman, *Guano and the Opening of the Pacific World*, 55.

50 Jordan, "Peruvian Guano Gospel in the Old South."

51 Navassa remains disputed territory today.

52 Lesher, "Load of Guano."

53 Lesher, 127.

54 *Kansas City Journal*, February 12, 1899, p. 17.

55 "The Navassa Riot," *Red Cloud Chief* (Webster County, NE), October 18, 1889, n.p.

56 See *Kansas City Journal*, February 12, 1899, p. 17, and Immerwahr, *How to Hide an Empire*.

57 There is considerable debate over when capitalist agriculture emerged, particularly in the United States. See, for example, Kulikoff, *Agrarian Origins of American Capitalism*; Bernstein, *Class Dynamics of Agrarian Change*.

58 See Foster and Clark, *Robbery of Nature*, 126.

59 For this reason, Marx referred to commodities as "fetishes." Marx, *Capital*, 163.

60 Crawford, "Cotton, Land, and Sustenance," 245; Stoll, *Larding the Lean Earth*, 189.

61 Melillo, "First Green Revolution."

62 Immerwahr, *How to Hide an Empire*.

63 Melillo.

64 Mcdonald, "Island with No Water"; Edwards, "Phosphate Mining and the Relocation of the Banabans to Northern Fiji in 1945." Even Navassa's rich fertilizer turned out to be mostly rock phosphate and not bird droppings.

65 Johnson, "Nitrogen Nation."

66 Cushman, *Guano and the Opening of the Pacific World*, 14.

67 Munita, "Guano"; Ybarra Zavala, "Guano Islands."

68 Wills, "Are We Entering a New Golden Age of Guano?"; Collyns, "Worldwide Fertiliser Shortage Prompts Peru to Turn to Bird Poo."

**CHAPTER THREE**

1 The famous opening line of President Roosevelt's speech before Congress declares "Yesterday, December 7, 1941—a date which will live in infamy—the United States of America was suddenly and deliberately attacked by the naval and air forces of the Empire of Japan."

2 Taylor, "People Nobody Wants," 66.

3 Krebs, "Bitter Harvest."

4 For this figure, see Huang, "Project to Be Unveiled in LA." The use of the term "concentration camp" is not intended to conflate the experiences of Japanese Americans with those of Holocaust victims and survivors or equate these very different histories. See Densho, "Terminology."

5 For historical overviews, see Chiang, *Nature Behind Barbed Wire*; Ng, *Japanese American Internment During World War II*. For personal accounts, see Hansen, *Japanese American World War II Evacuation Oral History Project*; Uchida, *Desert Exile*; Inada, *Only What We Could Carry*.

6 On July 2, 1948, President Truman signed the Japanese American Evacuation Claims Act, repaying a small fraction of estimated losses. On August 10, 1988, President Reagan signed HR 442, issuing an apology to the Japanese American community and offering $20,000 remuneration to surviving victims of wartime relocation.

7 I follow the terminological recommendations of Densho. As such, I use "forced removal" instead of "evacuation," "incarceration" instead of "internment," "Japanese American" instead of "Japanese," and "concentration camps" instead of "relocation centers." Roosevelt and other government officials used the term "concentration camps" in internal memos. See Densho, "Terminology." To properly reflect extant conditions, I refer to individuals in the process of removal as "evacuees" and those incarcerated at the camps as "inmates," not "internees."

8 See Fisher, *Exile of a Race*; Krebs, "Bitter Harvest," n.p.

9 Taylor, "People Nobody Wants," 66.

10 Commission on Wartime Relocation and Internment of Civilians, "Personal Justice Denied."

11 Lillquist puts the number at 43,775 out of approximately 97,000. See Lillquist, "Imprisoned in the Desert," 539.

12 Morehouse, "Farming Behind Barbed Wire."

13 Iwata, "Japanese Immigrants in California Agriculture," 28.

14 Bailey, "California, Japan, and the Alien Land Legislation of 1913."

15 Lee, "'Yellow Peril' and Asian Exclusion in the Americas."

16 See Fisher, *Exile of a Race*, 43.

17 Iwata, "Japanese Immigrants in California Agriculture."

18 Iwata, 30.

19 Iwata, 33.

20 Castillo, *Cross of Thorns*.

21 Taylor, "People Nobody Wants," 66.

22 Iwata, "Japanese Immigrants in California Agriculture," 31.

23 Iwata; Fisher, *Exile of a Race*.

24 McClatchy and Japanese Exclusion League of California, "Japanese Immigration and Colonization," 14.

25 Taylor, "People Nobody Wants," 66; McClatchy and Japanese Exclusion League of California, "Japanese Immigration and Colonization," 14–15. See the specific use of "colonization" on p. 15.

26 Iwata, "Japanese Immigrants in California Agriculture."

27 See McWilliams, *Factories in the Field*.

28 Iwata, "Japanese Immigrants in California Agriculture," 36.

29 Fisher, *Exile of a Race*, 33.

30 *San Francisco News*, "Jap Ban to Force Farm Adjustments."

31 Iwata, "Japanese Immigrants in California Agriculture."

32 Data from Tokunaga, "Japanese Internment as an Agricultural Labor Crisis."

33 Iwata, "Japanese Immigrants in California Agriculture."

34 Commission on Wartime Relocation and Internment of Civilians, "Personal Justice Denied," 122.

35 Taylor, "People Nobody Wants," 66.

36 Fisher, *Exile of a Race.*

37 Taylor, "People Nobody Wants," 66.

38 Tokunaga, "Japanese Internment as an Agricultural Labor Crisis," 80.

39 Fisher, *Exile of a Race*, 38.

40 US Army Western Defense Command, "Final Report," 34.

41 Tokunaga, "Japanese Internment as an Agricultural Labor Crisis," 93.

42 McLemore, "This Is War!"

43 Tokunaga, "Japanese Internment as an Agricultural Labor Crisis," 84.

44 Tokunaga.

45 *Bainbridge Island Review*, "Editorial."

46 Tokunaga, "Japanese Internment as an Agricultural Labor Crisis," 86.

47 Ten Broek et al., *Prejudice, War, and the Constitution.*

48 Tokunaga, "Japanese Internment as an Agricultural Labor Crisis," 94–95.

49 From USDA surveys of California's County Farm Labor subcommittees. See P. A. Minges, "Report on the Effects of the Japanese War on the Japanese Alien and Native-Born Vegetable Growers of California," December 29, 1941, Carton 2, Ralston Papers, Bancroft Library, University of California, Berkeley

50 See, for example, Evans, "Vacation for Victory"; Taylor, "People Nobody Wants."

51 Balderrama and Rodriguez, *Decade of Betrayal.*

52 Tokunaga, "Japanese Internment as an Agricultural Labor Crisis," 99.

53 Tokunaga, 99.

54 Fisher, *Exile of a Race*, 36.

55 Burton, *Confinement and Ethnicity.*

56 Lillquist, "Imprisoned in the Desert," 15.

57 Commission on Wartime Relocation and Internment of Civilians, "Personal Justice Denied," 121.

58 From a Los Angeles witness to evacuation, quoted in Commission on Wartime Relocation and Internment of Civilians, 123.

59 Commission on Wartime Relocation and Internment of Civilians, 126.

60 Commission on Wartime Relocation and Internment of Civilians, 126–27.

61 Chiang, *Nature Behind Barbed Wire*, 30.

62 This included orphans who were only one-half Japanese and living in white homes. See Lillquist, "Imprisoned in the Desert," 531.

63 Chiang, *Nature Behind Barbed Wire*, 30.

64 From the *Minidoka Irrigator*, September 10, 1942, cited in Chiang, 98.

65 Other criteria included distance from the West Coast and from military targets, availability of land, adequate water for irrigation and domestic use, and proximity to transportation infrastructure. See Lillquist, "Imprisoned in the Desert."

66 Chiang, *Nature Behind Barbed Wire*, 44.

67 Chiang, 45.

68 Lillquist, "Imprisoned in the Desert."

69 Chiang, *Nature Behind Barbed Wire*, 47.

70 Chiang.

71 Lillquist, "Imprisoned in the Desert," 412.

72 Chiang, *Nature Behind Barbed Wire*; Lillquist, "Imprisoned in the Desert."

73 Quoted in Morehouse, "Farming Behind Barbed Wire," n.p.

74 Chiang, *Nature Behind Barbed Wire*; Lillquist, "Imprisoned in the Desert."

75 For an excellent discussion of Japanese American labor as *prison* labor, see Hinnershitz, *Japanese American Incarceration*.

76 Chiang, *Nature Behind Barbed Wire*.

77 Chiang.

78 Morehouse, "Farming Behind Barbed Wire."

79 Taylor, *Jewel of the Desert*.

80 Lillquist, "Imprisoned in the Desert."

81 War Relocation Authority, "California Incidents of Terrorism Involving Persons of Japanese Ancestry."

82 Commission on Wartime Relocation and Internment of Civilians, "Personal Justice Denied," 124.

83 Hansen, "Japanese Internment."

84 See Harvey, *New Imperialism*.

85 Murphy in *Korematsu v. United States*.

86 Kurashige, *Two Faces of Exclusion*.

87 As a starting point for examination of racial capitalism, see Robinson, *Cedric J. Robinson*; Du Bois, "Inter-Racial Implications of the Ethiopian Crisis."

88 Robinson, *Black Marxism*; Frydenlund and Dunn, "Refugees and Racial Capitalism," 5.

## CHAPTER FOUR

1 *The Movement*, "Mississippi Negroes Occupy Air Base."

2 Mississippi Freedom Democratic Party, "Why We Are Here at the Greenville Air Force Base," 1.

3 US Bureau of the Census, "United States Census of Agriculture."

4 Cobb, "'Somebody Done Nailed Us on the Cross,'" 927.

5 Staff, "Poverty and the Federal Government," 18.

6 Dr. Raymond Wheeler, quoted in "Hunger in America, Part 1."

7 Washington County, Mississippi, infant mortality for Blacks in 1967 was 62.2 per 1,000 live births. Vietnam's estimated infant mortality rate for 1965–75 was 37.4. See US Senate, "Examination of the War on Poverty"; Savitz et al., "Vietnamese Infant and Childhood Mortality."

8 For mention of movie theaters, see Cochrane, *Curse of American Agricultural Abundance*; all other instances, see US House of Representatives, "Commodity Credit Corporation Grain Storage Activities."

9 USDA, "History of CCC Bin Storage Program."

10 For a list of private storage companies and earnings, see US House of Representatives, "Commodity Credit Corporation Grain Storage Activities," 36.

11 Phillips, "Price of Plenty."

12 Rose, *Powers of Freedom*, 58.

13 For example, see Smith, "Mississippi's War Against the War on Poverty."

14 Davidson, *Broken Heartland*, 8–9.

15 Franklin Delano Roosevelt, 1936, from Fireside Chat number 8, September 6, 1936.

16 McDonald, *Food Power*, 58.

17 Cochrane, *Curse of American Agricultural Abundance*, 11.

18 Cochrane, 11.
19 Cochrane, *Development of American Agriculture.*
20 Fitzgerald, *Every Farm a Factory*, 31.
21 Cochrane, *Curse of American Agricultural Abundance*, 29.
22 National Agricultural Statistics Service, *Farms and Land in Farms.*
23 Olmstead and Rhode, "Reshaping the Landscape."
24 Cobb, "'Somebody Done Nailed Us on the Cross.'"
25 Fitzgerald, *Every Farm a Factory.*
26 Blackmon, *Slavery by Another Name*; Davis, *Good and Faithful Labor.*
27 The USDA referred to this category as "croppers" whose landlords "furnish all work power," "land, management, equipment," and closely supervise the work. See US Bureau of the Census, "United States Census of Agriculture," 914.
28 Davidson, *Broken Heartland.*
29 National Bureau of Economic Research, Wholesale Price of Cotton Yarn, Mills for United States (M0400AUSM267NNBR) and Cotton Crop for United States (A01028USA558NNBR), from FRED, Federal Reserve Bank of St. Louis.
30 US Bureau of the Census, "United States Census of Agriculture."
31 For an example in this subject area, see Hippel, *Chemical Age.*
32 For example, see Mumford, *Technics and Civilization*; Latour, "Technology Is Society Made Durable."
33 Salisbury, *Manitou and Providence*; Schwartz, "Colonial New England Agriculture." In fact, this revolution built upon an even earlier crosspollination: the Columbian Exchange of genetic material from the New World to Europe in the fifteenth and sixteenth centuries. See Nunn and Qian, "Columbian Exchange."
34 Wallach, *How America Eats*, 71.
35 Olmstead and Rhode, "Reshaping the Landscape."
36 Olmstead and Rhode.
37 War Department, "Report on the Fixation and Utilization of Nitrogen," 201.
38 Jones, "Nitrogen," 391, 431.
39 Johnson, "Nitrogen Nation," 217.
40 Johnson.

41 Johnson, 211.

42 See Vail, *Chemical Lands.*

43 For more information on the evolution of seed technology, see Kloppenburg, *First the Seed*; Borowiak, "Farmers' Right."

44 See Latour, *Pasteurization of France.* For an Indigenous perspective, see Kimmerer, *Braiding Sweetgrass.*

45 See Tyner, "Violence, Surplus Production, and the Transformation of Nature."

46 Dunbar-Ortiz, *Indigenous Peoples' History of the United States.*

47 Olmstead and Rhode, "Reshaping the Landscape."

48 These statistics assume the worker is one of those mouths that must be fed. See Cochrane, *Curse of American Agricultural Abundance.*

49 Cochrane, 38–39.

50 Cochrane, 15.

51 Rasmussen et al., "Short History of Agricultural Adjustment."

52 USDA Economic Research Service.

53 Cobb, "'Somebody Done Nailed Us on the Cross.'"

54 Cobb.

55 See McDonald, *Food Power.*

56 See Hale, *Farm Chemurgic*; USDA, *Yearbook of Agriculture 1950–51.*

57 US Department of Energy, Alternative Fuels Data Center, data for 2022.

58 Zulauf et al., "Stock-to-Use Ratios"; Phillips, "Price of Plenty."

59 Phillips, "Price of Plenty."

60 Cobb, "'Somebody Done Nailed Us on the Cross'"; "Hunger in America, Part 1"; US Senate, "Examination of the War on Poverty."

61 US Senate, "Examination of the War on Poverty," 159.

62 Cochrane, *Curse of American Agricultural Abundance*, 15.

63 Phillips, "Price of Plenty," 606.

64 The 1966 amendment to the Fair Labor and Standards Act (FLSA) applied only to large farms, that is, those with sufficient capital to industrialize. See US Department of Labor, "History of Changes to the Minimum Wage Law"; Cobb, "'Somebody Done Nailed Us on the Cross,'" 918.

65 Cobb, "'Somebody Done Nailed Us on the Cross,'" 934; US Senate, "Examination of the War on Poverty."

66 See Fernandez-Cornejo et al., "Pesticide Use in US Agriculture," 70.

67 To be sure, Congress had humane intentions when it raised the minimum wage in 1967. Had landowners held onto their croppers—as minimum wage advocates hoped—higher pay may have helped croppers afford food stamps. In hindsight, it was foolish to raise the minimum wage when cheap alternatives to labor were already in hand.

68 Smith, *Uneven Development*; Harvey, "Globalization and the 'Spatial Fix.'"

69 Winders, *Politics of Food Supply.*

70 Kulikoff, *Agrarian Origins of American Capitalism.*

71 Schumpeter, *Capitalism, Socialism, and Democracy.*

72 Myrdal, *American Dilemma.*

73 US Senate, "Examination of the War on Poverty," 160.

74 The idea of restricting technology was first introduced by Willard Cochrane, chief USDA economist under President Kennedy. See Levins, *Willard Cochrane and the American Family Farm.*

75 Case in point, Southern sharecroppers who owned mules could usually negotiate better contract terms. See Cobb, "'Somebody Done Nailed Us on the Cross.'"

76 For an excellent account of how industrial advocates mobilized a discourse of agricultural "backwardness," see Fitzgerald, *Every Farm a Factory.*

77 USDA, "U.S. Public Agricultural Research," 2.

## CHAPTER FIVE

1 Communique from American Embassy, Guatemala to US Secretary of State, Washington, D.C., Subject: "Evidence of Continuing Human Rights Violations in Chimaltenango," November 10, 1982, GUATEM 08491. US Department of State Archive, Bureau of Public Affairs: Office of the Historian.

2 Communique from American Embassy.

3 Doyle et al., "Ex-Kaibil Officer Connected to Dos Erres Massacre."

4 Nolan-Ferrell, "Perspective."

5 I use "Third World" not to imply a particular "level" or "stage" of development, but to denote those colonies that gained independence following World War II that had to balance the influence of

free-market countries ("First World") and communist countries ("Second World").

6 As a starting point, see for Hawai'i: Siler, *Lost Kingdom*. Puerto Rico: Carro-Figueroa, "Agricultural Decline and Food Import Dependency in Puerto Rico." Colombia: Escobar, *Encountering Development*. Central America: Koeppel, *Banana*; Galeano, *Open Veins of Latin America*. Egypt: Hedley and Peacock, "Food for Peace." Cambodia: Rice, *Famine in the Remaking*. Sudan and Somalia: Hancock, *Lords of Poverty*. Bangladesh: McHenry and Bird, "Food Bungle in Bangladesh." Malawi: Zerbe, "Feeding the Famine?" Iraq: Gazdar, "Pre-modern, Modern and Post-modern Famine in Iraq."

7 As a starting point, see McDonald, *Food Power*; Hancock, *Lords of Poverty*; Thompson, "Food Aid and the Famine Relief Argument"; Hurt, *Green Revolution in the Global South*.

8 The annexation of Hawai'i was not a consequence of the Spanish-American War. US control of the cane market is based on the following production estimates for 1900 (all in tons): Hawaii (300,000), Puerto Rico (~200,000), Cuba (~200,000), Philippines (~250,000), global (~5.3 million).

9 Merleaux, *Sugar and Civilization*, 1.

10 Koivistoinen and Hyvönen, "Use of Sugar in Foods"; Mintz, *Sweetness and Power*.

11 Putzel, "Ladejinsky Model of Agrarian Reform."

12 Immerwahr, *How to Hide an Empire*.

13 Immerwahr.

14 Immerwahr, emphasis added.

15 William Howard Taft, speech given on May 2, 1910, at the Grant Day dinner of the Americus Club, Pittsburgh, PA, quoted in *The Independent* (New York) 68, no. 3206 (May 12, 1910).

16 Bernstein, *Class Dynamics of Agrarian Change*, 43.

17 Bernstein, 43. I use "peasant" to refer to small-scale farmers and tenants in colonial and postcolonial contexts.

18 Putzel, "Ladejinsky Model of Agrarian Reform."

19 Bucheli, *Bananas and Business*.

20 Galeano, *Open Veins of Latin America*.

21 Táíwò, "When the United Fruit Company Tried to Buy Guatemala."

22 Decree 2795. See Galeano, *Open Veins of Latin America.*

23 Under-Secretary of State Robert Olds, 1927 State Department memorandum, quoted in Tuathail, "Political Geography of Contemporary Events."

24 Butler, *War Is a Racket*, 10.

25 Schlesinger and Kinzer, *Bitter Fruit.*

26 Koeppel, *Banana.*

27 See Memorandum from the Chief of the Western Hemisphere Division, Central Intelligence Agency (King), to the Deputy Director for Plans of the Central Intelligence Agency (Wisner), Central Intelligence Agency, job 79-01025A, box 154, folder 2, Secret, drafted by King on July 22, 1954. US Department of State Archive, Bureau of Public Affairs: Office of the Historian.

28 Telegram from the Ambassador in Guatemala (Peurifoy) to the Department of State, Guatemala City, December 17, 1953, 7 p.m., National Archives and Records Administration, RG 59, Central File 611.14/12-1754.

29 See Memorandum from William Robertson of Operation PBSUCCESS to the Chief of the Project, Central Intelligence Agency, job 79-01025A, box 146, folder 4, Secret, RYBAT, PBSUCCESS. US Department of State Archive, Bureau of Public Affairs: Office of the Historian.

30 Gibbings and Vrana, *Out of the Shadow.*

31 Koeppel, *Banana.*

32 Galeano, *Open Veins of Latin America*, 107.

33 McDonald, *Food Power.*

34 Harry S. Truman, speech before a joint session of Congress recommending assistance to Greece and Turkey, March 12, 1947, now known as the Truman Doctrine.

35 Hubert Humphrey, testimony before the Senate Committee on Agriculture and Forestry, July 16, 1953, on considerations of S 2249 authorizing the use of surplus commodities for foreign assistance.

36 McDonald, *Food Power*, 49.

37 From a 1960 presidential campaign brochure for John F. Kennedy, "Every American Owes a Debt of Gratitude to the American Farmer."

38 McDonald, *Food Power*, 96.

39 See Hedley and Peacock, "Food for Peace."

40 The prohibition against trade with communist countries was removed by the 1970s. Purvis, "Some Observations on the Effects of PL 480 Wheat Sales."

41 Later amendments to the act created Title IV, which allowed the United States to provide loans for commercial food purchases. But these contributions were dwarfed by Title I.

42 For examples, see Bezuneh et al., "Food Aid Disincentives"; Abdulai et al., "Does Food Aid Really Have Disincentive Effects?"; Chabot and Dorosh, "Wheat Markets, Food Aid and Food Security in Afghanistan."

43 Hancock, *Lords of Poverty*, 169.

44 Hancock.

45 Hancock, 181.

46 Hancock.

47 Hancock; Economic Research Service, "12 Years of Achievement Under Public Law 480."

48 Zerbe, "Feeding the Famine?"; Clapp, "Financialization of Food."

49 Hedley and Peacock, "Food for Peace."

50 Carey, "Guatemala's Green Revolution."

51 Bates, "Recovery, Change and Development," 300.

52 Riding, "U.S. Food Aid Seen Hurting Guatemala."

53 Riding.

54 Trudeau, *Guatemalan Politics*.

55 Bates and Peacock, *Disaster and Social Change*, 324.

56 Gawronski and Olson, "Disasters as Crisis Triggers for Critical Junctures?"

57 See, for example, Orville Freeman's testimony at the Hearing Before the Select Committee on Hunger, 100th Congress, June 22, 1989.

58 Kloppenburg, *First the Seed*.

59 Hopkins, "Reform in the International Food Aid Regime."

60 For an overview of this transition in thinking, see Devereux, *New Famines*.

61 For an overview of how developmentalist narratives and policies emerged, see Latham, *The Right Kind of Revolution*.

62 See Baranski, *Globalization of Wheat.*

63 Escobar, *Encountering Development.*

64 Latham et al., *Hunger and Society.*

65 US Department of State, "2020 Investment Climate Statements"; Rice, "Madagascar's Famine Is More Than Climate Change."

66 Escobar, *Encountering Development*; Latham et al., *Hunger and Society.*

67 Webb, "When Projects Collapse."

68 Staudt, "Women's Politics and Capitalist Transformation in Subsaharan Africa."

69 Sachs, *Invisible Farmers*, 127.

70 Putzel, "Ladejinsky Model of Agrarian Reform," 185.

71 USDA, Economic Research Service.

72 Lappé and Collins, *Food First.*

73 Latham et al., *Hunger and Society.*

74 See, for example, "Letter from Thomas Jefferson to Henry Dearborn, 13 August 1802," Founders Online (original source: *The Papers of Thomas Jefferson*, vol. 38, *1 July–12 November 1802*, ed. Barbara B. Oberg [Princeton University Press, 2011], 209–210).

75 Escobar, *Encountering Development*, 166.

76 For these figures, see Galeano, *Open Veins of Latin America*; Carey, "Guatemala's Green Revolution."

77 Seay-Fleming, "Food Insecurity in the Northern Triangle," 12.

78 Carey, "Guatemala's Green Revolution," 297.

79 Seay-Fleming, "Food Insecurity in the Northern Triangle."

80 Von Braun et al., *Nontraditional Export Crops in Guatemala.*

81 Seay-Fleming, "Food Insecurity in the Northern Triangle," 9.

82 Grandia, "Poisonous Exports."

83 Carey, "Guatemala's Green Revolution," 298.

84 McDonald, *Food Power.*

85 See Phillips, "Price of Plenty."

86 Essex, *Development, Security, and Aid*, 82.

87 Essex, *Development, Security, and Aid.*

88 See Nketani and Wise, "Feeding Big Agribusiness, Starving Africans."

89 George, "More Food, More Hunger," 54.

90 Grandin, "Guatemalan Slaughter Was Part of Reagan's Hard Line."

91 Grandin.

92 Seay-Fleming, "Food Insecurity in the Northern Triangle";

93 Seay-Fleming, "Food Insecurity in the Northern Triangle."

**CHAPTER SIX**

1 Schwean-Lardner and Classen, "Lighting for Broilers." This 2010 study by Aviagen, Inc. recommends reducing daylight times from twenty-three hours per day to between seventeen and twenty. Twenty hours is now industry standard.

2 Schwean-Lardner and Classen.

3 According to the United Food and Commercial Workers (UFCW), in 2010 approximately 60 percent of plants failed to pay overtime to catchers. Cited in Stull and Broadway, *Slaughterhouse Blues.*

4 I use "animals" to refer to nonhuman animals and "humans" to refer to human animals. This convention is common in English but can reinforce anthropocentric and dualistic understandings of nature and society. See Castree, "Socializing Nature."

5 The idea that animal resistance is a primary force driving the evolution of technology in the meat industry follows the work of Dinesh Wadiwel. See Wadiwel, "Chicken Harvesting Machine."

6 For a detailed analysis of the global social and ecological costs of capitalist meat production, see Weis, *The Ecological Hoofprint.*

7 For example, the 2006 Animal Enterprise Terrorism Act. See Nibert, "Animals, Immigrants, and Profits."

8 See Specht, *Red Meat Republic.*

9 By the early 1980s, 95 percent of broilers were being raised under contract. See Stull and Broadway, *Slaughterhouse Blues.*

10 Statistic reported by UFCW, Stull and Broadway, 54.

11 Quandt et al., "3-D Jobs and Health Disparities"; Kirychuk et al., "Respiratory Symptoms and Lung Function."

12 Wallach, *How America Eats*, 197.

13 Musolin et al., "Health Hazard Evaluation Report." This report was produced by the CDC as part of the plant's request to increase line speeds.

14 Ken and León, "Necropolitical Governance and State-Corporate Harms."

15 Marx, *Capital*, 283.

16 Marx, 270.

17 Marx writes, "The value of labour-power is the value of the means of subsistence necessary for the maintenance of its owner." Marx, 274.

18 See, for example, Beldo, "Metabolic Labor"; Barua, "Lively Commodities and Encounter Value"; Barua, "Nonhuman Labour, Encounter Value, Spectacular Accumulation"; Hribal, "'Animals Are Part of the Working Class'"; Collard and Dempsey, "Capitalist Natures in Five Orientations"; Chhabra, "Animal Labourers and the Law in Colonial India."

19 Marx, *Capital*, 284.

20 Beldo, "Metabolic Labor"; Hribal, "'Animals Are Part of the Working Class.'"

21 See Tyner and Rice, "Agrarian Marxism, Animal Geographies, and Non-Human Labor."

22 Coulter, *Animals, Work, and the Promise of Interspecies Solidarity*; Wadiwel, "Chicken Harvesting Machine."

23 See Wadiwel, "Chicken Harvesting Machine," 530–31.

24 Beldo, "Metabolic Labor."

25 See Beldo.

26 Buller and Roe, *Food and Animal Welfare.*

27 Stull and Broadway, *Slaughterhouse Blues.*

28 Clapp, "The Financialization of Food: Who Is Being Fed," 2; Clapp, *Food.*

29 Fossum, "Precarity Capitalism and the Global Value Chain in Beef."

30 See Specht, *Red Meat Republic*; Wadiwel, "Chicken Harvesting Machine."

31 Dale et al., "Protein Feeds Coproduction in Biomass Conversion to Fuels and Chemicals."

32 Quoted in Smits, "Frontier Army and the Destruction of the Buffalo," 317.

33 Specht, *Red Meat Republic*, 17. See also Cunfer and Waiser, *Bison and People on the North American Great Plains*; Eichler, "Ecocide Is Genocide."

34 Smith, *Eating History.*

35 Cited in Horowitz, *Negro and White, Unite and Fight!*, 17.

36 Cronon, *Nature's Metropolis.*

37 Nibert, "Animals, Immigrants, and Profits."

38 Specht, *Red Meat Republic*; Russell, *Greatest Trust in the World*; Wallach, *How America Eats.*

39 Smith, *Eating History.*

40 Smith; Wallach, *How America Eats.*

41 Russell, *Greatest Trust in the World.*

42 Russell, 1–4.

43 Patel and Goodman, "Long New Deal."

44 As Specht notes, this argument continues today. See Specht, *Red Meat Republic*, 6.

45 Specht, 19.

46 Specht.

47 Wallach, *How America Eats*, 105.

48 Wallach, 105.

49 Sinclair, *The Jungle*, 40.

50 Sinclair, 130.

51 Sinclair, 40.

52 Arthur, *Radical Innocent.*

53 Wallach, *How America Eats.*

54 Quoted in Arthur, *Radical Innocent*, 83.

55 Horowitz, *Negro and White, Unite and Fight!*

56 Horowitz.

57 Horowitz.

58 Haedicke, "From Collective Bargaining to Social Justice Certification."

59 Patel and Goodman, "Long New Deal"; Jones, *Influenza 1918.*

60 Horowitz, *Negro and White, Unite and Fight!*, 23–24.

61 Horowitz, 20.

62 Horowitz, 20.

63 Nibert, "Animals, Immigrants, and Profits"; Halpern, *Down on the Killing Floor.*

64 In agriculture, see, for example, McWilliams, *Factories in the Field.*

65 Haedicke, "From Collective Bargaining to Social Justice Certification."

66 Horowitz, *Negro and White, Unite and Fight!*, 42.

67 Horowitz, 33.

68 Haedicke, "From Collective Bargaining to Social Justice Certification"; Davidson, *Broken Heartland.*

69 In "right-to-work" states, employees do not need to be union members to work in unionized workplaces. See Nibert, "Animals, Immigrants, and Profits."

70 Specht, *Red Meat Republic*, 10.

71 Frydenlund and Dunn, "Refugees and Racial Capitalism"; Freshour, "Poultry and Prisons."

72 Nibert, "Animals, Immigrants, and Profits."

73 Sittig and González, *Mayans Among Us.*

74 Nibert, "Animals, Immigrants, and Profits," 13.

75 Sittig and González, *Mayans Among Us*, 121–22.

76 Frydenlund and Dunn, "Refugees and Racial Capitalism," 3.

77 Fossum, "Precarity Capitalism and the Global Value Chain in Beef."

78 Ken and León, "Necropolitical Governance and State-Corporate Harms," 78.

79 Reported in Haedicke, "From Collective Bargaining to Social Justice Certification," 125.

80 Frydenlund and Dunn, "Refugees and Racial Capitalism," 3. TANF is the Temporary Assistance for Needy Families program.

81 Nibert, "Animals, Immigrants, and Profits," 10.

82 US Bureau of Labor Statistics; see also Luckstead et al., "US Workers' Willingness to Accept Meatpacking Jobs"; Nibert, "Animals, Immigrants, and Profits"; National Association of Manufacturers.

83 Sittig and González, *Mayans Among Us.*

84 Winders and Nibert, "Consuming the Surplus."

85 Wallach, *How America Eats.*

86 Stull and Broadway, *Slaughterhouse Blues.*

87 Stull and Broadway.

88 Emanuel, "Water in the Lumbee World."

89 Botheras, "Tail Docking of Dairy Cattle."

90 Guo et al., "Association of Common Zoonotic Pathogens."

91 Gilchrist et al., "Potential Role of Concentrated Animal Feeding Operations."

92 As described by a packing plant worker in Greeley, Colorado, quoted in Fossum, "Precarity Capitalism and the Global Value Chain in Beef."

93 See Ken and León, "Necropolitical Governance and State-Corporate Harms"; Fossum, "Precarity Capitalism and the Global Value Chain in Beef"; Jordan and Dickerson, "Poultry Worker's Death Highlights Spread of Coronavirus in Meat Plants."

94 Babic, "Shocking Underestimate of Deaths from COVID-19," n.p.

95 Beldo, "Metabolic Labor," 108.

96 Sinel and Weis, "Ventilation Shutdown"; Mackenzie, "Quarter of All Pigs Have Died." See also Rice and Tyner, "Revanchist 'Nature' and 21st Century Genocide."

97 Sinel and Weis, "Ventilation Shutdown," n.p.

98 Ken and León, "Necropolitical Governance and State-Corporate Harms."

99 Wadiwel, "Chicken Harvesting Machine," 528. See also Wadiwel, *Animals and Capital.*

100 For much more, see Vandenbosch and Lymbery, *Cultivated Meat to Secure Our Future*; Donaldson and Carter, *Future of Meat Without Animals*; Shapiro, *Clean Meat.*

101 *Washington Post*, "Upside Foods CEO on the Future of Cell-Cultivated Meats."

102 *Washington Post.*

103 Grossmann, "Sustainable Media Feedstocks for Cellular Agriculture."

104 UPSIDE Foods, "UPSIDE Foods: Company."

105 Aggregation and segregation can be found throughout the food system, not just meat packing. But it is in meat packing that these processes are most obvious.

106 For more on world systems theory, see Moore, *Capitalism in the Web of Life*; Wallerstein, "Land, Space, and People." For the body as a frontier, see Frydenlund and Dunn, "Refugees and Racial Capitalism." These ideas are part of broader discussions around biopolitics; cf. Braun,

"Biopolitics and the Molecularization of Life"; Rutherford and Rutherford, "Geography and Biopolitics"; Nally, "Biopolitics of Food Provisioning."

107 See Corfe et al., "Multifactorial Interplay of Diet, the Microbiome and Appetite Control."

108 Specht, *Red Meat Republic*, 3.

109 Fossum, "Precarity Capitalism and the Global Value Chain in Beef."

110 Felder, "How Oklahoma Tribes Are Reversing Years of Consolidation"; Reuters, "Explainer."

111 Ken and León, "Necropolitical Governance and State-Corporate Harms."

112 For much more detail on this point, see Specht, *Red Meat Republic*.

113 Yaffe-Bellany and Corkery, "Dumped Milk, Smashed Eggs, Plowed Vegetables."

114 Sittig and González, *Mayans Among Us*, 134.

115 Sittig and González, 136.

## CHAPTER SEVEN

1 See Flannery, "Fighting Pellagra in Alabama."

2 Leach, "Case of Pellagra"; Bollet, "Politics and Pellagra"; Searcy, "Epidemic of Acute Pellagra"; Randolph, "Some Notes on Pellgra and Pellagrins."

3 Akenhead, "Conquest of 'Hidden Hunger.'"

4 Bollet, "Politics and Pellagra"; Akenhead, "Conquest of 'Hidden Hunger.'" For an example of a suicidal patient, see Mason, "Pellagra," 345.

5 Searcy, "Epidemic of Acute Pellagra"; Marks, "Epidemiologists Explain Pellagra"; Grote, "Pellagra."

6 Flannery, "Fighting Pellagra in Alabama," 3.

7 CDC, "National Diabetes Statistics Report."

8 National Institute of Diabetes and Digestive and Kidney Diseases, "Overweight & Obesity Statistics."

9 Gray et al., "Association Between Residence in a Food Desert Census Tract and Adherence to Dietary Patterns."

10 Searcy, "Epidemic of Acute Pellagra," 391. See also Akenhead, "Conquest of 'Hidden Hunger.'"

11 Searcy, "Epidemic of Acute Pellagra," 391.
12 Leach, "Case of Pellagra," 446.
13 Cole, "Accidental Discovery of a Possible Cure for Pellagra," 406.
14 Dr. Steven F. Hale, quoted in Mason, "Pellagra," 347.
15 Bollet, "Politics and Pellagra"; Flannery, "Fighting Pellagra in Alabama."
16 Leach, "Case of Pellagra."
17 Moody, "Treatment of Pellagra"; Pruett, "Is Pellagra a Preventable Disease?"
18 Blosser, "Etiology of Pellagra," 406.
19 Blosser, 402.
20 Cole, "Accidental Discovery of a Possible Cure for Pellagra," 404.
21 Moody, "Treatment of Pellagra."
22 Siler et al., "Pellagra." See also Etheridge, *Butterfly Caste*.
23 See Marks, "Epidemiologists Explain Pellagra," 40.
24 Moody, "Treatment of Pellagra," 260.
25 Davenport, "Hereditary Factor in Pellagra," 6. "Nam Hollow" refers to a community of marginalized Native Americans in Upstate New York used by eugenicists to justify theories of racially determined behavior. "Cliffs of Newport" refers to the wealthy, white seaside communities of Newport, Rhode Island. See Jarvenpa, *Declared Defective*; Estabrook and Davenport, *Nam Family*.
26 Flannery, "Fighting Pellagra in Alabama," 9.
27 Bollet, "Politics and Pellagra." A century later, public hostility would target Chinese Americans during the COVID-19 pandemic.
28 Flannery, "Fighting Pellagra in Alabama," 10.
29 Leslie, "'Fighting an Unseen Enemy,'" 200.
30 Akenhead, "Conquest of 'Hidden Hunger'"; Bollet, "Politics and Pellagra."
31 Bollet, "Politics and Pellagra," 217. Pellagrin is the term used for a patient with pellagra.
32 Goldberger et al., "Study of the Relation of Diet to Pellagra Incidence"; Mooney et al., "Thompson-McFadden Commission and Joseph Goldberger."
33 Quoted in Akenhead, "Conquest of 'Hidden Hunger,'" 14.

34 Bollet, "Politics and Pellagra."

35 Marks, "Epidemiologists Explain Pellagra."

36 Akenhead, "Conquest of 'Hidden Hunger,'" 18.

37 Leach, "Case of Pellagra," 448. For that matter, corn (maize) was the cornerstone of Indigenous diets in the Americas but did not experience pellagra. Indigenous corn preparation involves nixtamalization, the soaking of kernels in alkaline solution (lime juice) before milling. This makes the tryptophan more available for absorption.

38 Grote, "Pellagra," 333.

39 Grote, 333.

40 Grote, 333.

41 Grote, 334.

42 Grote, 336.

43 Akenhead, "Conquest of 'Hidden Hunger.'"

44 Akenhead, 11–12.

45 Bollet, "Politics and Pellagra," 219.

46 Knecht-van Eekelen, "[History of Healing"]; Hardy, "Beriberi, Vitamin B1 and World Food Policy."

47 Chamberlain, "Eradication of Beriberi," 134.

48 Bishai and Nalubola, "History of Food Fortification in the United States"; Hardy, "Beriberi, Vitamin B1 and World Food Policy." For a fascinating look at the history of fortification programs, see also Bobrow-Strain, *White Bread*.

49 CDC, "Deaths and Mortality."

50 Milken Institute, "Americans' Obesity Weighs Down U.S. Economy."

51 Milken Institute.

52 De Master and Daniels, "Desert Wonderings."

53 Shannon, "Food Deserts," 252.

54 Though definitions vary. See Shannon, 248.

55 Eisenhauer, "In Poor Health."

56 See Redlining Map of the City of Birmingham, Alabama, 5/1933, Residential Security Maps, 1933–39, Records of the Federal Home Loan Bank Board, Record Group 195, National Archives at College Park, MD.

57 Pink, "Birmingham's 'Food Deserts' Have Been Shaped by Its Redlined Past."

58 Darrell O'Quinn, quoted in Pink.

59 Fountain Heights has a median household income of $14,700, Mountain Brook City $167,000. US census data for 2020, FARA data for 2019.

60 Eisenhauer, "In Poor Health," 128.

61 Eisenhauer, 128.

62 Shannon, "Food Deserts"; Cummins et al., "Understanding and Representing 'Place' in Health Research."

63 Reese, *Black Food Geographies.*

64 As a starting point, see Reese; Taylor and Ard, "Food Availability and the Food Desert Frame in Detroit"; Morton et al., "Solving the Problems of Iowa Food Deserts"; Widener et al., "Dynamic Urban Food Environments."

65 Brones, "Food Apartheid," n.p.

66 As a starting point, see Saletan, "Food Apartheid"; Sbicca, "Growing Food Justice by Planting an Anti-Oppression Foundation"; Alkon et al., *Recipe for Gentrification.*

67 Mayorga et al., "'I Hate That Food Lion.'"

68 Mayorga et al., 249, emphasis added.

69 Mayorga et al., 249–50.

70 For more on food apartheid, see Sbicca, "Growing Food Justice by Planting an Anti-Oppression Foundation"; Saletan, "Food Apartheid"; Brones, "Food Apartheid."

71 Shannon, "Food Deserts," 249.

72 See Williams-Forson, *Eating While Black.*

73 Guthman, *Weighing In*; Joassart-Marcelli, *$16 Taco*; Alkon et al., *Recipe for Gentrification.*

74 Pink, "Birmingham's 'Food Deserts' Have Been Shaped by Its Redlined Past."

75 USDA ERS, Food Access Research Atlas 2019, comparing data for 2015 and 2019 using low income and low access at one mile (urban) and ten miles (rural).

76 Zurawski, "Materializing the Metaphor."

77 Holt-Giménez, "Walmart's Food Deserts."

78 De Master and Daniels, "Desert Wonderings."

79 As a starting point, see Alkon et al., "Foodways of the Urban Poor"; Howerton and Trauger, "'Oh Honey, Don't You Know?'"; Cummins et al., "New Neighborhood Grocery Store"; Elbel et al., "Assessment of a Government-Subsidized Supermarket."

80 Perkins, "Why Are There So Few Black-Owned Grocery Stores?"

81 Cunningham, "Walmart Ends Free Shuttle Service."

82 As of February 2023. WIC is the Special Supplemental Nutrition Program for Women, Infants, and Children, a program that supports 40 percent of infants in the United States. See USDA, "Special Supplemental Nutrition Program for Women, Infants, and Children."

83 Perkins, "Why Are There So Few Black-Owned Grocery Stores?"

84 Zurawski, "Materializing the Metaphor," 287.

85 De Master and Daniels, "Desert Wonderings," 252; Hajer, *Politics of Environmental Discourse*, 22.

86 McDowell and Mason, "Prisoners in the US Are Part of a Hidden Workforce."

87 Shannon, "Food Deserts," 255.

88 De Master and Daniels, "Desert Wonderings."

89 Eisenhauer, "In Poor Health," 129.

90 Lambert, "Thousands of People Admitted to Hospital with Malnutrition."

91 Chen et al., "Consumption of Ultra-Processed Foods and Health Outcomes."

**CONCLUSION**

1 Civil Eats, "Food Policy Tracker."

2 See Hussain, *Could People in ICE Detention Centers Be Forced to Work in Slaughterhouses?*; Booth, "Ending Forced Labor in ICE Detention Centers."

3 Squire, "Farmers Face Whiplash as Trump Immigration Shifts Keep Ag in Limbo"; Civil Eats, "Food Policy Tracker."

4 Pollan, *In Defense of Food*.

5 Pollan, *Omnivore's Dilemma*, 245.

6 Buller and Roe, *Food and Animal Welfare*, 172.

7 Haedicke, "From Collective Bargaining to Social Justice Certification."

8 See Mayorga et al., "'I Hate That Food Lion.'"

9 For example, see Lappé and Collins, *Food First*; Lang and Heasman, *Food Wars*.

10 Hassanein, "Practicing Food Democracy," 79; Lang, "Food Policy for the 21st Century," 218.

11 Dahlberg, "Democratizing Society and Food Systems," 135.

12 For example, see Gates, *How to Avoid a Climate Disaster*; Monbiot, *Regenesis*; *Solving for Zero*.

13 As a starting point, see Hajer, *Politics of Environmental Discourse*.

14 Nketani and Wise, "Feeding Big Agribusiness, Starving Africans."

15 Armstrong, "'Megayachts' Are Environmentally Indefensible."

16 United Nations University, "UN Study Reveals the Hidden Environmental Impacts of Bitcoin."

17 Oxfam, "Climate Equality."

18 Wittman, "Food Sovereignty: A New Rights Framework for Food and Nature?," 87.

19 La Via Campesia, "About La Via Campesina | Via Campesina," np.

20 Sampson et al., "Food Sovereignty and Rights-Based Approaches Strengthen Food Security and Nutrition Across the Globe."

21 Patel and Goodman, "The Long New Deal," 449.

22 Mincyte and Dobernig, "Urban Farming in the North American Metropolis."

23 Mironesco, *Seeds of Change*.

24 For example, see Cordova Montes et al., "Constitutionalizing the Human Right to Food in Maine."

25 Kahiluoto, "Food Systems for Resilient Futures."

26 For a similar kind of deconstruction of "poverty," see Yapa, "What Causes Poverty?"

27 See Spataro, "Against a De-Politicized DIY Urbanism"; Wattnem, "Seed Laws, Certification and Standardization"; Ceryes and Heaney, "'Ag-Gag' Laws."

28 See Tyner, *Alienated Subject*.

29 See Davidson, *Broken Heartland*; Sodani and Kumar, "Cow Urine."

30 See Kropotkin, *Conquest of Bread.*

31 As an entry point into this conversation, see Liboiron, *Pollution Is Colonialism.*

32 Rockefeller Foundation, "True Cost of Food."

33 Food System Economics Commission, "Hidden Costs of Food."

34 Nketani and Wise, "Feeding Big Agribusiness, Starving Africans."

35 Alinsky, *Rules for Radicals*, 21.

36 Hoelscher, "Landmark Decision Recognizing Rights of Indigenous Peoples."

37 Iwata, "Japanese Immigrants in California Agriculture"; Sittig and González, *Mayans Among Us*; Crawford, "Cotton, Land, and Sustenance"; Horowitz, *Negro and White, Unite and Fight!*

# BIBLIOGRAPHY


Abdulai, Awudu, Christopher B. Barrett, and John Hoddinott. "Does Food Aid Really Have Disincentive Effects? New Evidence from Sub-Saharan Africa." *World Development* 33, no. 10 (2005): 1689–1704.

Akenhead, Linda. "Conquest of 'Hidden Hunger': The Work of Thomas Douglas Spies." Occasional paper, Jefferson County Medical Society / University of Alabama at Birmingham, July 1987.

Albritton, Robert. *Let Them Eat Junk: How Capitalism Creates Hunger and Obesity*. Pluto Press, 2009.

Alinsky, Saul. *Rules for Radicals: A Pragmatic Primer for Realistic Radicals*. Knopf Doubleday, 1989.

Alkon, Alison Hope, Daniel Block, Kelly Moore, Catherine Gillis, Nicole DiNuccio, and Noel Chavez. "Foodways of the Urban Poor." *Geoforum* 48 (August 1, 2013): 126–35. https://doi.org/10.1016/j.geoforum.2013.04.021.

Alkon, Alison Hope, Yuki Kato, and Joshua Sbicca. *A Recipe for Gentrification: Food, Power, and Resistance in the City*. New York University Press, 2020.

American Psychological Association. *Publication Manual of the American Psychological Association*. 7th ed. American Psychological Association, 2020.

Armstrong, Chris. "'Megayachts' Are Environmentally Indefensible. The World Must Ban Them." *The Guardian*, December 11, 2023, sec. Opinion. https://www.theguardian.com/commentisfree/2023/dec/11/megayachts-environment-carbon-emissions-ban.

Arthur, Anthony. *Radical Innocent: Upton Sinclair.* Random House, 2006.
Aubrun, Alex, Andrew Brown, and Joseph Grady. "Conceptualizing US Food Systems with Simplifying Models: Findings from TalkBack Testing." Frameworks Institute, April 2006.
Babic, Mary. "Shocking Underestimate of Deaths from COVID-19 in Meat and Poultry Industry." Oxfam America, October 2, 2023. https://politicsofpoverty.oxfamamerica.org/still-disposable-new-congressional-report-reveals-shocking-new-statistics-around-covid-19-in-meat-and-poultry-workforce/.
Bailey, Thomas A. "California, Japan, and the Alien Land Legislation of 1913." *Pacific Historical Review* 1, no. 1 (1932): 36–59. https://doi.org/10.2307/3633745.
*Bainbridge Island Review.* "Editorial: More Plain Talk." February 5, 1942.
Balderrama, Francisco E., and Raymond Rodriguez. *Decade of Betrayal: Mexican Repatriation in the 1930s.* University of New Mexico Press, 2006.
Baranski, Marci. *The Globalization of Wheat: A Critical History of the Green Revolution.* University of Pittsburgh Press, 2022.
Barua, Maan. "Lively Commodities and Encounter Value." *Environment and Planning D: Society and Space* 34, no. 4 (August 1, 2016): 725–44. https://doi.org/10.1177/0263775815626420.
———. "Nonhuman Labour, Encounter Value, Spectacular Accumulation: The Geographies of a Lively Commodity." *Transactions of the Institute of British Geographers* 42, no. 2 (2017): 274–88. https://doi.org/10.1111/tran.12170.
Bates, Frederick L. "Recovery, Change and Development: A Longitudinal Study of the 1976 Guatemalan Earthquake." Department of Sociology, University of Georgia, 1982.
Bates, Frederick L., and Walter G. Peacock. *Disaster and Social Change.* Franco Angeli, 1987.
Bear, Firman Edward. *Theory and Practice in the Use of Fertilizers.* Wiley, 1938.
Beldo, Les. "Metabolic Labor: Broiler Chickens and the Exploitation of Vitality." *Environmental Humanities* 9, no. 1 (2017): 108–28.
Bernstein, Henry. *Class Dynamics of Agrarian Change.* Kumarian Press, 2010.
Bezuneh, Mesfin, Brady Deaton, and Segu Zuhair. "Food Aid Disincentives: The Tunisian Experience." *Review of Development Economics* 7, no. 4 (2003): 609–21.

Bishai, David, and Ritu Nalubola. "The History of Food Fortification in the United States: Its Relevance for Current Fortification Efforts in Developing Countries." *Economic Development and Cultural Change* 51, no. 1 (2002): 37–53. https://doi.org/10.1086/345361.

Blaakman, Michael A. *Speculation Nation: Land Mania in the Revolutionary American Republic*. University of Pennsylvania Press, 2023.

Blackmon, Douglas A. *Slavery by Another Name: The Re-Enslavement of Black Americans from the Civil War to World War Two*. Icon Books, 2012.

Bloom, J. Dara, Joanna Lelekacs, Gretchen Hofing, Robyn Stout, Morgan Marshall, and Kristin Davis. "Integrating Food Systems and Local Food in Family and Consumer Sciences: Perspectives from the Pilot Extension Master Food Volunteer Program." *Journal of Agriculture, Food Systems, and Community Development* 9, no. 2 (2020): 197–220.

Blosser, Ray. "The Etiology of Pellagra with Especial Consideration of the Phenomena of Sensitization to Maize and Sugar-Cane Products." *Southern Medical Journal* 9, no. 5 (1916): 401–4.

Bobrow-Strain, Aaron. *White Bread: A Social History of the Store-Bought Loaf*. Beacon Press, 2012.

Bock, Brianna. "Planting the Seeds of Change." *Edible Vancouver Island*, December 27, 2019. https://ediblevancouverisland.ediblecommunities.com/food-thought/planting-seeds-change.

Bollet, A. J. "Politics and Pellagra: The Epidemic of Pellagra in the U.S. in the Early Twentieth Century." *Yale Journal of Biology and Medicine* 65, no. 3 (1992): 211–21.

Booth, Jonathan. "Ending Forced Labor in ICE Detention Centers." *Georgetown Immigration Law Journal* 34, no. 3 (2020): 373–612.

Borowiak, Craig. "Farmers' Right: Intellectual Property Regimes and the Struggle over Seeds." *Politics & Society* 32 (2004): 511–43.

Botheras, Naomi. "Tail Docking of Dairy Cattle: Is It Beneficial or a Welfare Issue?" Ohio Dairy Industry Resources Center, 2006. https://dairy.osu.edu/newsletter/buckeye-dairy-news/volume-8-issue-3/tail-docking-dairy-cattle-it-beneficial-or-welfare.

Braun, Bruce. "Biopolitics and the Molecularization of Life." *Cultural Geographies* 14, no. 1 (January 1, 2007): 6–28. https://doi.org/10.1177/1474474007072817.

Brones, Anna. "Food Apartheid: The Root of the Problem with America's Groceries." *The Guardian*, May 15, 2018, sec. Society. http://www

.theguardian.com/society/2018/may/15/food-apartheid-food-deserts-racism-inequality-america-karen-washington-interview.

Bucheli, Marcelo. *Bananas and Business: The United Fruit Company in Colombia, 1899–2000.* New York University Press, 2005.

Buller, Henry, and Emma Roe. *Food and Animal Welfare.* Bloomsbury, 2018.

Bureau of Indian Affairs. "Origin of Names of US States." January 4, 1974. https://www.bia.gov/as-ia/opa/online-press-release/origin-names-us-states.

Burton, Jeffery F. *Confinement and Ethnicity: An Overview of World War II Japanese American Relocation Sites.* University of Washington Press, 2011.

Butler, Smedley Darlington. *War Is a Racket.* Feral House, 2003.

Calo, Adam. "The Yeoman Myth: A Troubling Foundation of the Beginning Farmer Movement." *Gastronomica* 20, no. 2 (2020): 12–29.

Carey, David, Jr. "Guatemala's Green Revolution: Synthetic Fertilizer, Public Health, and Economic Autonomy in the Mayan Highland." *Agricultural History* 83, no. 3 (2009): 283–322.

Carro-Figueroa, Vivian. "Agricultural Decline and Food Import Dependency in Puerto Rico: A Historical Perspective on the Outcomes of Postwar Farm and Food Policies." *Caribbean Studies* 30, no. 2 (2002): 77–107.

Castillo, Elias. *A Cross of Thorns: The Enslavement of California's Indians by the Spanish Missions.* Craven Street Books, 2017.

Castree, Noel. "Socializing Nature: Theory, Practice, and Politics." In *Social Nature: Theory, Practice, and Politics*, edited by Noel Castree and Bruce Braun, 1–21. Wiley, 2001.

Centers for Disease Control and Prevention. "Deaths and Mortality." May 7, 2024. https://www.cdc.gov/nchs/fastats/deaths.htm.

———. "National Diabetes Statistics Report." June 6, 2024. https://www.cdc.gov/diabetes/php/data-research/index.html.

Ceryes, Caitlin A., and Christopher D. Heaney. "'Ag-Gag' Laws: Evolution, Resurgence, and Public Health Implications." *New Solutions* 28, no. 4 (February 1, 2019): 664–82. https://doi.org/10.1177/1048291118808788.

Chabot, Philippe, and Paul A. Dorosh. "Wheat Markets, Food Aid and Food Security in Afghanistan." *Food Policy* 32, no. 3 (2007): 334–53.

Chamberlain, Weston. "The Eradication of Beriberi from the Philippine (Native) Scouts by Means of a Simple Change in Their Dietary." *Philippine Journal of Science* 6 (1911): 133–46.

Chen, Xiaojia, Zhang Zhang, Huijie Yang, Peishan Qiu, Haizhou Wang, Fan Wang, Qiu Zhao, Jun Fang, and Jiayan Nie. "Consumption of

Ultra-Processed Foods and Health Outcomes: A Systematic Review of Epidemiological Studies." *Nutrition Journal* 19, no. 1 (August 20, 2020): 86. https://doi.org/10.1186/s12937-020-00604-1.

Chhabra, Heeral. "Animal Labourers and the Law in Colonial India." *South Asia Research* 39, no. 2 (2019): 166–83.

Chiang, Connie Y. *Nature Behind Barbed Wire: An Environmental History of the Japanese American Incarceration*. Oxford University Press, 2018.

City of Albuquerque. "The Albuquerque Indian School Cemetery at 4H Park." Accessed March 7, 2023. https://www.cabq.gov/office-of-equity-inclusion/about-office-of-equity-inclusion/native-american-affairs/4h-park-the-albuquerque-indian-school-cemetery.

Civil Eats. "Food Policy Tracker." *Civil Eats*, n.d. Accessed December 6, 2025. https://civileats.com/food-policy-tracker/.

Clapp, Jennifer. *Food*. John Wiley & Sons, 2016.

Clapp, Jennifer. "The Financialization of Food: Who Is Being Fed?" Paper, International Society for Ecological Economics, June 2012.

Cobb, James C. "'Somebody Done Nailed Us on the Cross': Federal Farm and Welfare Policy and the Civil Rights Movement in the Mississippi Delta." *Journal of American History* 77, no. 3 (1990): 912–36. https://doi.org/10.2307/2078991.

Cochrane, Willard Wesley. *The Curse of American Agricultural Abundance: A Sustainable Solution*. University of Nebraska Press, 2003.

———. *The Development of American Agriculture: A Historical Analysis*. University of Minnesota Press, 1993.

Cole, W. F. "Accidental Discovery of a Possible Cure for Pellagra." *Southern Medical Journal* 9, no. 5 (May 1916): 404–6.

Collard, Rosemary-Claire, and Jessica Dempsey. "Capitalist Natures in Five Orientations." *Capitalism Nature Socialism* 28, no. 1 (January 2, 2017): 78–97. https://doi.org/10.1080/10455752.2016.1202294.

Collyns, Dan. "Worldwide Fertiliser Shortage Prompts Peru to Turn to Bird Poo." *The Guardian*, October 4, 2022, sec. World news. https://www.theguardian.com/world/2022/oct/04/peru-guano-deposits-worldwide-fertiliser-shortage.

Columbia Broadcasting System News, dir. *Hunger in America, Part 1*. Aired May 1968, on CBS. 51:00.

Commission on Wartime Relocation and Internment of Civilians. "Personal Justice Denied: Report of the Commission on Wartime Relocation and Internment of Civilians." University of Washington Press, 2012.

Cordova Montes, R. Denisse, Heather Retberg, and Photini Kamvisseli Suarez. "Constitutionalizing the Human Right to Food in Maine: A People's Tool to Advance Food Sovereignty in the United States." *Maine Law Review* 76 (2024): 229.

Corfe, Bernard M., Charlotte J. Harden, Matthew Bull, and Iveta Garaiova. "The Multifactorial Interplay of Diet, the Microbiome and Appetite Control: Current Knowledge and Future Challenges." *Proceedings of the Nutrition Society* 74, no. 3 (August 2015): 235–44. https://doi.org/10.1017/S0029665114001670.

Coté, Charlotte. "Hishuk'ish Tsawalk—Everything Is One: Revitalizing Place-Based Indigenous Food Systems Through the Enactment of Food Sovereignty." *Journal of Agriculture, Food Systems, and Community Development* 9, no. A (2019): 37–48.

Coulter, Kendra. *Animals, Work, and the Promise of Interspecies Solidarity.* Springer, 2016.

Crawford, George B. "Cotton, Land, and Sustenance: Toward the Limits of Abundance in Late Antebellum Georgia." *Georgia Historical Quarterly* 72, no. 2 (1988): 215–47.

Cronon, William. *Nature's Metropolis: Chicago and the Great West.* Norton, 2009.

Crook, Martin, and Damien Short. "Marx, Lemkin and the Genocide–Ecocide Nexus." *International Journal of Human Rights* 18, no. 3 (April 3, 2014): 298–319. https://doi.org/10.1080/13642987.2014.914703.

Cummins, Steven, Sarah Curtis, Ana V. Diez-Roux, and Sally Macintyre. "Understanding and Representing 'Place' in Health Research: A Relational Approach." *Social Science & Medicine* 65, no. 9 (November 1, 2007): 1825–38. https://doi.org/10.1016/j.socscimed.2007.05.036.

Cummins, Steven, Ellen Flint, and Stephen A. Matthews. "New Neighborhood Grocery Store Increased Awareness of Food Access but Did Not Alter Dietary Habits or Obesity." *Health Affairs* 33, no. 2 (February 2014): 283–91. https://doi.org/10.1377/hlthaff.2013.0512.

Cunfer, Geoff, and Bill Waiser. *Bison and People on the North American Great Plains: A Deep Environmental History.* Texas A&M University Press, 2016.

Cunningham, Darren. "Walmart Ends Free Shuttle Service for Low-Income Residents." *Fox 17*, August 2, 2019. https://www.fox17online.com/2019/08/02/walmart-ends-free-shuttle-service-for-low-income-residents.

Cushman, Gregory T. *Guano and the Opening of the Pacific World: A Global Ecological History*. Cambridge University Press, 2013.

Dahlberg, Kenneth A. "Democratizing Society and Food Systems: Or How Do We Transform Modern Structures of Power?" *Agriculture and Human Values* 18, no. 2 (June 1, 2001): 135–51. https://doi.org/10.1023/A:1011175626010.

Dale, Bruce E., Michael S. Allen, Mark Laser, and Lee R. Lynd. "Protein Feeds Coproduction in Biomass Conversion to Fuels and Chemicals." *Biofuels, Bioproducts and Biorefining* 3, no. 2 (2009): 219–30.

Davenport, C. B. "The Hereditary Factor in Pellagra." Robert M. Thompson Pellagra Commission, Eugenic Record Office, March 9, 1916.

Davidson, Osha Gray. *Broken Heartland: The Rise of America's Rural Ghetto*. University of Iowa Press, 1996.

Davis, Ronald L. F. *Good and Faithful Labor: From Slavery to Sharecropping in the Natchez District, 1860–1890*. Bloomsbury, 1982.

De Master, Kathryn Teigen, and Jess Daniels. "Desert Wonderings: Reimagining Food Access Mapping." *Agriculture and Human Values* 36, no. 2 (June 1, 2019): 241–56. https://doi.org/10.1007/s10460-019-09914-5.

Densho. "Terminology." Densho: Japanese American Incarceration and Japanese Internment. Accessed December 28, 2022. https://densho.org/terminology/.

Devereux, Stephen. *The New Famines: Why Famines Persist in an Era of Globalization*. Taylor & Francis, 2006.

Diné Policy Institute. "Diné Food Sovereignty: A Report on the Navajo Nation Food System and the Case to Rebuild a Self-Sufficient Food System for the Diné People." Diné Policy Institute, 2014.

Donaldson, Brianne, and Christopher Carter. *The Future of Meat Without Animals*. Rowman & Littlefield, 2016.

Doyle, Kate, Jesse Franzblau, and Emily Willard. "Ex-Kaibil Officer Connected to Dos Erres Massacre Arrested in Alberta, Canada." National Security Archive, George Washington University, May 7, 2020. https://nsarchive2.gwu.edu/NSAEBB/NSAEBB316/index.htm.

Du Bois, W. E. B. "Inter-Racial Implications of the Ethiopian Crisis—A Negro View." *Foreign Affairs* 14 (1935): 82–92.

Duell, Rebecca. "Is 'Local Food' Sustainable? Localism, Social Justice, Equity and Sustainable Food Futures." *New Zealand Sociology* 28, no. 4 (2013): 123–44.

Dunbar-Ortiz, Roxanne. *An Indigenous Peoples' History of the United States.* Beacon, 2015.

Economic Research Service. "12 Years of Achievement Under Public Law 480." ERS Foreign, US Department of Agriculture, November 1967.

Edwards, Judith, and Nicki Peter Petrikowski. *The Relocation of Native Peoples of North America*. Enslow, 2015.

Edwards, Julia B. "Phosphate Mining and the Relocation of the Banabans to Northern Fiji in 1945: Lessons for Climate Change-Forced Displacement." *Journal de La Société Des Océanistes*, no. 138–139 (December 15, 2014): 121–36. https://doi.org/10.4000/jso.7100.

Egan, Sophie. *How to Be a Conscious Eater: Making Food Choices That Are Good for You, Others, and the Planet*. Workman, 2020.

Eichler, Lauren. "Ecocide Is Genocide: Decolonizing the Definition of Genocide." *Genocide Studies and Prevention* 14, no. 2 (September 4, 2020). https://doi.org/10.5038/1911-9933.14.2.1720.

Eisenhauer, Elizabeth. "In Poor Health: Supermarket Redlining and Urban Nutrition." *GeoJournal* 53, no. 2 (February 1, 2001): 125–33. https://doi.org/10.1023/A:1015772503007.

Elbel, Brian, Alyssa Moran, L. Beth Dixon, Kamila Kiszko, Jonathan Cantor, Courtney Abrams, and Tod Mijanovich. "Assessment of a Government-Subsidized Supermarket in a High-Need Area on Household Food Availability and Children's Dietary Intakes." *Public Health Nutrition* 18, no. 15 (October 2015): 2881–90. https://doi.org/10.1017/S1368980015000282.

Emanuel, Ryan E. "Water in the Lumbee World: A River and Its People in a Time of Change." *Environmental History* 24, no. 1 (January 2019): 25–51. https://doi.org/10.1093/envhis/emy129.

Environmental Protection Agency. "Symptoms of Exposure to Pesticide Products Used on Minor Colorado Crops." Worker Protection Standards, Environmental Protection Agency Region 8, 2002.

Escobar, Arturo. *Encountering Development: The Making and Unmaking of the Third World*. Princeton University Press, 2001.

Essex, Jamey. *Development, Security, and Aid: Geopolitics and Geoeconomics at the U.S. Agency for International Development*. University of Georgia Press, 2013.

Estabrook, Arthur Howard, and Charles Benedict Davenport. *The Nam Family: A Study in Cacogenics*. New Era, 1912.

Etheridge, Elizabeth. *The Butterfly Caste: A Social History of Pellagra in the South*. Bloomsbury, 1972.

Evans, Clara. "Vacation for Victory." *Elementary English Review* 19, no. 5 (1942): 175–86.

Felder, Ben. "How Oklahoma Tribes Are Reversing Years of Consolidation with Their Own Meat Processing Plants." *The Oklahoman*, May 5, 2024. https://www.oklahoman.com/story/news/2024/05/05/oklahoma-tribal-nations-sage-cherokee-muscogee-quapaw-meat-processing/73555125007/.

Fernandez-Cornejo, Jorge, Richard F. Nehring, Craig Osteen, Seth Wechsler, Andrew Martin, and Alex Vialou. "Pesticide Use in US Agriculture: 21 Selected Crops, 1960–2008." *USDA-ERS Economic Information Bulletin*, no. 124 (2014).

Fisher, Anne Reeploeg. *Exile of a Race: A History of the Forcible Removal and Imprisonment by the Army of the 115,000 Citizens and Alien Japanese Who Were Living on the West Coast in the Spring of 1942*. B.C., 1965.

Fitzgerald, Deborah Kay. *Every Farm a Factory: The Industrial Ideal in American Agriculture*. Yale University Press, 2008.

Flannery, Michael. "Fighting Pellagra in Alabama: Carl A. Grote's Forgotten Contribution to Conquering a Southern Scourge." Reynolds-Finley Library, University of Alabama at Birmingham, 2009. https://library.uab.edu/locations/reynolds/collections/regional-history/pellagra.

Food System Economics Commission. "The Hidden Costs of Food." November 29, 2022. https://foodsystemeconomics.org/hidden-food-costs/.

Fossum, John. "Precarity Capitalism and the Global Value Chain in Beef: The Plight of Meatpacking Workers at JBS Greeley." Working paper, Rapoport Center, 2022.

Foster, John Bellamy, and Brett Clark. *The Robbery of Nature: Capitalism and the Ecological Rift*. New York University Press, 2020.

Founders Online. National Archives, n.d. https://founders.archives.gov.

Freshour, Carrie. "Poultry and Prisons." *Monthly Review* 72, no. 3 (2020): 32–47.

Frydenlund, Shae, and Elizabeth Cullen Dunn. "Refugees and Racial Capitalism: Meatpacking and the Primitive Accumulation of Labor." *Political Geography* 95 (2022): 102575.

Galeano, Eduardo. *Open Veins of Latin America: Five Centuries of the Pillage of a Continent*. New York University Press, 1997.

Galtung, Johan. "Violence, Peace, and Peace Research." *Journal of Peace Research* 6, no. 3 (1969): 167–92.

Garth, Hanna, and Ashanté M. Reese. *Black Food Matters: Racial Justice in the Wake of Food Justice*. University of Minnesota Press, 2020.

Gates, Bill. *How to Avoid a Climate Disaster: The Solutions We Have and the Breakthroughs We Need*. Knopf Doubleday Publishing Group, 2021.

Gawronski, Vincent T., and Richard Stuart Olson. "Disasters as Crisis Triggers for Critical Junctures? The 1976 Guatemala Case." *Latin American Politics and Society* 55, no. 2 (2013): 133–49.

Gazdar, Haris. "Pre-modern, Modern and Post-modern Famine in Iraq." *IDS Bulletin* 33, no. 4 (2002): 63–69.

Genovese, Eugene D. *The Political Economy of Slavery: Studies in the Economy and Society of the Slave South*. Wesleyan University Press, 1989.

George, Susan. "More Food, More Hunger: Development." *Seeds of Change* 1, no. 2 (1986): 53–63.

Gibbings, Julie, and Heather Vrana. *Out of the Shadow: Revisiting the Revolution from Post-Peace Guatemala*. University of Texas Press, 2020.

Gilchrist, Mary J., Christina Greko, David B. Wallinga, George W. Beran, David G. Riley, and Peter S. Thorne. "The Potential Role of Concentrated Animal Feeding Operations in Infectious Disease Epidemics and Antibiotic Resistance." *Environmental Health Perspectives* 115, no. 2 (February 2007): 313–16. https://doi.org/10.1289/ehp.8837.

Goldberger, Joseph, G. A. Wheeler, and Edgar Sydenstricker. "A Study of the Relation of Diet to Pellagra Incidence in Seven Textile-Mill Communities of South Carolina in 1916." *Public Health Reports* 35, no. 12 (1920): 648–713. https://doi.org/10.2307/4575517.

Grandia, Liza. "Poisonous Exports: Pesticides, Peasants, and Conservation Paradigms in Guatemala." *Latin American Perspectives* 49, no. 6 (2022): 124–52. https://doi.org/10.1177/0094582X221124535.

Grandin, Greg. "Guatemalan Slaughter Was Part of Reagan's Hard Line." *New York Times*, May 21, 2013, sec. Room for Debate.

Gray, Marquita S., Sindhu Lakkur, Virginia J. Howard, Keith Pearson, James M. Shikany, Monika Safford, Orlando M. Gutiérrez, Natalie Colabianchi, and Suzanne E. Judd. "The Association Between Residence in a Food Desert Census Tract and Adherence to Dietary Patterns in the REGARDS Cohort." *Food and Public Health* 8, no. 4 (2018): 79–85.

Grey, Sam, and Raj Patel. "Food Sovereignty as Decolonization: Some Contributions from Indigenous Movements to Food System and

Development Politics." *Agriculture and Human Values* 32, no. 3 (September 1, 2015): 431–44. https://doi.org/10.1007/s10460-014-9548-9.

Grossmann, Lutz. "Sustainable Media Feedstocks for Cellular Agriculture." *Biotechnology Advances* 73 (July 1, 2024): 108367. https://doi.org/10.1016/j.biotechadv.2024.108367.

Grote, Carl. "Pellagra: From the View Point of a County Health Officer." In *Transactions of the Medical Association of the State of Alabama*. Brown, 1916.

Guo, Yaqiong, Una Ryan, Yaoyu Feng, and Lihua Xiao. "Association of Common Zoonotic Pathogens with Concentrated Animal Feeding Operations." *Frontiers in Microbiology* 12 (January 10, 2022). https://doi.org/10.3389/fmicb.2021.810142.

Guthman, Julie. *Weighing In: Obesity, Food Justice, and the Limits of Capitalism*. University of California Press, 2011.

Haedicke, Michael A. "From Collective Bargaining to Social Justice Certification: Workers' Rights in the American Meatpacking Industry." *Sociological Focus* 46, no. 2 (2013): 119–37.

Hajer, Maarten A. *The Politics of Environmental Discourse: Ecological Modernization and the Policy Process*. Clarendon, 1995.

Hale, William Jay. *The Farm Chemurgic: Farmward the Star of Destiny Lights Our Way*. Stratford, 1934.

Halpern, Rick. *Down on the Killing Floor: Black and White Workers in Chicago's Packinghouses, 1904–54*. University of Illinois Press, 1997.

Hancock, Graham. *Lords of Poverty: The Power, Prestige, and Corruption of the International Aid Business*. Atlantic Monthly Press, 1989.

Hansen, Arthur A. *Japanese American World War II Evacuation Oral History Project*. Meckler, 1991.

Hansen, Gladys. "Japanese Internment—War Hits California Farm Lands—1942." Museum of the City of San Francisco, 1998. http://www.sfmuseum.org/hist9/harvest.html.

Hardy, Anne. "Beriberi, Vitamin B1 and World Food Policy, 1925–1970." *Medical History* 39, no. 1 (January 1995): 61–77. https://doi.org/10.1017/S0025727300059482.

Harvey, David. "Globalization and the 'Spatial Fix.'" *Geographische Revue* 2 (2001): 23–30.

———. *The New Imperialism*. Oxford University Press, 2005.

———. "The Spatial Fix—Hegel, Von Thunen, and Marx." *Antipode* 13, no. 3 (1981): 1–12. https://doi.org/10.1111/j.1467-8330.1981.tb00312.x.

Hassanein, Neva. "Practicing Food Democracy: A Pragmatic Politics of Transformation." *Journal of Rural Studies* 19, no. 1 (January 1, 2003): 77–86. https://doi.org/10.1016/S0743-0167(02)00041-4.

Hedley, Douglas, and David Peacock. "Food for Peace, PL: 480 and American Agriculture." *Agricultural Economics Report* 14 (1970): 156.

Higgins, James. *Lima: A Cultural History*. Oxford University Press, 2005.

Hinnershitz, Stephanie. *Japanese American Incarceration: The Camps and Coerced Labor During World War II*. University of Pennsylvania Press, 2021.

Hippel, Frank A. von. *The Chemical Age: How Chemists Fought Famine and Disease, Killed Millions, and Changed Our Relationship with the Earth*. University of Chicago Press, 2020.

Hoelscher, Megan. "Landmark Decision Recognizing Rights of Indigenous Peoples." *Lockridge Grindal Nauen PLLP* (blog), February 13, 2024. https://www.locklaw.com/landmark-decision-recognizing-rights-of-indigenous-peoples/.

Holt-Giménez, Eric. "Walmart's Food Deserts: Greening the Bottom Line." *Common Dreams*, August 2, 2011. https://www.commondreams.org/views/2011/08/02/walmarts-food-deserts-greening-bottom-line.

Hopkins, Raymond F. "Reform in the International Food Aid Regime: The Role of Consensual Knowledge." *International Organization* 46, no. 1 (1992): 225–64.

Hormel, Leontina M., and Kari M. Norgaard. "Bring the Salmon Home! Karuk Challenges to Capitalist Incorporation." *Critical Sociology* 35, no. 3 (May 1, 2009): 343–66. https://doi.org/10.1177/0896920508101502.

Horowitz, Roger. *Negro and White, Unite and Fight! A Social History of Industrial Unionism in Meatpacking, 1930–90*. University of Illinois Press, 1997.

Howerton, Gloria, and Amy Trauger. "'Oh Honey, Don't You Know?' The Social Construction of Food Access in a Food Desert." *ACME: An International Journal for Critical Geographies* 16, no. 4 (December 20, 2017): 740–60.

Hribal, Jason. "'Animals Are Part of the Working Class': A Challenge to Labor History." *Labor History* 44 (2003): 435–53. https://doi.org/10.1080/0023656032000170069.

Huang, Josie. "Project to Be Unveiled in LA Names Every Japanese American Incarcerated During WWII." *LAist*, September 23, 2022. https://laist.com/news/wwii-monument-project-japanese-american-incarceration.

Hugo, Victor. *Les Miserables*. Vol. 5. Graphic Arts Books, 2021.

Hurt, R. Douglas. *American Agriculture: A Brief History*. Purdue University Press, 2002.

———. *The Green Revolution in the Global South: Science, Politics, and Unintended Consequences*. University of Alabama Press, 2020.

Hussain, Grace. *Could People in ICE Detention Centers Be Forced to Work in Slaughterhouses?* Food. April 16, 2025. https://sentientmedia.org/people-in-ice-detention-forced-to-work/.

Immerwahr, Daniel. *How to Hide an Empire: A History of the Greater United States*. Farrar, Straus and Giroux, 2019.

Inada, Lawson Fusao. *Only What We Could Carry: The Japanese American Internment Experience*. Heyday Books, 2000.

Indian Land Tenure Foundation. "Land Tenure History." Accessed December 23, 2023. https://iltf.org/land-issues/history/.

Institute for Government Research. "The Problem of Indian Administration." Johnson Reprint, 1971.

Iwata, Masakazu. "The Japanese Immigrants in California Agriculture." *Agricultural History* 36, no. 1 (1962): 25–37.

James, Jennifer C. "'Buried in Guano': Race, Labor, and Sustainability." *American Literary History* 24, no. 1 (2012): 115–42.

Jarvenpa, Robert. *Declared Defective: Native Americans, Eugenics, and the Myth of Nam Hollow*. University of Nebraska Press, 2018.

Joassart-Marcelli, Pascale. *The $16 Taco: Contested Geographies of Food, Ethnicity, and Gentrification*. University of Washington Press, 2021.

Johnson, Timothy. "Nitrogen Nation: The Legacy of World War I and the Politics of Chemical Agriculture in the United States, 1916–1933." *Agricultural History* 90, no. 2 (2016): 209–29.

Jones, Esyllt W. *Influenza 1918: Disease, Death, and Struggle in Winnipeg*. Vol. 31. University of Toronto Press, 2007.

Jones, Grinnell. "Nitrogen: Its Fixation, Its Uses in Peace and War." *Quarterly Journal of Economics* 34, no. 3 (1920): 391–431.

Jordan, Miriam, and Caitlin Dickerson. "Poultry Worker's Death Highlights Spread of Coronavirus in Meat Plants." *New York Times*, April 9, 2020, sec. U.S. https://www.nytimes.com/2020/04/09/us/coronavirus-chicken-meat-processing-plants-immigrants.html.

Jordan, Weymouth T. "The Peruvian Guano Gospel in the Old South." *Agricultural History* 24, no. 4 (1950): 211–21.

Kahiluoto, Helena. "Food Systems for Resilient Futures." *Food Security* 12, no. 4 (August 1, 2020): 853–57. https://doi.org/10.1007/s12571-020-01070-7.

Kapayou, Derrick G., Emma M. Herrighty, Christina Gish Hill, V. Cano Camacho, Ajay Nair, Donna M. Winham, and Marshall D. McDaniel. "Reuniting the Three Sisters: Collaborative Science with Native Growers to Improve Soil and Community Health." *Agriculture and Human Values* 40, no. 1 (March 1, 2023): 65–82. https://doi.org/10.1007/s10460-022-10336-z.

Kappler, Charles Joseph. *Indian Affairs: Laws and Treaties*. Government Printing Office, 1975. https://catalog.hathitrust.org/Record/008319763.

Ken, Ivy, and Kenneth Sebastian León. "Necropolitical Governance and State-Corporate Harms: COVID-19 and the US Pork Packing Industry." *Journal of White Collar and Corporate Crime* 3, no. 2 (2022): 76–89.

Kenner, Robert, dir. *Food, Inc.* Magnolia Pictures, 2008.

Kiernan, Ben. *Blood and Soil: A World History of Genocide and Extermination from Sparta to Darfur*. Yale University Press, 2008.

Kimmerer, Robin Wall. *Braiding Sweetgrass: Indigenous Wisdom, Scientific Knowledge, and the Teachings of Plants*. Milkweed Editions, 2020.

Kirychuk, Shelley P., Ambikaipakan Senthilselvan, James A. Dosman, Victor Juorio, John J. R. Feddes, Philip Willson, Henry Classen, Stephen J. Reynolds, Wilhelm Guenter, and Thomas S. Hurst. "Respiratory Symptoms and Lung Function in Poultry Confinement Workers in Western Canada." *Canadian Respiratory Journal* 10, no. 7 (October 2003): 375–80. https://doi.org/10.1155/2003/109679.

Kloppenburg, Jack Ralph. *First the Seed: The Political Economy of Plant Biotechnology*. University of Wisconsin Press, 2005.

Kluger, Richard. *Seizing Destiny: The Relentless Expansion of American Territory*. Knopf Doubleday, 2008.

Knecht-van Eekelen, A. de. "[The History of Healing; Beriberi: 'Kind of Paralysis']." *Nederlands Tijdschrift Voor Geneeskunde* 141, no. 24 (June 14, 1997): 1199–1203.

Koeppel, Dan. *Banana: The Fate of the Fruit That Changed the World*. Penguin, 2008.

Koivistoinen, P., and L. Hyvönen. "The Use of Sugar in Foods." *International Dental Journal* 35, no. 3 (September 1985): 175–79.

*Korematsu v. United States*, 323 US 214 (1944).

Krebs, A. V. "Bitter Harvest." *Washington Post*, February 2, 1992.

Kropotkin, Peter. *The Conquest of Bread*. Dialectics annot. ed. Vanguard, 2012.

Kulikoff, Allan. *The Agrarian Origins of American Capitalism*. University of Virginia Press, 1992.

Kurashige, Lon. *Two Faces of Exclusion: The Untold History of Anti-Asian Racism in the United States*. UNC Press Books, 2016.

La Via Campesia. "About La Via Campesina | Via Campesina." *La Via Campesina - EN*, n.d. Accessed December 1, 2025. https://viacampesina.org/en/international-peasants-voice/.

Lambert, Georgia. "Thousands of People Admitted to Hospital with Malnutrition." *The Times*, July 13, 2024. https://www.thetimes.com/uk/healthcare/article/times-health-commission-thousands-of-people-admitted-to-hospital-suffering-from-malnutrition-n23hqgzjr.

Lang, Tim. "Food Policy for the 21st Century: Can It Be Both Radical and Reasonable?" In *For Hunger-Proof Cities: Sustainable Urban Food Systems*, edited by Mustafa Koc, Rod MacRae, Luc J. A. Mougeot, and Jennifer Welsh, 216–24. IDRC, 1999.

Lang, Tim, and Michael Heasman. *Food Wars: The Global Battle for Mouths, Minds and Markets*. Routledge, 2015.

Lappé, Frances Moore, and Joseph Collins. *Food First: The Myth of Scarcity*. Souvenir Press, 1980.

Latham, Michael C., Lars Bondestam, and Urban Jonsson. *Hunger and Society: An Examination of Country Cases*. Cornell University, Program in International Nutrition, 1988.

Latham, Michael E. *The Right Kind of Revolution: Modernization, Development, and U.S. Foreign Policy from the Cold War to the Present*. Cornell University Press, 2011.

Latour, Bruno. *The Pasteurization of France*. Harvard University Press, 1993.

———. "Technology Is Society Made Durable." *Sociological Review* 38, no. S1 (1990): 103–31.

Laudan, Rachel. "Words Matter. Declaring 'The Food System Is (or Isn't) Broken' Does Not Help Understand or Improve the Food Supply." *Rachel Laudan* (blog), May 5, 2020. https://www.rachellaudan.com/2020/05/words-matter-declaring-the-food-system-is-or-isnt-broken-does-not-help-understand-or-improve-the-food-supply.html.

Lawrence, Felicity. "The Pigs' Revenge." *The Guardian*, May 1, 2009, sec. World news. https://www.theguardian.com/world/2009/may/02/swine-flu-pandemic-mexico-pig-farming.

Leach, Sydney. "A Case of Pellagra." In *Transactions of the Medical Association of the State of Alabama*. Brown, 1908.

Lee, Erika. "The 'Yellow Peril' and Asian Exclusion in the Americas." *Pacific Historical Review* 76, no. 4 (November 1, 2007): 537–62. https://doi.org/10.1525/phr.2007.76.4.537.

Lemkin, Raphael. *Axis Rule in Occupied Europe: Laws of Occupation, Analysis of Government, Proposals for Redress*. Lawbook Exchange, 2005.

Leonard, Julia Platt. "The Whole Food System in Britain Needs to Change to Achieve Real Sustainability #20pledges." *The Independent*, November 27, 2020. https://www.independent.co.uk/20-pledges/feeding-britain-food-sustainability-b1762970.html.

Lesher, Pete. "A Load of Guano: Baltimore and the Fertilizer Trade in the Nineteenth Century." *The Northern Mariner / Le Marin Du Nord* 18, no. 3–4 (2008): 121–28.

Leslie, Chris. "'Fighting an Unseen Enemy': The Infectious Paradigm in the Conquest of Pellagra." *Journal of Medical Humanities* 23, no. 3 (December 1, 2002): 187–202. https://doi.org/10.1023/A:1016842301517.

Levins, Richard A. *Willard Cochrane and the American Family Farm*. University of Nebraska Press, 2016.

Liboiron, Max. *Pollution Is Colonialism*. Duke University Press, 2021.

Lillquist, Karl. "Imprisoned in the Desert: The Geography of World War II–Era, Japanese American Relocation Centers in the Western United States." Central Washington University, Geography and Land Studies Department, 2007.

Lorr, Benjamin. *The Secret Life of Groceries: The Dark Miracle of the American Supermarket*. Penguin, 2021.

Luckstead, Jeff, Rodolfo M. Nayga, and Heather Snell. "US Workers' Willingness to Accept Meatpacking Jobs amid the COVID-19 Pandemic." *Journal of the Agricultural and Applied Economics Association* 1, no. 1 (April 2022): 47–60. https://doi.org/10.1002/jaa2.8.

Mackenzie, Debora. "A Quarter of All Pigs Have Died This Year Due to African Swine Fever." *New Scientist*, May 11, 2019. https://www.newscientist.com/article/2222501-a-quarter-of-all-pigs-have-died-this-year-due-to-african-swine-fever/.

Madley, Benjamin. *An American Genocide: The United States and the California Indian Catastrophe, 1846–1873*. Yale University Press, 2016.

———. "Reexamining the American Genocide Debate: Meaning, Historiography, and New Methods." *American Historical Review* 120, no. 1 (February 1, 2015): 98–139. https://doi.org/10.1093/ahr/120.1.98.

Mann, Charles C. *1493: Uncovering the New World Columbus Created*. Vintage Books, 2012.

Marks, Harry M. "Epidemiologists Explain Pellagra: Gender, Race, and Political Economy in the Work of Edgar Sydenstricker." *Journal of the History of Medicine and Allied Sciences* 58, no. 1 (January 1, 2003): 34–55. https://doi.org/10.1093/jhmas/58.1.34.

Marx, Karl. *Capital: A Critique of Political Economy*. Vol. 1. Vintage Books, 1977.

———. *Grundrisse: Foundations of the Critique of Political Economy*. Penguin, 2005.

Mason, E. Marvin. "Pellagra." In *Transactions of the Medical Association of the State of Alabama*. Brown, 1910.

Mayorga, Sarah, Megan Underhill, and Lauren Crosser. "'I Hate That Food Lion': Grocery Shopping, Racial Capitalism, and Everyday Disinvestment." *City & Community* 21, no. 3 (September 1, 2022): 238–55. https://doi.org/10.1177/15356841221091811.

McClatchy, V. S., and Japanese Exclusion League of California. "Japanese Immigration and Colonization: Skeleton Brief." Government Printing Office, 1921. https://catalog.hathitrust.org/Record/006557654.

McDonald, Bryan L. *Food Power: The Rise and Fall of the Postwar American Food System*. Oxford University Press, 2017.

Mcdonald, Joshua. "The Island with No Water: How Foreign Mining Destroyed Banaba." *The Guardian*, June 8, 2021, sec. World news. https://www.theguardian.com/world/2021/jun/09/the-island-with-no-water-how-foreign-mining-destroyed-banaba.

McDonnell, Janet A. *The Dispossession of the American Indian, 1887–1934*. Indiana University Press, 1991.

McDowell, Robin, and Margie Mason. "Prisoners in the US Are Part of a Hidden Workforce Linked to Hundreds of Popular Food Brands." *AP News*, January 29, 2024. https://apnews.com/article/prison-to-plate-inmate-labor-investigation-c6f0eb4747963283316e494eadf08c4e.

McEwan, Barbara. *Thomas Jefferson, Farmer*. McFarland, 1991.

McHenry, Donald F., and Kai Bird. "Food Bungle in Bangladesh." *Foreign Policy*, no. 27 (1977): 72–88.

McIlwaine, Cathy. "Geography and Development: Violence and Crime as Development Issues." *Progress in Human Geography* 23, no. 3 (September 1, 1999): 453–63. https://doi.org/10.1177/030913259902300309.

McKinney, Lillie. "History of the Albuquerque Indian School." *New Mexico Historical Review* 20, no. 2 (1945).

McLemore, Henry. "This Is War! Stop Worrying About Hurting Jap Feelings." *Seattle Times*, January 30, 1942.

McWilliams, Carey. *Factories in the Field.* 1939. Peregrine, 1971.

Melillo, Edward D. "The First Green Revolution: Debt Peonage and the Making of the Nitrogen Fertilizer Trade, 1840–1930." *American Historical Review* 117, no. 4 (2012): 1028–60.

Merleaux, April. *Sugar and Civilization: American Empire and the Cultural Politics of Sweetness.* UNC Press Books, 2015.

Midtrød, Tom Arne. *The Memory of All Ancient Customs: Native American Diplomacy in the Colonial Hudson Valley.* Cornell University Press, 2012.

Milken Institute. "Americans' Obesity Weighs Down U.S. Economy by $1.4 Trillion." November 29, 2016. https://milkeninstitute.org/article/americans-obesity-weighs-down-us-economy-14-trillion.

Mincyte, Diana, and Karin Dobernig. "Urban Farming in the North American Metropolis: Rethinking Work and Distance in Alternative Food Networks." *Environment and Planning A: Economy and Space* 48, no. 9 (2016): 1767–86. https://doi.org/10.1177/0308518X16651444.

Mintz, Sidney W. *Sweetness and Power: The Place of Sugar in Modern History.* Penguin, 1986.

Mironesco, *Monique. Seeds of Change: Women Cultivating Food Sovereignty in Hawaiʻi and Aotearoa.* Common Ground Research Networks, 2025.

Mississippi Freedom Democratic Party. "Why We Are Here at the Greenville Air Force Base." January 31, 1966. Wisconsin Historical Society Library Microforms Room, micro 788, reel 2, segment 2, pt. 1.

Monbiot, George. *Regenesis: Feeding the World Without Devouring the Planet.* Penguin, 2022.

Moody, Earle. "The Treatment of Pellagra." In *Transactions of the Medical Association of the State of Alabama.* Brown, 1914.

Mooney, Stephen J., Justin Knox, and Alfredo Morabia. "The Thompson-McFadden Commission and Joseph Goldberger: Contrasting 2 Historical Investigations of Pellagra in Cotton Mill Villages in South Carolina." *American Journal of Epidemiology* 180, no. 3 (August 1, 2014): 235–44. https://doi.org/10.1093/aje/kwu134.

Moore, Jason W. *Capitalism in the Web of Life: Ecology and the Accumulation of Capital.* Verso Books, 2015.

Morehouse, Lisa. "Farming Behind Barbed Wire: Japanese-Americans Remember WWII Incarceration." *NPR*, February 19, 2017, sec. Food for Thought. https://www.npr.org/sections/thesalt/2017/02/19/515822019/farming-behind-barbed-wire-japanese-americans-remember-wwii-incarceration.

Morton, Lois Wright, Ella Annette Bitto, Mary Jane Oakland, and Mary Sand. "Solving the Problems of Iowa Food Deserts: Food Insecurity and Civic Structure." *Rural Sociology* 70, no. 1 (2005): 94–112. https://doi.org/10.1526/0036011053294628.

*The Movement.* "Mississippi Negroes Occupy Air Base." 2, no. 2 (February 1966).

Muir, Cameron. *The Broken Promise of Agricultural Progress: An Environmental History.* Routledge, 2014.

Mumford, Lewis. *Technics and Civilization.* University of Chicago Press, 2010.

Munita, Tomás. "Guano: The Bird's Bounty." *Save Our Seas Magazine*, December 7, 2018. https://saveourseasmagazine.com/guano-the-birds-bounty/.

Musolin, Kristen, Jessica G. Ramsey, James T. Wassell, David L. Hard, and Charles A. Mueller. "Health Hazard Evaluation Report: Evaluation of Musculoskeletal Disorders and Traumatic Injuries Among Employees at a Poultry Processing Plant." Health Hazard Evaluation Program, March 2014. www.cdc.gov/niosh/hhe/reports/pdfs/2012-0125-3204.pdf.

Myrdal, Gunnar. *An American Dilemma*, vol. 1: *The Negro Problem and Modern Democracy.* Transaction, 1996.

Nally, David. "The Biopolitics of Food Provisioning." *Transactions of the Institute of British Geographers* 36, no. 1 (2011): 37–53.

Nash, J. A., and M. P. Parish, eds. *Plough, the Loom and the Anvil: An American Farmers' Magazine.* J. S. Skinner, 1857.

National Agricultural Statistics Service. *Farms and Land in Farms: 2024 Summary.* US Department of Agriculture, 2025.

National Association of Manufacturers. "Facts About Manufacturing." *NAM*, n.d. Accessed February 25, 2026. https://nam.org/mfgdata/facts-about-manufacturing-expanded/.

National Congress of American Indians. "NCAI Response to Usage of the Term, 'Indian Country.'" 2019. https://www.ncai.org/news/ncai-response-to-usage-of-the-term-indian-country.

National Institute of Diabetes and Digestive and Kidney Diseases. "Overweight & Obesity Statistics." 2024. https://www.niddk.nih.gov/health-information/health-statistics/overweight-obesity.

National Park Service History, Electronic Library and Archive. "A History of the Daniel Boone National Forest 1770–1970 (Appendix A)." April 7, 2010. http://npshistory.com/publications/usfs/region/8/daniel-boone/history/appa.htm.

Native Governance Center. "Blood Quantum and Sovereignty: A Guide." *Native Governance Center* (blog), 2022. https://nativegov.org/resources/blood-quantum-and-sovereignty-a-guide/.

Nesheim, Malden C., Maria Oria, and Peggy Tsai Yih, eds. *Overview of the U.S. Food System: A Framework for Assessing Effects of the Food System.* National Academies Press, 2015.

Nesterak, Max. "The 1950s Plan to Erase Indian Country." *APM Reports,* November 1, 2019. https://www.apmreports.org/episode/2019/11/01/uprooted-the-1950s-plan-to-erase-indian-country.

Ng, Wendy L. *Japanese American Internment During World War II: A History and Reference Guide.* Greenwood, 2002.

Nibert, David. "Animals, Immigrants, and Profits: Slaughterhouses and the Political Economy." In *Critical Animal Studies: Thinking the Unthinkable,* edited by John Sorenson, 3–17. Canadian Scholars' Press, 2014.

Nixon, Rob. *Slow Violence and the Environmentalism of the Poor.* Harvard University Press, 2011.

Nketani, Mutinta, and Timothy Wise. "Feeding Big Agribusiness, Starving Africans." *Al Jazeera,* September 4, 2023.

Nolan-Ferrell, Catherine. "Perspective: Past U.S. Policies Have Made Life Worse for Guatemalans." *Washington Post,* June 21, 2021. https://www.washingtonpost.com/outlook/2021/06/21/past-us-policies-have-made-life-worse-guatemalans/.

Norgaard, Kari M., Ron Reed, and Carolina Van Horn. "A Continuing Legacy: Institutional Racism, Hunger, and Nutritional Justice on the Klamath." In *Cultivating Food Justice: Race, Class, and Sustainability,* edited by Alison Hope Alkon and Julian Agyeman. MIT Press, 2011.

Nunn, Nathan, and Nancy Qian. "The Columbian Exchange: A History of Disease, Food, and Ideas." *Journal of Economic Perspectives* 24, no. 2 (2010): 163–88.

Olmstead, Alan L., and Paul W. Rhode. "Reshaping the Landscape: The Impact and Diffusion of the Tractor in American Agriculture, 1910–1960." *Journal of Economic History* 61, no. 3 (2001): 663–98.

O'Neill, Eilís. "Inmates Will Soon Be Growing Wine Grapes, Hops Inside the Washington State Penitentiary." July 19, 2019. https://www.kuow.org/stories/inmates-will-soon-be-growing-wine-grapes-hops-inside-the-washington-state-penitentiary.

Oshinsky, David M. *Worse Than Slavery: Parchman Farm and the Ordeal of Jim Crow Justice*. Free Press, 1996.

Ouden, Amy E. Den. *Beyond Conquest: Native Peoples and the Struggle for History in New England*. University of Nebraska Press, 2005.

Oxfam. "Climate Equality: A Planet for the 99%." 2023. https://policy-practice.oxfam.org/resources/climate-equality-a-planet-for-the-99-621551/.

Patel, Raj. *Stuffed and Starved*. HarperCollins, 2007.

Patel, Raj, and Jim Goodman. "The Long New Deal." *Journal of Peasant Studies* 47, no. 3 (2020): 431–63.

Pelzer, Karl Josef. *Planters Against Peasants: The Agrarian Struggle in East Sumatra 1947–1958*. Nijhoff, 1982.

Perkins, Tom. "Why Are There So Few Black-Owned Grocery Stores?" *Civil Eats*, January 8, 2018. https://civileats.com/2018/01/08/why-are-there-so-few-black-owned-grocery-stores/.

Phillips, James. "American Agrarianism's Answers to the Nation's (In) Securities." *Connecticut Public Interest Law Journal* 9, no. 2 (2010): 343–56.

Phillips, Sarah T. "The Price of Plenty: Getting Farm Policy Right in the 1960s." *Journal of American History* 109, no. 3 (December 1, 2022): 596–620. https://doi.org/10.1093/jahist/jaac346.

Pink, Randi. "Birmingham's 'Food Deserts' Have Been Shaped by Its Redlined Past." *Civil Eats*, September 26, 2018. https://civileats.com/2018/09/26/birminghams-food-deserts-have-been-shaped-by-its-redlined-past/.

Polanyi, Karl. *The Great Transformation: The Political and Economic Origins of Our Time*. Beacon, 2001.

Pollan, Michael. *In Defence of Food: The Myth of Nutrition and the Pleasures of Eating*. Penguin, 2008.

———. *In Defense of Food: An Eater's Manifesto*. Penguin, 2009.

———. *The Omnivore's Dilemma: A Natural History of Four Meals*. Penguin, 2007.

Prucha, Francis Paul. *American Indian Treaties: The History of a Political Anomaly*. University of California Press, 2023.

Pruett, T. J. "Is Pellagra a Preventable Disease?" In *Transactions of the Medical Association of the State of Alabama*. Brown, 1914.

Purvis, Malcolm. "Some Observations on the Effects of PL 480 Wheat Sales." *Journal of Farm Economics* 45, no. 4 (1963): 858–62.

Putzel, James. "The Ladejinsky Model of Agrarian Reform: The Philippine Experience." MA thesis, McGill University, 1986.

Quandt, Sara A., Alice E. Arcury-Quandt, Emma J. Lawlor, Lourdes Carrillo, Antonio J. Marín, Joseph G. Grzywacz, and Thomas A. Arcury. "3-D Jobs and Health Disparities: The Health Implications of Latino Chicken Catchers' Working Conditions." *American Journal of Industrial Medicine* 56, no. 2 (2013): 206–15.

Randolph, James. "Some Notes on Pellgra and Pellagrins, with Report of Cases." *Mobile Medical and Surgical Journal* 14, no. 2 (February 1909): 65–87.

Rasmussen, Wayne, Gladys Baker, and James Ward. "A Short History of Agricultural Adjustment, 1933–75." Agricultural Information Bulletin, US Department of Agriculture, Economic Research Service, n.d.

Reese, Ashanté M. *Black Food Geographies: Race, Self-Reliance, and Food Access in Washington, D.C.* UNC Press Books, 2019.

Renwick, Melissa. "Healing the Land and Spirits Through a Community Garden." *Ha-Shilth-Sa*, June 3, 2020. https://hashilthsa.com/news/2020-06-03/healing-land-and-spirits-through-community-garden.

Reuters. "Explainer: How Four Big Companies Control the U.S. Beef Industry." June 17, 2021, sec. Business. https://www.reuters.com/business/how-four-big-companies-control-us-beef-industry-2021-06-17/.

Rice, Stian. "Divide and Cultivate: The Role of Prisons and Indian Reservations in U.S. Agricultural Imperialism." *Food and Foodways* 30 (2022): 16–37.

———. *Famine in the Remaking: Food System Change and Mass Starvation in Hawaii, Madagascar, and Cambodia*. West Virginia University Press, 2020.

———. "Genocide." In *International Encyclopedia of Human Geography*, edited by Audrey Kobayashi, 2nd ed., 5:327–34. Elsevier, 2020.

———. "Madagascar's Famine Is More Than Climate Change." *Georgetown Journal of International Affairs* 23, no. 2 (January 24, 2022).

Rice, Stian, and James A. Tyner. "Revanchist 'Nature' and 21st Century Genocide." *Space and Polity* 25 (2021): 347–52. https://doi.org/10.1080/13562576.2021.1991785.

Riding, Alan. "U.S. Food Aid Seen Hurting Guatemala." *New York Times*, November 6, 1977, sec. Archives. https://www.nytimes.com/1977/11/06/archives/us-food-aid-seen-hurting-guatemala-donations-sent-after-1976-quake.html.

Robinson, Cedric J. *Black Marxism: The Making of the Black Radical Tradition*. University of North Carolina Press, 2000.

———. *Cedric J. Robinson: On Racial Capitalism, Black Internationalism, and Cultures of Resistance*. Pluto Press, 2019.

Rockefeller Foundation. "True Cost of Food: Measuring What Matters to Transform the U.S. Food System." Accessed August 24, 2024. https://www.rockefellerfoundation.org/report/true-cost-of-food-measuring-what-matters-to-transform-the-u-s-food-system/.

Rose, Nikolas. *Powers of Freedom: Reframing Political Thought*. Cambridge University Press, 1999.

Russell, Charles Edward. *The Greatest Trust in the World*. Ridgway-Thayer Company, 1905.

Rutherford, Stephanie, and Paul Rutherford. "Geography and Biopolitics." *Geography Compass* 7, no. 6 (June 1, 2013): 423–34. https://doi.org/10.1111/gec3.12047.

Sachs, Carolyn. *The Invisible Farmers: Women in Agricultural Production*. Rowman & Allanheld, 1983.

Saletan, William. "Food Apartheid." *Slate*, July 31, 2008. https://slate.com/technology/2008/07/banning-fast-food-in-poor-neighborhoods.html.

Salisbury, Neal. *Manitou and Providence: Indians, Europeans, and the Making of New England, 1500–1643*. Oxford University Press, 1995.

Sampson, Devon, Marcela Cely-Santos, Barbara Gemmill-Herren, et al. "Food Sovereignty and Rights-Based Approaches Strengthen Food Security and Nutrition Across the Globe: A Systematic Review." *Frontiers in Sustainable Food Systems* 5 (September 2021). https://doi.org/10.3389/fsufs.2021.686492.

*San Francisco News*. "Jap Ban to Force Farm Adjustments." March 4, 1942.

Sato-Parker, Julianne, and Rob Finch, dirs. *Solving for Zero*. Blue Chalk Media, 2022. 69 mins.

Savitz, David A., Minh Thang Nguyen, Ingrid Elizabeth Swenson, and Erika M. Stone. "Vietnamese Infant and Childhood Mortality in Relation to the Vietnam War." *American Journal of Public Health* 83, no. 8 (August 1993): 1134–38.

Sbicca, Joshua. "Growing Food Justice by Planting an Anti-Oppression Foundation: Opportunities and Obstacles for a Budding Social Movement." *Agriculture and Human Values* 29, no. 4 (December 1, 2012): 455–66. https://doi.org/10.1007/s10460-012-9363-0.

Schlesinger, Stephen, and Stephen Kinzer. *Bitter Fruit: The Story of the American Coup in Guatemala*. Rev. ed. Harvard University Press, 2005.

Schlosser, Eric. *Fast Food Nation: The Dark Side of the All-American Meal*. Houghton Mifflin Harcourt, 2012.

Schumpeter, Joseph A. *Capitalism, Socialism, and Democracy*. 1942. Routledge, 2013.

Schwartz, Amy D. "Colonial New England Agriculture: Old Visions, New Directions." *Agricultural History* 69, no. 3 (1995): 454–81.

Schwean-Lardner, Karen, and Hank Classen. "Lighting for Broilers." Huntsville, AL: Aviagen, 2010.

Searcy, George. "An Epidemic of Acute Pellagra." In *Transactions of the Medical Association of the State of Alabama*. Brown, 1907.

Seay-Fleming, Carrie. "Food Insecurity in the Northern Triangle: Leveraging Agricultural Policies and Programs for the Benefit of Smallholders." Latin America's Environmental Policies in Global Perspective. Wilson Center, October 2021.

Shannon, Jerry. "Food Deserts: Governing Obesity in the Neoliberal City." *Progress in Human Geography* 38, no. 2 (April 1, 2014): 248–66. https://doi.org/10.1177/0309132513484378.

Shapiro, Paul. *Clean Meat: How Growing Meat Without Animals Will Revolutionize Dinner and the World*. Simon & Schuster, 2018.

Siler, Julia Flynn. *Lost Kingdom: Hawaii's Last Queen, the Sugar Kings, and America's First Imperial Adventure*. Grove/Atlantic, 2012.

Siler, Joseph Franklin, P. E. Garrison, and W. J. MacNeal. "Pellagra: A Summary of the First Progress Report of the Thompson-McFadden Commission." *Journal of the American Medical Association* 62 (January 3, 1914): 8–12.

Sims, Jonathan. "Indian School Graves Rediscovered Under City Park." *The Paper*, July 27, 2021. http://abq.news/2021/07/indianschools/.

Sinclair, Upton. *The Jungle*. Doubleday, 1906.

Sinel, Martin, and Tony Weis. "Ventilation Shutdown and the Breath-Taking Violence of Infectious Disease Emergency Management in Industrial Livestock Production." *Environment and Planning E: Nature and Space* 7 (2024): 1076–97, https://doi.org/10.1177/25148486241229012.

Sittig, Ann L., and Martha Florinda González. *The Mayans Among Us: Migrant Women and Meatpacking on the Great Plains*. University of Nebraska Press, 2016.

Skaggs, Jimmy M. *The Great Guano Rush: Entrepreneurs and American Overseas Expansion*. St. Martin's Press, 1994.

Smith, Andrew F. *Eating History: 30 Turning Points in the Making of American Cuisine*. Columbia University Press, 2009.

Smith, Bobby J., II. "Mississippi's War Against the War on Poverty: Food Power, Hunger, and White Supremacy." *eGROVE*, July 1, 2019. https://egrove.olemiss.edu/studythesouth/11/.

Smith, Neil. *Uneven Development: Nature, Capital, and the Production of Space*. University of Georgia Press, 2008.

Smith, Walter Buckingham, and Arthur Harrison Cole. *Fluctuations in American Business, 1790–1860*. Russell & Russell, 1962.

Smits, David D. "The Frontier Army and the Destruction of the Buffalo: 1865–1883." *Western Historical Quarterly* 25, no. 3 (1994): 312–38.

Sodani, Rekha, and Sunil Kumar. "Cow Urine: A Boon for Sustainable Agriculture." *International Journal of Current Microbiology and Applied Sciences* 6, no. 2 (2017): 1824–29.

Spataro, David. "Against a De-Politicized DIY Urbanism: Food Not Bombs and the Struggle over Public Space." *Journal of Urbanism* 9, no. 2 (April 2, 2016): 185–201. https://doi.org/10.1080/17549175.2015.1056208.

Specht, Joshua. *Red Meat Republic: A Hoof-to-Table History of How Beef Changed America*. Princeton University Press, 2020.

Squire, Mariah. "Farmers Face Whiplash as Trump Immigration Shifts Keep Ag in Limbo." Successful Farming, 2025. https://www.agriculture.com/from-raids-to-reprieves-tracking-trump-s-impact-on-ag-labor-11791862.

Staff. "Poverty and the Federal Government: Five Documents from Greenville Mississippi." *Liberation: An Independent Monthly*, March 1966.

Staudt, Kathleen A. "Women's Politics and Capitalist Transformation in Subsaharan Africa." Working paper, Office of Women in International Development, Michigan State University, 1984.

Steinbeck, John. *The Grapes of Wrath*. 1939. Penguin, 2006.

———. *In Dubious Battle*. 1936. Penguin, 2006.

Stern, Eliana. "The Navajo Nation: A Case Study on Food Colonialism and Environmental Justice." *Earth Refuge*, March 4, 2021. https://earthrefuge.org/the-navajo-nation-a-case-study-on-food-colonialism-and-environmental-justice/.

Stewart, Watt. *Chinese Bondage in Peru: A History of the Chinese Coolie in Peru, 1849–1874*. Pickle Partners, 2018.

Stoll, Steven. *Larding the Lean Earth: Soil and Society in Nineteenth-Century America*. Farrar, Straus and Giroux, 2003.

Stull, Donald D., and Michael J. Broadway. *Slaughterhouse Blues: The Meat and Poultry Industry in North America*. Cengage, 2012.

Svensson, Måns, Rustamjon Urinboyev, Anders Wigerfelt Svensson, Peter Lundqvist, Margareta Littorin, and Maria Albin. "Migrant Agricultural Workers and Their Socio-Economic, Occupational and Health Conditions—A Literature Review." SSRN, July 26, 2013. http://dx.doi.org/10.2139/ssrn.2297559.

Táíwò, Olúfémi O. "When the United Fruit Company Tried to Buy Guatemala." *The Nation*, December 7, 2021. https://www.thenation.com/article/economy/united-fruit-guatemala/.

Tajima, Kayo. "The Marketing of Urban Human Waste in the Early Modern Edo/Tokyo Metropolitan Area." *Environnement Urbain / Urban Environment* 1 (September 9, 2007). https://journals.openedition.org/eue/1039?lang=en.

Taylor, Dorceta E., and Kerry J. Ard. "Food Availability and the Food Desert Frame in Detroit: An Overview of the City's Food System." *Environmental Practice* 17, no. 2 (June 2015): 102–33. https://doi.org/10.1017/S1466046614000544.

Taylor, Frank. "The People Nobody Wants." *Saturday Evening Post*, May 9, 1942.

Taylor, Sandra C. *Jewel of the Desert: Japanese American Internment at Topaz*. University of California Press, 1993.

Teixeira, Fabio. "Picked by Slaves: Coffee Crisis Brews in Brazil." Reuters, December 12, 2019, sec. World. https://www.reuters.com/article/world/picked-by-slaves-coffee-crisis-brews-in-brazil-idUSKBN1YG13D/.

Ten Broek, Jacobus, Edward N. Barnhart, and Floyd W. Matson. *Prejudice, War, and the Constitution: Causes and Consequences of the Evacuation of the Japanese Americans in World War II*. University of California Press, 1954.

Thompson, Paul B. "Food Aid and the Famine Relief Argument (Brief Return)." *Journal of Agricultural and Environmental Ethics* 23, no. 3 (June 1, 2010): 209–27. https://doi.org/10.1007/s10806-009-9181-6.

Timler, Kelsey, and Dancing Water Sandy. "Gardening in Ashes: The Possibilities and Limitations of Gardening to Support Indigenous Health and Well-Being in the Context of Wildfires and Colonialism." *International Journal of Environmental Research and Public Health* 17, no. 9 (2020): 3273.

Tokunaga, Yu. "Japanese Internment as an Agricultural Labor Crisis: Wartime Debates over Food Security Versus Military Necessity." *Southern California Quarterly* 101, no. 1 (2019): 79–113.

Trudeau, Robert H. *Guatemalan Politics: The Popular Struggle for Democracy.* Lynne Rienner, 1993.

Tuathail, Gearóid Ó. "Political Geography of Contemporary Events VIII: The Language and Nature of the 'New Geopolitics'—The Case of US-El Salvador Relations." *Political Geography Quarterly* 5, no. 1 (January 1, 1986): 73–85. https://doi.org/10.1016/0260-9827(86)90014-5.

Tuck, Eve, and K. Wayne Yang. "R-Words: Refusing Research." In *Humanizing Research: Decolonizing Qualitative Inquiry with Youth and Communities*, edited by Django Paris and Maisha T. Winn, 223–248. Sage, 2014.

Tyner, James A. *The Alienated Subject: On the Capacity to Hurt.* University of Minnesota Press, 2022.

———. *Violence in Capitalism: Devaluing Life in an Age of Responsibility.* University of Nebraska Press, 2016.

———. "Violence, Surplus Production, and the Transformation of Nature During the Cambodian Genocide." *Rethinking Marxism* 26, no. 4 (October 2, 2014): 490–506. https://doi.org/10.1080/08935696.2014.947756.

Tyner, James, and Joshua Inwood. "Violence as Fetish: Geography, Marxism, and Dialectics." *Progress in Human Geography* 38, no. 6 (December 1, 2014): 771–84. https://doi.org/10.1177/0309132513516177.

Tyner, James, and Stian Rice. "Agrarian Marxism, Animal Geographies, and Non-Human Labor in Democratic Kampuchea." *Environment and Planning E: Nature and Space* 5, no. 3 (2022): 1252–72.

Uchida, Yoshiko. *Desert Exile: The Uprooting of a Japanese American Family.* University of Washington Press, 2015.

United Nations University. "UN Study Reveals the Hidden Environmental Impacts of Bitcoin: Carbon Is Not the Only Harmful By-Product." October 24, 2023. https://unu.edu/press-release/un-study-reveals-hidden-environmental-impacts-bitcoin-carbon-not-only-harmful-product.

United Nations FAO. *Agricultural Production Statistics 2010–2023.* FAOSTAT Analytical Briefs, No. 96. FAO, 2024. https://openknowledge.fao.org/handle/20.500.14283/cd3755en.

UPSIDE Foods. "UPSIDE Foods: Company." Accessed June 10, 2024. https://upsidefoods.com/company.
US Army Western Defense Command. "Final Report: Japanese Evacuation from the West Coast, 1942." Government Printing Office, 1943.
US Bureau of Labor Statistics. Occupational Employment and Wage Statistics Profiles. Accessed September 1, 2025. https://data.bls.gov/oesprofile/
US Bureau of the Census. "United States Census of Agriculture: 1950, Volume 2, General Report, Statistics by Subjects." Government Printing Office, 1952.
US Department of Agriculture (USDA). "History of CCC Bin Storage Program." Agricultural Stabilization and Conservation Service, 1975.
———. "Special Supplemental Nutrition Program for Women, Infants, and Children." Food and Nutrition Service. Accessed July 29, 2024. https://www.fns.usda.gov/wic.
———. "U.S. Public Agricultural Research: Changes in Funding Sources and Shifts in Emphasis, 1980–2005." Economic Information Bulletin. Economic Research Service, March 2009.
———. *The Yearbook of Agriculture 1950–51: Crops in Peace and War.* Forest Service, 1950.
US Department of Labor. "History of Changes to the Minimum Wage Law." Accessed July 4, 2023. http://www.dol.gov/agencies/whd/minimum-wage/history.
US Department of State. "2020 Investment Climate Statements: Madagascar." 2020 Investment Climate Statements. US Department of State, Bureau of Economic and Business Affairs, n.d.
US House of Representatives. "Commodity Credit Corporation Grain Storage Activities." Committee on Government Operations, 85th Cong. 2d Session, August 31, 1960.
US Senate. "Examination of the War on Poverty: First Session on Hunger and Malnutrition in America." Subcommittee on Employment, Manpower, and Poverty of the Committee on Labor and Public Welfare, 90 Cong., 1 sess., July 1967.
Vail, David D. *Chemical Lands: Pesticides, Aerial Spraying, and Health in North America's Grasslands Since 1945.* University of Alabama Press, 2018.
Vandenbosch, Michel, and Philip Lymbery. *Cultivated Meat to Secure Our Future: Hope for Animals, Food Security, and the Environment.* Lantern Books, 2023.

Vantrease, Dana. "Commod Bods and Frybread Power: Government Food Aid in American Indian Culture." *Journal of American Folklore* 126, no. 499 (2013): 55–69.

Vigne, Godfrey Thomas. *Travels in Mexico, South America, Etc. Etc.* Vol. 2. W. H. Allen, 1863.

Villarejo, Don. "The Health of US Hired Farm Workers." *Annual Review of Public Health* 24, no. 1 (2003): 175–93.

Vizenor, Gerald. "Aesthetics of Survivance: Literary Theory and Practice." In *Survivance: Narratives of Native Presence*, edited by Gerald Vizenor, 1–23. University of Nebraska Press, 2008.

Von Braun, Joachim, David Hotchkiss, and Maarten Dirk Cornelis Immink. *Nontraditional Export Crops in Guatemala: Effects on Production, Income, and Nutrition*. Vol. 73. International Food Policy Research Institute, 1989.

Wadiwel, Dinesh Joseph. *Animals and Capital*. Edinburgh University Press, 2023.

———. "Chicken Harvesting Machine: Animal Labor, Resistance, and the Time of Production." *South Atlantic Quarterly* 117, no. 3 (July 1, 2018): 527–49. https://doi.org/10.1215/00382876-6942135.

Wageningen University. "New Course Shows Bigger Picture of the Food System." July 17, 2018. https://www.wur.nl/en/newsarticle/new-course-shows-bigger-picture-of-the-food-system.htm.

Wakeham, Pauline. "The Slow Violence of Settler Colonialism: Genocide, Attrition, and the Long Emergency of Invasion." *Journal of Genocide Research* 24, no. 3 (2022): 337–56.

Wallach, Jennifer Jensen. *How America Eats: A Social History of U.S. Food and Culture*. Rowman & Littlefield, 2013.

Wallerstein, Immanuel. "Land, Space, and People: Constraints of the Capitalist World-Economy." *Journal of World-Systems Research* 18, no. 1 (2012): 6–14. https://doi.org/10.5195/jwsr.2012.485.

Walls, Melissa L., and Les B. Whitbeck. "The Intergenerational Effects of Relocation Policies on Indigenous Families." *Journal of Family Issues* 33, no. 9 (September 1, 2012): 1272–93. https://doi.org/10.1177/0192513X12447178.

War Department. "Report on the Fixation and Utilization of Nitrogen." Nitrate Division, Ordnance Office, 1922.

War Relocation Authority. "California Incidents of Terrorism Involving Persons of Japanese Ancestry." 1945.

*Washington Post*. "Upside Foods CEO on the Future of Cell-Cultivated Meats." August 16, 2023. https://www.washingtonpost.com/washington-post-live/2023/08/15/upside-foods-ceo-future-lab-grown-meats/.

Wattnem, Tamara. "Seed Laws, Certification and Standardization: Outlawing Informal Seed Systems in the Global South." *Journal of Peasant Studies* 43, no. 4 (July 3, 2016): 850–67. https://doi.org/10.1080/03066150.2015.1130702.

Webb, Patrick. "When Projects Collapse: Irrigation Failure in the Gambia from a Household Perspective." *Journal of International Development* 3, no. 3 (1991): 339–53.

Weis, Tony. *The Ecological Hoofprint: The Global Burden of Industrial Livestock*. Bloomsbury Publishing, 2013.

Wetherington, Mark V. *American Agriculture: From Farm Families to Agribusiness*. Rowman & Littlefield, 2021.

Widener, Michael J., Sara S. Metcalf, and Yaneer Bar-Yam. "Dynamic Urban Food Environments: A Temporal Analysis of Access to Healthy Foods." *American Journal of Preventive Medicine* 41, no. 4 (October 1, 2011): 439–41. https://doi.org/10.1016/j.amepre.2011.06.034.

Williams, Robert A. *Savage Anxieties: The Invention of Western Civilization*. Macmillan, 2012.

Williams-Forson, Psyche A. *Eating While Black: Food Shaming and Race in America*. UNC Press Books, 2022.

Wills, Matthew. "Are We Entering a New Golden Age of Guano?" *JSTOR Daily*, May 4, 2016. https://daily.jstor.org/golden-age-of-guano/.

Winders, Bill. *The Politics of Food Supply: U.S. Agricultural Policy in the World Economy*. Yale University Press, 2009.

Winders, Bill, and David Nibert. "Consuming the Surplus: Expanding 'Meat' Consumption and Animal Oppression." *International Journal of Sociology and Social Policy* 24, no. 9 (January 1, 2004): 76–96. https://doi.org/10.1108/01443330410790786.

Wittman, Hannah. "Food Sovereignty: A New Rights Framework for Food and Nature?" *Environment and Society* 2, no. 1 (2011): 87–105.

Wolfe, Patrick. "Settler Colonialism and the Elimination of the Native." *Journal of Genocide Research* 8, no. 4 (December 1, 2006): 387–409. https://doi.org/10.1080/14623520601056240.

Yaffe-Bellany, David, and Michael Corkery. "Dumped Milk, Smashed Eggs, Plowed Vegetables: Food Waste of the Pandemic." *New York Times*, April 11, 2020, sec. Business. https://www.nytimes.com/2020/04/11/business/coronavirus-destroying-food.html.

Yapa, Lakshman. "What Causes Poverty? A Postmodern View." *Annals of the Association of American Geographers* 86, no. 4 (1996): 12.

Ybarra Zavala, Alvaro. "The Guano Islands." *Alvaro Ybarra Zavala* (blog). Accessed November 18, 2023. https://alvaroybarra.com/project/the-guano-islands/.

Younging, Gregory. *Elements of Indigenous Style: A Guide for Writing By and About Indigenous Peoples*. Brush Education, 2018.

Zerbe, Noah. "Feeding the Famine? American Food Aid and the GMO Debate in Southern Africa." *Food Policy* 29, no. 6 (2004): 593–608. https://doi.org/10.1016/j.foodpol.2004.09.002.

Zulauf, Carl, Gary Schnitkey, Krista Swanson, and Nick Paulson. "Stock-to-Use Ratios of US Corn, Soybeans, and Wheat Since 1960." *Farmdoc Daily* 11, no. 92 (June 14, 2021). https://farmdocdaily.illinois.edu/2021/06/stock-to-use-ratios-of-us-corn-soybeans-and-wheat-since-1960.html.

Zurawski, Erica. "Materializing the Metaphor: Theorizing the Food Desert as a Sociospatial–Legal Instrument in the Production of Space." *Human Geography* 16, no. 3 (November 1, 2023): 286–98. https://doi.org/10.1177/19427786231173631.

# INDEX

accumulation, 10, 102, 167, 177, 184, 292n18
  through dispossession, 100, 246
  of flesh, 203
  frontiers of, 210
  of guano, 64
activist, ix, 1, 14, 137, 148, 209, 242
Adams, John Quincy, 28
Africa, 18, 151, 155, 172, 200
  East, 58, 266
  North, 105
  South, 237
  sub-Saharan, 159, 161, 173, 252, 290n68
African swine fever (ASF), 206
Africans, 82, 156, 290n88, 301n14
Agency for International Development (USAID), 13, 153, 155–57, 165–68, 172–74
aggregation
  of animals, 202–4
  of farms, 108
  social benefits of, 196, 198
  spatial strategy of, 13, 180, 184–85, 209–10, 295n105
  of wealth, 78
agrarianism, 21, 50, 272n13
agribusinesses, 173, 186, 199, 209, 248, 257
  government regulation of, xii, 124, 140
  government support for, 101, 134, 167, 261
  and labor, 103, 159, 212
  multinational, ix, 131, 149, 252,
  US, 2, 50, 79, 108, 144, 169, 178
  white-owned, 78, 83, 93, 100, 102
Agricultural Adjustment Act (1933), 126
Agricultural Trade Development and Assistance Act (1954). *See* PL-480
agriculture
  American, 7, 18, 38, 45, 47–48, 65, 271n5
  capitalist, 55, 57, 71, 140, 184, 257, 278n57
  carceral, 99
  cellular, 207–9
  community-supported, ix, 255
  and crisis, 18, 62, 72, 121, 134, 150

agriculture (*continued*)
  and depression, 38, 44
  and development. *See* development
  economic sector of, 38–39, 46, 90, 111, 121, 128, 135, 169
  and education, 31, 36, 130, 166, 266
  efficiency of, 112, 160–62, 170, 173
  and empire, 2, 70–71, 73
  environmental impacts of, 252
  growth of, 7, 37, 140, 271n23
  history of, 93, 100, 187
  Indigenous, 26, 32, 36, 40
  industrial, ix, 14, 49, 129, 131–32, 186, 218, 246
  intensive, 54, 129
  inputs to, 9, 59–60, 74, 271n30
  and land, 34, 64, 77, 94, 96, 100–3
  plantation. *See* plantation
  and policies, 106, 121, 124, 129, 134
  and production, 7, 60, 112–13, 155, 164, 187, 259, 271n23
  and race, 81–82
  row, 183, 204
  and settlers, 7, 28, 31–32, 42, 46–47, 50, 62, 118
  and smallholders, 63, 161, 252
  and soil. *See* soil
  success of, 54, 109, 171
  and surplus. *See* surplus
  systems of, 107, 151, 254
  and technology, 13, 45, 107, 118, 126, 130, 133, 139, 251–53
  and transition, 107, 110, 117, 122–23, 166, 227, 243
  and treadmill. *See* treadmill
  and violence, 257
  and workers, 78–80, 88–89, 98, 123, 214, 247
Agriculture Department. *See* United States Department of Agriculture
Aguinaldo, Emilio, 141
aid
  consequences of, 155, 159, 167, 172
  domestic, 106, 127, 227
  to Guatemala, 138, 156–57, 174
  motivations for, 151, 153–55, 158
  and PL-480. *See* PL-480
  to reservations, 30, 32
  stipulations on, 152, 172
  as surplus disposal, 122, 139, 152, 156–58, 160, 169–71
  tied, 152
  as trap, 153–54
  *See also* commodity foods; food stamps
Alabama (US state), 215, 229, 266, 298n56
  medical societies of, 219–20, 223
Alaska (US state), 18, 49
Albuquerque Indian School (AIS), vii, 36–37
Albuquerque, New Mexico, 16, 36, 275n62
Algonquin, 49
Alien Land bill (1913), 80
Alinsky, Saul, 262
allotment, 30, 33–40, 45
alternative food networks, 255
American Guano Company, 68
American Legion, 76
ammonia, 55, 66, 74, 120, 178

ammunition. *See* munitions
Andrews, Garnett, 72
anemia, 107
animals
  as commodities, 13, 71, 112, 175–76, 182, 186, 188–90, 209–11
  culling of, 206–7
  labor of, 178–79, 181–83, 207
  needs of, 187
  relationships with, 49–50, 55, 122, 183, 212
  resistance of, 176–77, 184–85, 207, 209
  as sources of nutrients, 56–58, 60
  as traction, 123, 181
  violence against, xi, 13, 179–80, 193–94, 204–5, 254
  welfare of, 3, 14, 184, 208, 211, 214
  in the wild, 29
  *See also* concentrated animal feeding operation; agriculture, cellular
Anson, Austin E., 76–79, 82, 84–85, 87, 93, 104, 149, 245, 262
antibiotics, 205–7
Apache, 16, 37
Apartheid, 237–38, 241, 299n70
Arapaho, 29, 187
Árbenz, Jacobo, 147–48, 160, 173, 258
Archer Daniels Midland (ADM), 108
Arevalo, José, 147–48
Argentina, 167
Arizona (US state), 96, 274n39
Arkansas (US state), 63
Armas, Carlos Castillo, 148
Armour, 189, 198
army, 45, 53, 86, 90, 147, 156
  of Guatemala, 136–37, 148
  of the US. *See* US Army
Asiatic Exclusion Law (1924), 80
Asiatic Exclusion League, 80
assimilation, 19, 24, 36–39, 43, 119
Associated Farmers, 76, 89
asylum, 222, 225, 227, 229
Atlantic, 2, 23, 26, 47, 123
Australia, 74
automation, 177, 182, 199, 207
avian flu (H5N1), 206

bacon, 32, 197, 228
bacteria, 55, 57, 119, 178, 265
Bainbridge Island, Washington, 88, 91
Baltimore, Maryland, 52–54, 65, 68–70
banana, 59, 138–40, 145–47, 156, 162, 169, 258
bankruptcy, 1, 7, 38, 93, 167, 177, 185, 253
banks, 6, 38, 65, 144, 167, 186, 210, 242
Baptiste, Miska Jean, 205
beans, 32–33, 46, 58, 97, 154–57, 168, 172, 219
beef
  canned, 192
  production of, 61, 180, 188–89, 193, 203, 211–12
  rations, 32–33, 274n51
  slaughterhouses, 179
  *See also* ranching
Beef Trust, 190–94
Beldo, Les, 182

beriberi, 231–32
Beveridge, Albert Jeremiah, 141–42
Big Four, 190–97, 209, 211, 234, 253
biodiversity, 3, 251
Birmingham, Alabama, 228, 235, 239, 298n56
bison, 29–32, 46, 118, 187, 210
Black, 12, 54, 106, 129, 132–33, 145, 242
  children, 13, 107, 283n7
  consumers, 102, 234, 240
  croppers, 9, 102, 105–6, 108, 125, 246
  entrepreneurship, 235, 240
  guano workers, 54, 70, 102
  and pellagra, 216, 218, 223
  meatpackers, 127, 196–99
blindness, 1, 66
blood quantum, 34–35, 40
Blosser, Ray, 221
boarding schools, 10, 36, 40, 47, 49
body, 56, 182, 184, 209–11, 233
bones, 37, 55–56, 60–61, 68, 74, 188, 203
boom, 42, 120–21, 177, 185, 230
border, 10, 71, 91, 140, 169–71, 199, 214, 234
Bosch, Carl, 74, 120
boss, 80–81
Boston, Massachusetts, 189, 191
bottleneck, 185–86, 211, 213,
Bovine spongiform encephalopathy (BSE), 206
Bracero Program, 91, 100–101
bran, 3, 232
Brazil, xi
Britain, 54, 60–61, 65–66, 69, 71, 73
British Phosphate Commission, 74
broiler, 175–79, 182–83, 186, 203, 108, 213, 291n9
  *See also* chicken
Buchanan, James, 69
buffalo. *See* Bison
Buffalo, New York, 31
Bureau of Indian Affairs (BIA), 34, 36, 40–43, 45, 47, 91
Bureau of Reclamation, 94
Burke Act (1906), 35
butchers, 60, 179, 189–92, 194–95, 207, 209
Butler, Smedley, 145

cacao, xi, 140
CAFO. *See* concentrated animal feeding operation
California, 2, 29, 76, 79, 81, 82, 87–91, 99, 190–91
  agriculture of, 13, 77–79, 84–85, 96, 100–1, 174
  government of, 80–81, 103
  and gold, 67
  Northern, 41
California Farm Bureau, 76, 89
California Plan, 89, 91, 103
calorie, 30, 57, 123, 192, 204, 215, 220
campesino, x, xi, 202, 254
Canada, 173, 206
canned goods, 40, 69, 127, 229
  *See also* beef, canned
capital
  agricultural, 22, 128, 140, 164, 166, 173, 199, 278n57, 285n64
  challenges for, 22, 171, 200, 206
  finance, 257, 261
  fixed, 185, 210

foreign, 161, 163
industrial, 120, 124, 126, 129–30, 134–35, 156
and labor, 11, 79, 101, 196, 198, 257
meatpacking, 181–84, 187, 192, 208, 212, 291n6
movement of, 10, 102, 144, 152, 158, 211
perspective of, 9, 100, 218, 256–57, 260–62
plantation, 22, 25, 28
retail, 242
scientific, 209
surplus, 117
stagnation of, 160
*See also* capitalism; plantation; spatial fix

capitalism
definition of, 6–7, 9
characteristics of, 7, 34, 83, 180, 210, 247
and food production, 6–7, 12, 48, 55, 57, 71, 102, 132, 183–84, 214
and growth, 132, 142, 259
history of, 8, 57, 278n57
and inequality, 102, 246
as political-economic system, 151, 160, 174
racial, 13, 101–2, 283n87
and regulation, 194, 255
and violence, 11, 247
*See also* overproduction

carbon dioxide, 55
carceral fixes, 12–13, 45–46, 101, 131, 247, 275n80
Cargill, 108, 201, 209
Caribbean, 12, 51–53, 66, 68–69, 74, 140, 144–45, 192
Carson, Kit, 16–17, 20, 47
cash crops, 40, 123, 143, 168–69
and development, 160–65, 171
Castro, Fidel, 149
Catholic Relief Services, 157
cattle. *See* cows
cavalry, 17, 25, 29–30, 36, 44
cell, 55–56, 181, 207–9, 212
cellular agriculture. *See* agriculture, cellular
census, 218, 234, 236, 239, 241, 258, 299n59
Central America, 12–13, 17, 143–45, 148, 168
Central Intelligence Agency (CIA), 13, 138, 147–50
centralization, 207, 211, 257
Chapin, Frederick, 136–37, 157, 167, 173
charity, 36
cheese, 40, 108, 127
chef, 1, 237, 239, 258
chemical, 1, 3–4, 64, 122, 128–29, 158, 168–69
companies, 120–21, 134, 150, 159, 161, 171–73
engineering, 122, 207
fertilizer, 107, 120–21, 202
innovations, 13, 123, 134
inputs, 42, 56, 121, 125, 163, 251
manufacturing of, 130, 133
and metabolism, 55–58
pesticide, 107, 112, 118, 121, 128, 202
*See also* Green Revolution
chemurgy, 126
Cherokee, 23, 25–26, 273n23
Cheyenne, 29, 187

Chicago, Illinois, 31, 85, 188, 191, 194, 208–9, 262
  and beef, 189, 192
  and labor, 189, 195–96
  and meatpacking, 2, 188–89, 193–94, 198, 210
  and neighborhoods, 196–97
  *See also* Big Four
Chickasaw, 26–27, 273n33
chicken
  behavior, 176–77
  bodies, 175, 180, 201, 204, 207–8
  culling, 206
  debeaking. *See* debeaking
  disassembly, 178
  houses, 176–77, 186, 203–4
  industry, 180, 186, 203, 213
  labor of, 177–78, 181–82, 209–11
  sleep, xi, 175–76, 180, 212, 269n2
  treatment of, 2, 178–79
  workers, 177
  *See also* broiler; domestication
children
  and disease, 4, 37, 107, 230, 263
  and education, 36, 47
  and hunger, 8, 13, 106, 108–9, 127, 130, 133–35, 169
  and incarceration, 94, 99, 247
  and inheritance, 63, 76, 80, 213
  killing of, 25–26, 29, 136–37, 189, 199
  and labor, 199
  and malnourishment, x, 4, 40, 174, 270n10, 300n82
  *See also* boarding schools; nisei
China, 58, 67–68, 142, 173, 278n43
  *See also* coolie
Chincha Islands, 64–65
Choctaw, 23, 26–27, 39, 273n25, 273n34
cholera, 221, 241
CIA. *See* Central Intelligence Agency (CIA)
Cincinnati, Ohio, 31, 188
citizen, 25, 30–31, 43, 52, 63, 90
  and democracy, 250
  and incarceration, 76, 79, 86, 88
  and ideology, 21
  and rights, 80, 86, 104, 251
civil rights, 106, 109
civil servants, 154, 172
civil war, 137–38, 148, 156, 170, 174, 199–200, 278n43
  *See also* war
Civil War, 29, 64, 114–15, 132, 187, 189, 228
Clapp, Jennifer, 186
class, xii, 28, 70, 229, 237, 241
  conflict, 12, 21, 23, 31, 44–48, 197
  and elite. *See* elite
  and landowning. *See* landowning class
  and lower. *See* lower class
  and middle. *See* middle class
  prejudice, 13
  structure, 19, 130–31, 160, 164, 228
  and working. *See* working class
  and yeoman. *See* yeoman
climate, 128, 159, 221, 251, 261
coal, 56, 97, 108, 181, 229–30
coalition, 1, 132, 254
Cochrane, Willard, 38, 109, 111, 113, 128, 272n16

coffee, xi, 140, 143, 162, 169, 194, 269n2
plantations, 156, 168
Cold War, 109, 151
Collier, John, 39
colonialism, 16, 18, 46, 143, 148, 173, 273n20
*See also* settler, colonialism
colony, 62, 142
Colorado (US state), 29, 39, 98, 200–1, 274n39, 295n92
Colorado River Indian Reservation, 96
Columbus, Christopher, 51–53, 62, 64, 72, 253
commissioner, 33, 36, 39, 42, 274n51
commodity, 6, 32, 40, 56, 72, 111, 128, 180
and overproduction. *See* overproduction
exports, 152–55, 288n35
production, 59, 71, 112, 116, 125, 210
surplus, 19, 39, 131, 134, 154, 169
and tropical, 13, 140, 143–44
commodity foods, 10, 33, 40–41, 46–47, 106–7, 127
Commodity Credit Corporation (CCC), 126, 283n10
communal ownership, 34
communism, 140, 149, 151, 156, 159–60
fight against, 9, 138, 147–49, 155
Communist Party, 196
community, x, xi, 99, 199, 216, 236, 242, 245
Black, 235, 240
Indigenous, 17, 37, 48–49
Japanese American, 93, 104 279n6, 297n25
community supported agriculture (CSA), ix, 255
companies
chemical, 120–21, 126, 130, 134, 158–59, 161, 168, 172–73
food retail, 240, 242
government relationship with, 70
grain, 234, 240, 243, 283n10
in guano trade, 53, 65–70, 72–75
in Guatemala, 144–45, 147, 164
influence of, 13, 148–49, 252, 262
and imperialism, 139, 142, 144, 146, 149, 166
and labor, 90, 195–202, 210, 216
meatpacking, 179–86, 188–91, 195–205, 207–9, 211–13, 248, 252
and monopolization, 22, 164
of the New South, 217, 226, 229–33, 236
partnerships between, 78
in the Philippines, 142
railroad, 5, 13, 31, 34, 188, 190
seed, 121, 158
subsidies to, 108, 120–21, 150, 152
technology, 251
transfer of wealth to, 131, 230
*See also* agribusinesses; Big Four; meatpacking; plantations; Sugar Trust; United Fruit
company store, 69, 216, 226, 229, 231, 236
competition, 39, 42, 79, 82–83, 139–40, 155, 169–71, 197

concentrated animal feeding operation (CAFO), 183, 186, 207, 210–11, 213, 249, 252
and conditions, 204, 208
and environmental costs, 183, 203–4, 208
and disease, 185, 203, 205–6
concentration camps, 86, 96–99, 102, 138, 141, 279n4, 279n7
conditions at, 94
construction of, 91
removal to, 17, 76–77, 103, 174
site selection, 94
closure of, 79
Confederacy, 32, 220
Congress, 32–33, 40, 87, 120, 126, 129, 147, 151
acts of, 24, 29–31, 36, 44–45, 68, 124, 141
battles in, 28, 141, 174
hearings before, 127, 158
investigations by, 108
members of, 28, 76, 133, 191
conquistadors, 18
consumer, xii, 42, 122, 154, 202, 249, 257
contagion, 206, 221, 223–25
contracts, 33, 93, 115, 189–90
cropper, 127, 286n75
food aid, 154–55
guano labor, 67, 69, 72
Japanese American labor, 80–81, 83
in poultry, 176–77, 186, 207–8, 291n9
union, 198, 202, 249
contradiction, 2, 37, 47, 109, 134
convicts, xi, 2, 53–54, 66, 75, 225
coolies, 67–68, 72–73, 75, 79, 278n43
Cooper, Edward K., 52, 68
cooperatives, 81, 83, 236, 262
corn, 40, 62–64, 71, 117, 126, 132, 170
in commodity foods, 33
export of, 138, 155–56, 172–74
and feed, 183, 187, 202, 210
in Indigenous foodways, 40, 45, 49, 58, 298n37
milling of, 231, 242–44
prices of, 7, 32, 111–12, 140
production of, 44, 107, 113–15, 123, 128–29, 270n14
in Southern diet, 223, 228, 231
spoiled, 215–16, 219–20, 228
yields of, 150
Corn Laws, 54, 61
corn syrup, xi, 244
cornbread, 49, 216, 219, 221, 224–25, 227, 231, 238
corporations. *See* companies
Coté, Charlotte, 49
cotton, 105, 112, 128, 131–33, 156, 231, 270n14
expansion of, 23, 62–63, 226, 262
exports of, 62, 71–72, 143, 162
industrialization of, 106, 114, 125, 129
milling, 216–17, 223, 284n29
production of, 72, 116–17, 124, 243
soil impacts of, 63
switch to, 22, 117, 231
counterinsurgency, 137–38
countryside, 2, 46, 54, 59, 136–37, 199, 202
coup, 147–49, 151, 164, 262

courts, 17, 19, 45, 54, 70, 190
*See also* Supreme Court
COVID-19, 186, 205–6
cows, 179, 182, 184–87, 190–91, 203–4, 206–7, 211, 229–30
cowboys, 187, 191
Crawford, George, 62–63
credit, 6, 115, 167–68, 229
Creek, 23, 25–26
crime, 5, 8, 99, 193
criminal, 10, 25
crop dusting, 121
croppers, 115–16, 124–25, 129, 133–34, 284n27, 286n67
*See also* sharecropping
crops, 22, 42, 55, 59, 143
alternative, 128, 140
destruction of, 7
displacement of, 63
failure of, 119
feed, 95
growing of, 58, 63, 71, 92
harvested, 187
high-value, 84
marketing of, 81, 83
metabolism of, 56
planting of, 58, 62, 95
selection of, 82, 117
subsistence, 21, 115, 117, 127, 143, 230, 270n14
*See also* cash crops; staple crops; high-yielding varieties (HYVs)
Crow Creek Reservation, 30
Cuba, 140, 145, 149, 192, 287n8
culling. *See animals, culling of*
cultivated meat, 207–9, 251
cultivation, 58–59, 85, 92, 110, 114, 162, 165
intensive, 71, 119
cultivars, 117, 122, 160
culture, 18, 36, 154, 205, 246, 257, 260
Cushman, Gregory, 74, 276n10

Dahlberg, Kenneth, 251
dairy, 107, 125, 224, 226, 228, 230, 258
farms, 204
industry, 40
Dakota, 30–31
dams, 39–41, 49, 105
Daniels, Roger, 89
Davenport, C.B., 223–24, 238
Davidson, Osha Gray, 110
Dávila, Miguel, 147
Dawes Act (1887), 34, 36
Dead Indian Act (1904), 35
Dearborn, Henry, 26, 273nn30–32, 290n74
death
of agricultural workers, 145, 169
of animals, 175, 184, 206
of croppers, 107
from disease, 216–17, 227–28
of guano workers, 66–67, 70
of Guatemalans, 137–38, 145, 157, 167, 169, 200
of Indigenous people, 26, 30, 37
debeaking, 213
debt
government, 27, 44, 62, 74, 108
of farms and families, 38, 48, 115–16, 125, 167, 177
of Indigenous communities, 27–28, 45, 167
peonage, 67–68, 71, 144

deficit, 236, 238
demand
elasticity of, 111–13
for farm products, 7, 38, 101, 107, 112–13, 153, 186, 192
for inputs, 53, 60–61, 65, 158, 163, 166, 171
for labor, 42, 80, 90, 98, 123, 126, 246
for land, 22
for organic, 75, 249
and price, 7, 111–13, 128
and supply, 100, 102, 111, 128, 209, 249, 256
for imported commodities, 140, 151, 160, 162, 169–70, 246, 258
democracy, 40, 148, 150–51,170, 174, 251, 254
*See also* food democracy
Denver, Colorado, 29, 274n58
Department of State, 145, 147, 153, 155, 158, 170, 245
Department of the Interior, 33, 35, 41
dependence, 26, 32, 169–70, 172, 243, 261
deportation, 26, 90, 100, 200
depression
economic, 38–39, 42, 44, 109, 140
psychological, 43, 244
*See also* Great Depression
Desert Land Act (1877), 30
Detroit, Michigan, 149, 240
development, 43, 172, 238, 234
agricultural, 138–39, 226
economic, 225, 261, 286n5
and gender, 165
as ideology, 160, 289n61
rural, 161, 171, 174, 199
and treadmill. *See* treadmill, development
*See also* Agency for International Development (USAID); Green Revolution; PL-480
DeWitt, John, 86
diabetes, 1, 41, 217–18, 233
Dichlorodiphenyltrichloroethane (DDT), 121
dictator, 9, 136, 148–49, 172, 174
diet, ix, 168, 211, 215, 228
changes in, 41, 154
and disease, 4, 217
interventions in, 36
and malnourishment, 97, 225, 238
and pellagra, 216, 219–21, 223, 225–27, 230
and policy, 218
Southern, 219, 224, 226, 231
traditional, 1, 36, 58, 154, 187, 298n37
Diné. *See* Navajo
disassembly, 178, 183, 188–89, 210–11, 250, 262–63
discrimination, 11, 43, 83–84, 101–4, 147, 262
disease, 13, 39, 215–21, 227–28, 232, 240–41
in animals, 185, 202, 204–6
benefit of, 244
burden of, 10, 233, 244
chronic, 4, 210
emerging, 3, 218
heart, 1, 41, 217–18, 233, 244
outbreaks of, 206, 233
and starvation, 142, 187

transmission of, 5, 196, 204–6, 221, 223
*See also* pellagra
disincentive, 153, 155
dispossession, 105, 246
of Indigenous people, 18–19, 24, 28, 44, 47
of Japanese Americans, 89, 100
dissent, 101, 145, 160
DNA, 47
dogs, 181, 198, 217, 220, 227–28, 243
dollar store, 236
domestic rearrangements, 13, 131–33, 247
domestication, 175, 207, 211
Dominican Republic, 146
Dominican Republic-Central America Free Trade Agreement (DR-CAFTA), 168
Dos Erres, Guatemala, 137, 157, 174
Douglass, Frederick, 3
Dow Chemical, 134
Dulles, Allen, 147
Dulles, John Foster, 147
Dust Bowl, 39, 54, 100
Dutch East Indies, 232

East (US region), 12, 17, 26, 28–29, 62, 187–89, 191
East Africa, 58, 200, 266
East Asia, 67, 140, 149–50, 232
East Coast (US region), 22, 118
eating, xi, 4, 208, 212, 219, 228, 246, 250, 252
economy, 8–9, 41, 80, 87–89, 115, 118, 134, 233
and excess, 124, 142
local, 96
settler, 38–39, 42, 44, 45–46,
state, 78, 220
ecosystem, 3, 56, 60, 71, 73, 187, 256
education, 36, 39, 166, 218, 233, 251
efficiency, 9, 160–63, 170–71, 181, 193, 211
of land and labor, 112, 114, 119, 123
and technology, 28, 113–14
Egypt, 153, 206
Eisenhower, Dwight D., 126, 147, 152
Eisenhower, Milton, 94
El Salvador, 138
electricity, 41–42, 56
elites, 18, 31, 139, 167, 249, 253
and land, 20–23, 47, 141–43, 156, 160
and politics, 22, 142, 146, 171, 257
Emancipation, 2
embassy, 136, 138, 148, 286nn1–2
empire, 16, 279n1
and agriculture, 2, 76, 104
and Europe, 18
overseas, 9, 12, 52
employment, 79, 90, 115, 166
enclosure, 241–43, 247
energy, 55–57, 74, 107, 208, 253
engineering, 207, 209
England, 61, 65, 71, 244
epidemic, 13, 37, 156, 217, 221, 232–33, 240
equity, 237–38, 251, 260
erosion, 19, 39–40
Essex, Jamey, 172, 266
Ethiopia, 201
ethnicity, 201, 214, 258
eugenics, 217, 223, 297n25
Euro-Americans, 11, 20–23, 47, 123

Europe, 12, 47, 52, 71, 173, 215–16
and agriculture, 47, 58–62, 65, 118, 122, 150–51, 253
arrivals from, 18, 82, 100, 109, 188, 196
and colonization, 11, 18, 22, 81, 104, 142–43
and conflict, 20, 23, 38, 123, 192
and reforms, 10–11, 23, 59
and settlers, xii, 19
European Union, 155
executive order, 76, 86, 91, 104
excess, 117, 131, 153, 194
consumption of, 40
disposal of, 125, 128, 131, 134–35, 150
making of, 105, 122, 124, 129
management of, 107, 108, 124, 126, 130
and violence, 48, 117, 122, 170, 257, 259
excrement, 2, 52–53, 59, 64, 68–69, 74–75, 197, 203, 225
exploitation, 10, 19, 173, 213
of labor, 2, 73, 79, 104, 192
exports, 147, 155, 162–64, 168, 173, 202
from the US, 62, 72, 74, 126, 150–52, 172, 174
expropriation, 14, 54, 59, 71–73, 104, 246–48, 253–56, 258
extermination, 19, 24, 28, 47
extraction, 10, 48, 50, 66, 75, 143, 181, 243
extraterritorial fixes, 13, 71–73, 170, 247

factory, 56, 130, 177–78, 183, 185, 188, 197–99
jobs, 31, 119
workers, 13, 131, 179, 181, 183, 191, 196
fallow, 22, 58, 106
family, 16, 84, 93, 97–99, 137, 191, 266–67
and diet, 216, 228, 250, 260
*See also* farm, family
famine, 65, 107, 153, 155, 159
farms
aggregation of, 108
challenges facing, 88, 110, 130, 253, 258
equipment of, 78, 92, 99–100, 119, 122
family, 46, 110, 128, 131, 134
large, 107, 134
loss of, 109, 117, 156, 177, 231, 247
and owners/operators, 4, 81–82, 117
and prices, 33, 37–39, 113, 124, 130, 153, 168, 170
and products, 7, 38, 44, 126, 151, 155, 159
*See also* farmworkers
Farm Bill, 248
Farm Service Administration (FSA), 93, 100–1
farm-to-table, 248
farmers, 251
farmer's markets, ix, x, 1, 249, 255
farmworkers, xi, 107, 218, 90, 258
displaced, 10, 106, 129–31, 247
fat
animal, 175–76, 183, 189, 201
in children, 107
in diets, 41, 48, 211, 220

fear, 4, 31, 77, 79–80, 98–99, 148, 164, 201–2
public, 85–89, 137, 156, 192, 216, 250
feces. *See* excrement
Federal Bureau of Investigation (FBI), 93, 86
feed (for livestock), 95, 177, 183, 186–87, 190, 203–4, 210
feedlots, 180, 202, 208
fertilizer, 10, 12, 162, 252, 278n64
application of, 117
chemical production of, 119–21, 186
demand for, 61, 158, 166
purchase of, 60, 163, 170
synthetic, 60, 75, 150, 168–69
use of, 158, 166
*See also* companies, chemical
Fiji, 74
Filipino, 81, 103, 142, 170, 232
Fillmore, Millard, 53, 68, 276n3
finance, 148, 167, 257, 261
fines, 69, 98–99
First World, 139, 286n5
fish, 2, 8, 18, 32, 41, 52, 64, 75
fishers, 6, 52, 254
fix. *See* spatial fixes
Fixed Nitrogen Research Laboratory (FNRL), 120
Fixico, Donald, 43
fixing (of nitrogen), 58, 119–20
flesh, 6
accumulation of, 180, 182–83, 203–4, 207, 213
animal, 178, 183–84, 188, 207–9, 211, 262
Florida (US state), 174, 190, 216
flour, 32–33, 40, 48, 127, 154, 274n51
fodder, 41, 58, 63, 187
Food Access Research Atlas (FARA), 234, 236, 299n59
Food and Agriculture Organization (FAO), 260
Food and Drug Administration (FDA), 232
food apartheid, 238
food democracy, 251, 254
food deserts, 13, 218, 238–39, 241–42
and criticism, 236–38
and definition, 233–34
and mapping, 242–43
and metaphor, 238–39, 241–42, 252
and supermarkets, 236–39
food sovereignty, 254–55
food stamps, 106, 127–28, 131, 286n67
food system, 11–15, 48, 54, 60, 77–78, 88, 170
brokenness of, 1–2, 8–9, 14, 248
crisis in, 103, 108, 130
change in, xii, 11, 79, 85, 100, 109, 131, 217, 232, 246
definition of, 6–7
fixing, 249–253, 255–61
foreign, 170–72
Indigenous, 32, 41, 46, 50, 123
purpose of, 9, 109, 212–13
settler, 18–19, 21, 24, 44–46, 50, 123
structure of, xi, 87, 102, 186, 211
violence of, x, xi, 3–5, 10–11, 15, 117, 179, 214, 246–48
and wealth accumulation, 11, 14, 134, 139

foodways, 36, 43, 47, 49, 154
Ford Foundation, 160
Ford, Henry, 188
Ford, Leland, 86
Fordson, 92, 119
foreclosure, 35, 93, 100, 218
foreign exchange, 162–63
forest, 1, 162, 251, 288n35
Forrest, Nathan Bedford, 220
Fort Confederation, 27
Fort Jackson, 24
Fort Laramie, 30
fortification, 232, 298n48
fraud, 17
free market, 6, 100, 109–11, 128, 171, 230, 247, 250, 286n5
free trade agreements, 10, 138, 168
freemen, 115
freight, 188, 190
French Indochina, 232
frontier
  of accumulation. *See* accumulation, frontiers of
  advancement of, 20, 140
  closing of, 33–34, 37, 44, 140, 169
  and Indian country, 17, 23, 30, 46
  and plantations, 30, 46, 110
  settlement of, 21–22,31, 123
  tropical, 140–41, 150
fructose, xi, 244
fruit, 8, 40, 190–91, 199
  and vegetables, 78, 270n14
funding, 39, 118, 134, 203, 220, 225, 241, 247

Galeano, Eduardo, xii
gardens, ix, x, 10, 59, 223, 236
  organic, 49–50
Garrison, Philip, 221
Gates, Bill, 209, 251–52
gender, xii, 13, 36, 223–24, 237
  and development, 165–66, 185,
  and labor, 196, 198, 201, 210
Genetically modified organism (GMO), 155, 251
genocide, 20, 46–47, 187, 201, 246
  cultural, 36, 275n59
  in Guatemala, 137–38, 143
  and settler colonialism, 2, 17, 271n2
gentrification, 238–39
geography, x, xii, 14, 19, 138, 176–77, 218, 257
  as answer, 142, 158, 179
  and crisis, 23, 48
  uneven, 10, 12, 217
geopolitics, 4, 139, 142, 151, 158, 172
George, Susan, 173
Georgia (US state), 22, 25–26, 62–63, 72, 216
Georgia State Asylum, 227
germ, 3, 221, 223, 229, 231
Germany, 74
Gersony, Robert, 157
Ghana, xi, 132, 266
Gibbs, Antony, 65–66, 68
Gila River, 95–96
globalization, 192
gold, 52, 59, 67, 140, 143
Goldberger, Joseph, 224–29, 233
Golden Age, 38, 134
González, Martha, 200
government, 60, 68, 85, 99, 120, 126, 199, 261
  debt. *See* debt, government

food aid, 30, 32–33, 42, 107, 127, 167
of Guatemala, 148–50, 156–57, 167–69, 262
and imperialism, 53, 140
institutions of, 11
and land, 27, 34–35, 45, 62, 96, 134
legislation, 11, 26, 90
market controls, 7, 107, 111, 115, 125–26, 130
policies, 124, 243–44
regulation, 100, 128–29, 177, 194, 207, 211
of Southern states, 109
subsidies. *See* subsidies
and territory, 20, 22, 24, 29, 44
and trade, 10, 47
treaties, 25
and tribal. *See* tribal government
grain, 2, 10, 13, 28
as aid, 32, 153, 157, 172
as animal feed, 203, 204, 211
in diet, 228, 232
movement of, 61, 65, 153, 226, 231
production of, 31, 45, 58, 165, 234, 240, 243
surpluses, 105, 108, 150, 152, 160, 170, 202, 211
*See also* disincentive
granary, 130, 141
Grand Rapids, Michigan, 239–40
Granger movement, 132
grassland, 29
graves, 29, 37, 56, 60–61, 68, 177, 233
grazing, 35, 40, 187, 210
Great Depression, 39, 42, 100
conditions, 90, 96, 145, 197
policies, 7, 106, 124, 126, 246
Great Migration, 106
Great Plains (US region), 28, 187, 189
Green Revolution, 138, 159–60
consequences of, 161, 171
new, 252
technologies, 166, 171–72, 253
Greenville AFB, Mississippi, 105–7, 125, 135, 262
Greenville, Indiana, 24
grits, 216, 219, 225
grocer, 234–35, 237
grocery stores, x, xii, 237–38, 240, 242–43, 247, 249
and food deserts, 242
and white flight, 234, 242–43
Gross Domestic Product (GDP), 27
Grote, Carl, 229–30, 233, 236, 240
groundwater, 3
grower-shippers, 77–78, 83, 90, 93, 100, 103
guano, 51–52, 65, 79, 178–79, 252
companies. *See* companies, in guano trade
craze, 65, 67
deposits, 52
harm, 66–68, 252
islands, 60, 70, 74
mining, 53, 66, 69
and nutrients, 59, 65, 72, 204
Peruvian, 53, 61, 66, 72–73, 75
and Quechua, 64, 75
trade, 66–68, 72–74
workers, 54–55, 67–68
*See also* Chincha Islands; coolies; Guano Islands Act (1856)

Guano Islands Act (1856), 52–54, 68, 70, 72–73
Guatemala, 136, 139, 164, 286n1, 288n28
agriculture in, 166
business interests in, 144–45
civil war in, 156, 200
democracy in, 173, 258
earthquake in, 157
land holdings in, 145, 147, 149, 160
migrants from, 174, 200, 202, 213–14
military intervention in, 13, 148–49
threat of communism in, 147
US exports to, 138, 168
*See also* Central Intelligence Agency (CIA); Maya; plantations; United Fruit Company (UFC)
Guatemala City, 144, 288n28
Guerilla Army of the Poor (EGP), 156
Guevara, Ernesto "Che", 149

Haber, Fritz, 74, 120
Haber-Bosch process, 74
Haiti, 52, 69–70, 145, 153,
Hajer, Maarten, 241
Hale, Stephen, 220
Hancock, Graham, 153–54, 287nn6–7
harassment, 102, 197, 200
Harris, Seale, 226
Harrison, Benjamin, 70
harvest, 7, 115, 157, 168, 187
of cotton, 114
of fish and wildlife, 29, 75
of grain, 114, 153, 164
and labor, 49, 98, 125, 129
mechanization of, 114, 119, 248, 259–60
and rotation, 58
seasonal, 92
of vegetables, 99
Harvey, David, 100
Hassanein, Neva, 250
Hawai'i, 68–69, 73, 85, 140, 255, 287n6, 287n8
Healthy Food Financing Initiative (HFFI), 239, 241
heart disease. *See* disease, heart
Heart Mountain, 94, 96, 102
hen. *See* broiler
herbicide, 121, 123, 129
herds, 29, 41, 45–46, 62
Hewes, Laurence I., 93
high-yielding varieties (HYV), 158–63, 171
Hillman Hospital, 228
Hispanic, 9, 199, 201, 218, 234, 242
Hispaniola, 51
history, xii, 35, 47, 49, 244, 247, 257
and agriculture, 93, 100, 121, 187
and challenges, 41, 108
economic, 8, 234
and Guatemala, 138, 147
and innovation, 118, 260, 265
and meatpacking, 180, 200, 209, 211
and narrative, 24, 77, 118
and the South, 220, 232
and violence, 10, 30, 77, 91, 255
world, 1, 259
Hitomi, Lucille, 97

hogs. *See* pigs
homes, 17, 62, 92, 94, 138, 157, 179, 202
homestead, 28, 31, 34–35,
homesteaders, 17, 20, 24, 52, 58, 101, 110
homestead acts, 30, 44, 46
Honduras, 138, 146, 148
Hoover Dam, 40–41
Hopi, 42
Hormel, Leontina, 41
hormones, 2, 211
Horowitz, Roger, 198
hospitals, 42, 90, 96, 205, 233
Hugo, Victor, 59, 203
von Humboldt, Alexander, 64–65
Humphrey, Hubert, 151, 156, 288n35
hunger, 54, 77, 149, 153, 161, 174, 215
  benefits of, 256
  and death, 8, 20
  as economic problem, 159–60, 169, 244
  for land, 19–21, 48
  and malnutrition, 3, 50, 173
  and politics, 151, 156, 159, 171, 241
  rural, 109
  widespread, 30, 156, 159, 173
Hwéeldi, 17
hybrid, 113, 118, 122, 125

Ibo, 23
ideology, 21, 102, 142, 156, 170, 263
illness, 5, 178–79, 189, 210
  *See also* disease
Immerwahr, Daniel, 73, 272n9
immigrants, 188–89, 193, 195–97
  European, 100, 201, 224
  Filipino, 81
  Japanese, 76–80, 84
  Mayan, 200
  Mexican, 81
immigration, 80, 185, 196, 200–1, 212
Immigration and Customs Enforcement (ICE), 200, 212
imperialism, 12, 28, 51, 54, 70–73, 83, 192
implements, 32, 119, 128
import, 61, 71, 139, 151, 155, 162, 243
incarceration, 76–77, 79, 86, 88, 93–94, 99–104, 245–47
incentives, 6, 72, 124, 153–54
  *See also* disincentives
indenture, 18, 22–23, 53
independence, 16, 23, 47, 141–43, 148, 173, 192, 286n5
India, 132, 142, 260
Indians. *See* Indigenous people
Indian Affairs. *See* Bureau of Indian Affairs (BIA)
Indian Appropriations Act (1851), 29
Indian country, 28, 38, 40, 42–46, 49, 271n3
  definition of, 17–18, 24, 33, 46
  and food system, 32, 48
Indian New Deal, 39–40
Indian Relocation Act (1956), 31, 42–43, 91
Indian Removal Act (1830), 26, 28, 44, 46
Indian Reorganization Act (1934), 40, 44

Indigenous people, 2, 4, 17–18, 20, 81, 123
assimilation of, 34, 36, 43, 45
definition of, 11–12, 102
and food, 30, 58
removal of, 24, 26, 28, 91, 245
indigo, 22
Indonesia, xi
industrial agriculture, ix, 14, 49, 186
Industrial Revolution, 59, 259
industry, 118
and agriculture, 117, 122, 133–35
and crisis, 120–22, 158
and growth, 65, 88
and overproduction, 130
public funding for, 134
*See also* meatpacking
inequality, 161, 171, 230–31, 254–55
influenza, 37
infrastructure, 48, 235
for agriculture, 39, 150, 154, 212
in Guatemala, 144
for Indian reservations, 39
for irrigation, 163
for metabolism, 188, 208
for transportation, 180, 184, 194, 282n65
injuries, 196, 210, 217, 243
inmates, 2, 77, 99, 216
and agriculture, 97
conditions facing, 94, 96
labor of, 97–98, 101
resistance of, 98, 99
*See also* convicts
inner cities, xi, 218, 235, 238, 258
innovation, 118–19, 188–89, 213, 259–60
in chemistry, 13
and cross-pollination, 118, 121
in geography, 198
in industry, 117, 132
mechanical, 119
in technology, 4, 130, 132, 231, 252
inputs, 6, 9, 81, 83, 129–31, 159–60, 258
chemical. *See* chemical inputs
cost of, 2, 110, 113, 115, 134, 167–69
markets for, 134, 171
mechanical, 42
packages of, 159–63
for metabolism, 56, 208
subsidization of, 133
*See also* Green Revolution
inspection, 178
*See also* Meat Inspection Act (1906)
insurance, 93, 235
insurgents, 136, 138
*See also* counterinsurgency
integrator, 177–78, 186
intensification, 71, 84, 271n23
International Center for Tropical Agriculture (CIAT), 161
International Maize and Wheat Improvement Center (CIMMYT), 161
International Monetary Fund (IMF), 161, 163–64, 168
International Rice Research Institute (IRRI), 160, 166
International Union of All Workers (IUAW), 198
International Workers of the World (IWW), 198

internment, 77, 95, 279n7
  *See also* incarceration
Inuit, 49
invasion, 23–24, 138–40, 151
  of Guatemala, 148–49
investment, 100, 180, 219, 237, 242, 258
  in foreign countries, 139, 142, 161, 165
  in infrastructure, 39
  in research and education, 166, 233
  return on, 27, 158, 166
Iowa (US state), 199–200
Iowa Beef Packers (IBP), 199, 202, 209, 211
Ireland, 61
Irish, 61, 65, 196, 198, 201
Iroquois, 24, 26
irrigation
  and CAFOs, 204
  in concentration camps, 94, 96, 282n65
  and Green Revolution, 159, 163, 165, 168
  in Indian reservations, 40
issei, 79, 104
  bosses, 80–81
  businesses, 85, 87
  cooperatives, 81, 83
  discrimination against, 80, 82–83
  farming, 84, 87
  incarceration of, 88–89, 91, 93, 102
  land improvements, 82, 85
  land leasing, 81
  land ownership, 81, 83
  property
  workers, 81, 90
  *See also* nisei
Italian, 196, 198, 215–16, 224
Italy, 216, 224

Jackson, Andrew, 16, 25–26, 28, 47
Jamaica, 51, 72
Jamestown, Virginia, 18
Japan, 59, 79, 82, 104, 173, 259
Japanese Americans, 12–13, 104, 279n4
  and agriculture, 78–80, 84–85, 87, 98
  and businesses, 81, 83
  discrimination against, 82–86, 102, 104
  and families
  incarceration of, 77, 94–96, 99, 103, 246
  and property, 78–79, 89, 93
  removal of, 76, 87–89, 91–93, 100
  labor of, 90–91
  *See also* issei; nissei
Japanese Exclusion League, 82
Jay, John, 21
JBS, 186, 200–1
Jefferson, Thomas, 21, 26–27, 46–47, 50, 83, 134, 167, 257
Jim Crow, 235
John Deere, 28
Johnson, Lyndon, 105–7, 126–27, 130, 135, 154
Jones v. United States, 54,

Kansas (US state), 31, 39, 70, 274n39
Kansas City, Missouri, 188
Karuk, 41, 49
Kennedy, John F., 126, 151, 286n74, 288n37

Kent, Ohio, ix, 14, 245, 266
Kentucky (US state), 23
kidnapping, 67–68, 148, 156
killing, 19, 68, 70, 98, 145, 175
  mass, 36, 102, 137, 217
killing floor, 180, 183–84, 188, 195–96, 210, 213
Kiowa, 187
Kiribati, 74
Klamath, 41, 49
Korematsu v. United States, 101
Ku Klux Klan, 220
Kulikoff, Allan, 132, 278n57

La Via Campesina (LVC), 254–55
laborers, 1, 11, 44–45, 69, 188
  agricultural, 79, 88–90, 103, 112, 114, 135, 231
  animal. *See* animals, labor of
  hourly, 106, 125
  impoverished, 73, 216
  migrant, 90, 254
  resistance from, 184
  violence against, 70, 73, 181, 210
  white, 90, 100,
Lake Mead, 40, 45
Lakota, 29, 187, 189
landlords, 35, 81, 115, 117, 128, 134, 284n27
landowners, 48, 258, 35
  in California, 71, 83
  in Europe and its colonies, 22, 61, 143
  in Guatemala, 160
  in the US South, 106, 115–17, 127–31, 133, 135, 286n67
language, 14, 18, 36, 185, 214, 258
Laotian, 199
latifundios, 142–43, 148–50, 174
Latin America, 12, 149, 161, 174, 200, 254
  economic change in, 132
  labor in, 149, 167, 210
  US companies in, 70, 144, 150
Latina/Latino, 12, 201
layoffs, 184, 202, 214
Leach, Sydney, 228
lease, 81, 92–93, 149, 164, 167, 247
Lee, Robert E., 220
legislators, 1, 5, 103, 191
legislature, 68, 78, 80, 190
legume, 58
Lend Lease, 99
lending, 1–2
Leslie, Chris, 225
license, 21, 121
von Liebig, Justus, 61, 65
life expectancy, 42
Lincoln, Abraham, 30
Little Big Horn, 24
livestock, 32, 56, 58, 97, 188, 202, 230
  confinement of, 176, 185
  grazing of, 40, 63
  nutrients from, 72
  reductions in, 41
lobbyist, 1, 141
localism, ix, x, xii, 250
locavore, ix, xi, 14, 245–46, 253, 266
Lombroso, Cesare, 216, 219
loop, xi, 59, 72,
Los Angeles, California, 85–87, 90, 92, 281n58
Lost Cause narrative, 220
Louisiana Purchase, 26
lower class, 11, 223, 230–31
lynching, 4, 197, 220

machinery, 183–84, 187, 193
*See also* mechanization
Madagascar, 164
maize. *See* corn
Malawi, 155
malnourishment, x, 1, 4, 8, 109, 238, 246
malnutrition, 1, 3, 8, 133, 216–18, 243–44, 247
*See also* nutrition
Manifest Destiny, 20, 48
manufacturer, 2, 119, 121, 134, 140, 159, 162
manure, 57–59, 61–62, 65, 69
Manzanar, 94–95, 102
marasmus, 107
markets
and commodities, 72, 126
for farm products, 21, 32, 157
for farmland, 89, 96
for exports 162–63
free, 100, 109–11, 128, 171, 230, 247–49
for frontier land, 22–23, 27, 47
for fruits and vegetables, 78, 87, 101
for meat, 189, 194
for inputs, 120, 168, 204
and overproduction, 33, 44, 99, 107, 112
and prices, 112, 117
*See also* overproduction
Marshall Plan, 126
Marx, Karl, 57, 60–61, 180–82, 278n59, 292n17
Maryland (US state), 22, 62, 65, 68, 203, 205, 266
Masagana 99, 167
Massachusetts (US state), 49
Maya, 156, 169
communities, 136–37, 157, 167
genocide against, 138, 157, 174
labor migration, 168, 174, 200, 202
smallholders, 218
women, 200, 213–14
McClatchy, V.S., 82–84, 280n24
McCormick, 28, 114
McDonald, Bryan, 111
McFadden, Henry. *See* Thompson-McFadden Commission
McKinley, William, 141
McLemore, Henry, 86, 88
McWilliams, Carrey, 2
measles, 37
meat, 63, 97, 175, 248–49, 291nn5–6, 295n105
canned, 40
and diet, 125, 214, 219, 224, 228, 230, 236, 256
hunting for, 29
public demand for, 2, 107
*See also* beef, canned; meatpacking
Meat Inspection Act (1906), 194, 253
meatpacking, 179–80, 183, 202, 209, 248
centralization in, 180, 186, 211
and labor, 179, 181, 191, 197, 246
and mechanization, 187–88, 211, 262
and migrants, 193, 200–1, 247
and monopolies, 189, 199, 194–95
and unions, 198–99, 202, 245
and working conditions, 201, 205, 214
*See also* Big Four

mechanization, 40, 114, 123, 131, 212
media, 85, 90, 205
Mediterranean, 215, 221
Meriam Report, 39
mestizo, 143, 145
metabolic rift, 63, 71, 74–75
  as crisis,12, 58–60, 247
  definition, 57
  solutions to, 73, 255
metabolism, 56–57, 71, 182, 207
  *See also* metabolic rift
metaphors, 6, 13, 142, 182
  in public health, 218, 233, 236–39, 241–42, 247, 263
  in settler colonialism, 19, 36, 272n11
Mexicans
  deportation of, 90–91, 100
  labor of, 81, 90–91, 199, 103
Mexico, 90, 161, 174, 199, 234
  and economic development, 132, 149, 167
  and US imperialism, 29, 145, 274n39
microbe, 58, 184, 205
Mid-Atlantic (US region), 22, 62, 199
middle class, 232, 234, 237–39
Middle East, 74, 151
Midwest (US region), 117, 188, 196, 206, 220, 231
migration, 13, 39, 108, 131, 174
  *See also* Great Migration; immigration
Miles, Nelson A., 192
military, 13, 17, 18, 271n3, 274n58, 282n65
  of Guatemala, 136, 148, 156–57
  in the western US, 29–30, 187
  in World War II, 77, 84–86, 88–89
  overseas, 106, 142, 174, 232
militia, 25, 29
milk, 40, 194, 213, 225, 228
mill towns, 10, 223, 226, 229, 231, 241, 244
milling, 165, 216, 220, 231–32, 240, 298n37
milpa, 58
miners, 13, 39, 229–30, 239
Minges, P.A, 88, 281n49
Minidoka, 94, 97, 99
minimum wage, 129, 147, 253, 286n67
mining, 229–30, 240
Minnesota (US state), 30, 198, 200
Mississippi (US state), 23, 62, 106, 127–29, 133, 225
Mississippi Delta, 105–6, 117, 124, 135, 227, 234, 258
Mississippi River, 17, 26–29, 47, 245
Missouri (US state), 30
Modernization, 166, 211, 232, 259
Monbiot, George, 251
monoculture, 50–51, 143, 252–53, 256
  growth in, 31–32, 46, 54
  impact of, 48–49, 74, 150, 170
  and industrialization, 114, 158, 162
  and overproduction, 44, 151
monopoly, 190, 211, 261
  in guano, 53, 65, 68,
  and Guatemala, 138, 145, 147

Monsanto, 166
Montgomery Street, 77–78, 82, 84, 89, 103
Morrill Act (1862), 31
Mount Vernon, Alabama, 215
movements, xii, 9, 14, 141, 164, 249, 254
Mozambique, 153, 155
munitions, 119–20, 138, 150
murder, 8, 51, 54, 70, 137, 156
Murphy, Frank, 101
Muskogee, 25
Muslim, 201
mutual aid, 255
Myami, 31
Myanmar, 201
Myer, Dillon S., 42, 91
myths, 19–20

Nam Hollow, 224, 238, 243, 297n25
narrative, 1, 11, 77, 109, 244
  and agrarianism, 19, 110
  and development, 122, 149, 160, 169, 173, 259, 289n61
  and settler colonialism, 19, 33, 36, 46, 50
Nash, Maurice, 67
National Association for the Advancement of Colored People (NAACP), 127
National Congress of American Indians (NCAI), 17
National Defense Act (1916), 120
National Labor Relations Board (NLRB), 197
national security, 48, 53, 147
Native. *See* Indigenous people
natural resource, 21, 254
  and crisis, 2, 12, 19, 38, 44, 54
Navajo, 1, 17, 20, 32, 37, 40–45, 47, 49, 218
Navassa, 53–54, 60, 68–70, 72–75, 262, 278n51, 278n64
Navassa Phosphate Company, 53, 68
navy, 76, 86, 88
Nebraska (US state), 31, 200, 202, 213–14, 262
networks, 157, 202, 236, 256–58, 262–63
  for communication, 144
  of enterprises, 83, 194–95, 211
  and food system definition, 2, 5, 6, 14, 270n21
  for transportation, 29, 187, 198
  *See also* alternative food networks
New Deal, 3, 33, 39–41, 91, 127, 133
  *See also* Indian New Deal
New Echota, 24, 25
New Mexico (US state), 16, 32, 37, 274n39
New Orleans, Louisiana, 144, 146
New South, 216, 219, 226
New World, 10–11, 62–64, 82, 110, 118, 143, 216, 253
New York (US state), 49, 297n25
New York City, New York, 63, 144, 189, 191
New Zealand, 74, 255
niacin, 228, 231–32, 258
Nicaragua, 138, 146, 148

nisei, 79–80
  businesses, 85, 87
  citizenship, 80, 88
  farming, 84, 87–88
  incarceration of, 89, 91, 93, 104
  land improvements, 85
  land ownership, 81
  property, 102
  workers, 90
  *See also* issei; Japanese Americans
nitrate, 73, 120–21, 135
  in guano, 53
  in soil, 56, 119
nitrogen, 55, 59–60, 65, 119, 174, 204, 208
  cycle of, 59, 74–75, 204
  fixing of, 55, 58, 74, 119–20
  and industry, 158, 168
  for plant growth, 54–55, 71, 120, 276n6
nogyo kumiai, 81
nonviolence, 255, 257
Norgaard, Kari, 41
North (US region), 106, 125, 217, 220, 238, 240, 242–43
North America, 10–12, 109, 122–23, 154, 166, 216, 253
  *See also* Indigenous people
North Carolina (US state), 204, 237
North Vietnam, 105, 135
Nuevo México, 16
nurses, 107, 216, 219
nutrients, 64, 71–72, 246
  and expropriation, 54, 59–60
  lack of, 53, 58, 62
  for metabolism, 55–58, 207, 227–32
  replacement of, 58, 61
nutrition, 233, 253
  *See also* malnutrition

Obama, Michelle, 239
obesity
  condition of, 1, 218, 233, 238, 240, 244
  epidemic of, 13, 217, 233, 238
Ohio (US state), 24, 237
oil, 56, 74, 108, 145, 159, 194, 246, 252
Okamoto, Shoichi James, 99
Oklahoma (US state), 39, 274n39
Old World, xii, 64, 110
Olds, Robert, 145, 288n23
Olson, Culbert, 78, 89, 91
Oneida, 49
Onondaga, 26, 273n28
O'Quinn, Darrell, 235
orchards, 2, 17, 94
Oregon (US state), 29, 41, 76, 108
organic, x, 10, 14, 49, 249
  fertilizers, 75, 159, 166
orphans, 93, 138, 173, 225, 282n62
orphanages, 216, 225
overproduction, 12, 48, 124, 130, 134
  causes of, 7, 37, 113
  consequences of, 7, 32, 140, 152
  control of, 13, 33, 125–26, 128, 131
  violence of, 10, 128, 170, 247
overseer, 53, 70, 72
overtime, 176, 178, 291n3

Pacific, 78–79, 84, 87, 101, 104
  and guano trade, 52, 64, 67–69, 72–74

Pacific Railway Acts (1862, 1864), 31
PacificCorp, 49
Packers and Stockyards Act (1921), 195
packing plant, 1, 82, 186, 199–205, 213–14, 295n92
Pakistan, 253
palm oil, xi
panacea, 255–56
paradox, 3, 113, 130, 132, 179, 214, 246
Parliament (UK), 61
pastoralists, 18
pasture, 22, 58, 188, xii
patents, 120, 231, 260
pathogens, 204
  in animals, 182, 205–6
  in humans, 219, 227–28, 233
  *See also* disease
Pearl Harbor, 76, 78, 85, 103–4
peasant, 10, 47, 137, 141–43, 147, 160–61, 171
Peasant's Way, The. *See* La Via Campesina (LVC)
pellagra, 215–34, 236, 238–39, 241–44, 298n37
pellagrin, 225, 297n31
peonage, 67–68, 115
perpetrators, 3, 5, 138
Peru, 53, 64–69, 73, 278n43
pesticide, xi, 117, 163, 167, 169, 202, 249
  and chemical, 7, 107, 112, 114, 118, 121, 161–62
petroleum, 74, 122, 158, 166, 172, 252
  *See also* oil
Philippines, 143, 232, 258
  development of, 160, 164, 166–67
  US imperialism in, 70, 140–43, 149–50, 173, 192
phosphate, 53, 55–56, 60, 65, 72, 74, 278n64
phosphorus, 54–57, 59–61, 71, 208, 246, 276n6
pickers, xi, 2, 114, 125, 266
Pierce, Franklin, 68
pigs, 2, 8, 149, 182, 193–94, 203–4, 206–7, 213
Piscataway, 23
PL-480, 152–58, 160, 163, 245, 253
plant. *See* factory
plantations, xi, 19, 25, 47, 62, 73
  and bananas, 144–45, 156
  and capital, 25, 28
  cotton, 115–16, 131, 133, 226, 243
  and labor, 67, 115, 160, 162, 168, 174
  and monocropping, 12, 22, 27
  and ownership, 21, 24
  and slavery, 2, 23, 82
  sugar, 141–43
  yeoman conflict with, 22–24, 31–32, 46, 110
  *See also* United Fruit Company (UFC)
planters, 23, 62, 71, 114–16, 125–29, 133
plow, 7, 22, 28, 85, 92–93, 95
  and tractor, 112, 114, 119
pneumonia, 189, 216
poison, 169, 221, 263
politicians, 31, 110, 128, 170, 186, 192
  California, 76, 86
  Southern, 220, 223, 243

Pollan, Michael, 249, 261
pollution, 235
populism, 9
pork, 32, 63, 180, 219, 193, 203
Poston, 96, 98
potatoes, 8, 61, 64, 219
poultry, 175, 180, 199, 203–4, 209
  houses, 176
  workers, 179, 205
poverty, 109–10, 170, 174, 246
  and development, 160, 252
  and diet, 220, 226–27, 230
  on Indian reservations, 39, 43
  and migration, 199, 203, 213
  rural, 43, 133, 227, 234
  *See also* War on Poverty
Pratt, Richard Henry, 36, 43, 47, 274n58
press, 1, 87, 151, 243
prison, 44, 70, 99, 216, 225, 242, 282n75
prison camps, xi, 48
prisoners, 25, 30, 138, 141, 217, 225
privy, 223–24, 229
problem closure, 241–42
problem enclosure, 241–42, 247
processed foods, 1, 107, 126, 140, 235
  ultra-, 244, 249,
processor, 2, 134, 140, 159, 168, 171, 211, 232
production, 6, 117, 143, 149, 180, 250–52, 255
  and agriculture, 60, 62, 85, 112–14, 156, 162–64
  of commodities, 46, 71, 123–25, 140, 145, 162–65
  controls, 131
  crises of, 103
  of crops, 11–12, 58, 63, 88–89, 104, 159, 168
  disincentives, 153–55
  ecological, 56–57
  of farm equipment, 119
  increase in, 7, 39, 57, 101, 113, 118
  of inputs, 119–21
  of meat, 177, 182–85, 187, 189–90, 193–94, 202, 208–13
  wartime, 95, 111
property, 93, 99–100, 121, 144, 208, 255, 260
  confiscation of, 93
  destruction of, 78, 93, 102
  maintenance of, 93
  protection of, 89
  rights to, 143
  sale of, 78, 92
  of the state, 141
  taxed, 230
  theft of, 77–78
propaganda, 80, 86, 102, 148, 151
public health, 3, 109, 217, 232–33
  *See also* United States Public Health Service (PHS)
Public Law 480. *See* PL-480
Pueblo, 16, 37
Puerto Rico, 140, 192, 287n6, 287n8
purchasing power, 169, 235, 242, 244, 249, 258
Pure Food and Drug Act (1906), 196

Quechua, 64, 75
Queen Victoria, 244

race, xii, 12, 197, 235, 237–38
  and classification, 81, 201, 223
  and ideology, 86, 102

*See also* white supremacy
racial capitalism. *See* capitalism, racial
racism, 2, 20, 77, 101–2, 248
railroads, 67, 108, 147, 189, 202, 217
companies, 31, 110, 140, 144, 190–91, 194, 198, 230
and government support, 31, 44, 132, 185
and meatpacking, 185, 187–88, 194, 198
rain, 9, 55, 64, 165, 204
ranchers, 6, 21, 35, 40–41, 187, 191, 194
ranching, 29–31, 35, 37, 40, 49, 192
Rapanui (Easter Island), 68
rape, 17, 51, 137
rash, 215, 220
rations, 69, 97, 227, 232
and reservations, 30, 32–33, 36
Reagan, Ronald, 174, 279n6
reaper, 28, 114
rebels, 53, 137–38, 145, 148, 278n43
recession, 38, 133
Reconstruction, 54, 115, 133, 135, 145, 220
reconstruction committees, 157, 214
redlining, 102, 235, 237, 243, 298n56
Reese, Ashanté, 236, 265
Reeseman, Era, 95
reformists, 34, 36, 179
refrigeration, 184–85, 189, 192
refugees, 106, 136–38, 199, 200–2, 211, 213, 250
regulation, xii, 5, 84, 175, 195, 202, 211, 233
relief, 152, 158
religion, 19, 21, 36, 142, 196
relocation
in meatpacking, 202, 214
of Japanese Americans, 77, 91
of Indigenous people, 19, 31, 43
relocation centers, 91, 279n7
remittances, 202
removal, 49, 206, 208
of Japanese Americans. *See* Japanese Americans, removal of
of Indigenous people. *See* Indigenous people, removal of
rent, 4, 81, 93, 107, 127, 129, 202
researchers, 121, 216–17, 220, 227, 233, 241, 243
reservations, xi, 11, 16, 49, 73, 79, 96, 154
and agriculture, 40–41
and allotment, 30, 34–35, 44
and commodity foods, 33, 40
conditions on, 39, 42, 258
creation of, 24, 28–29, 32
elimination of, 43
and Indian country, 17, 43
management of, 42
and surpluses, 19
resistance, xii, 19, 106, 180, 185, 206
animals, 184, 186, 209, 291n5
Indigenous, 123, 187
responses to, 209, 213
worker, 184, 186, 198, 209
resources
retail flight, 235, 237
retailers, 2, 83, 87, 185–86, 189, 239–41, 252

revolution, 103, 179, 208–9, 252
agricultural, 23, 117–18, 122–23
political, 19, 135, 143, 149, 164
social, 157, 160, 194
in technology, 3, 107, 109, 118, 123–25, 129, 208
*See also* Green Revolution; Industrial Revolution, Revolutionary War
Revolutionary War (US), 12, 27
Rhode Island (US state), 27, 297n25
rice, 22, 82, 105, 155, 164–65, 167–68, 232
rift. *See* metabolic rift
rights, 19, 65, 80, 104, 143, 157, 255
*See also* civil rights
right-to-work, 199, 202, 294n69
Ríos Montt, José Efraín, 136–38, 174
riots, 9, 197
Robinson, Solon, 72
Rockefeller Foundation, 160, 260
Roosevelt, Franklin Delano (FDR), 279n1, 279n7
and Japanese American incarceration, 76, 78, 83, 86–87, 91, 100, 103
and New Deal, 39–40, 110, 124, 134
Roosevelt, Theodore, 83, 141, 192
and meat industry, 193–94, 211–12
Rose, Nikolas, 108
rotation, 58, 62
Round Valley Reservation, 30
runoff, 3, 41, 74
Russell, Charles Edward, 190–1

sabotage, 86–87, 92–93
Sachs, Carolyn, 166
Sacramento, California, 84, 103
Salinas Valley, 76, 86
Salinas Valley Vegetable Grower-Shipper Association, 76, 89, 149
salmon, 41, 49
salt, 32–33, 189, 219
San Francisco, California, 77, 208
San Joaquin Valley, 84, 96
Sand Creek, 29
scarcity, 7–8, 10–11, 107, 130–31, 169, 258
Schofield, John M., 29
Schultz, Theodore, 128
Schumpeter, Joseph, 132
science, 226, 233, 260
seabird, 52, 75
Searcy, George, 215–16, 219, 233
Seattle, Washington, 87
Secretary of the Interior, 35
seeds, 49, 114, 121, 125, 160, 168, 251
*See also* high-yielding varieties (HYV); hybrids
segregation
and race, 133, 235
spatial strategy of, 13, 180, 185, 197–98, 209–10, 247, 249
self-sufficiency, 32, 38, 72, 95, 151, 154, 170–72
Seminole, 26
settlers, xii, 18, 29, 30
and agriculture, 21, 31, 46, 62, 122
and class conflict
and Indigenous conflict, 19, 24–25, 27–28, 33–34, 38–39
and land, 17, 20, 28, 38, 47, 143

and westward expansion, 12, 23, 25, 32, 187
*See also* agrarianism
settler colonialism, 11, 18–19, 24, 26, 47–48, 262, 273n20
sharecroppers. *See* croppers
sharecropping, 115–17, 124, 127, 133–34, 144
treadmill. *See* treadmill, sharecropping
sheep, 40, 49
Sherman, William T., 187
ships, 51–53, 65–67, 85, 108, 148–49, 192
shortage
of food, 63, 87, 89–90, 160, 169, 174
of grocery stores, 237
of labor, 4, 98, 101
of nutrients, 55
sick leave, xi, 5, 205
sickness, 69, 238
Siler, Joseph, 221, 223–24, 228, 236, 241
Sinclair, Upton, 2, 192–94
Sioux, 30–31, 35
Sittig, Ann, 200, 202, 213–14
slaughter, 6, 159, 184, 188, 211
slaughterhouse, 97, 175, 179, 182–83, 187–88, 200–2, 208
slave, 2, 17, 22, 25, 28, 67–68, 70, 198
slavery, 3, 23, 28, 44, 66–67, 75, 220
sleep-deprivation, 175–76
smallholders, 21, 160–1, 252, 254, 259
and development treadmill, 164, 167
and free trade, 10
in Guatemala, 9, 168, 172, 218, 253
and plantations, 24
and precarity, 13, 139, 149, 153
and wage labor, 156, 231
snake, 2, 3, 14, 75, 213, 253
socialism, 194
sodium nitrate, 73
soil, 17, 49, 52, 61, 128, 171, 223
conservation, 39, 64, 71–72
erosion, 39–40
fertility, 119, 214
fertility loss, 3, 12, 54, 62, 65, 169
nutrients, 55–59, 246
*See also* nutrients
Soil Conservation Service (SCS), 39
Somalia, 153, 201, 287n6
South (US region), 4, 63, 107, 130, 197, 238
and agriculture, 64, 71, 114–17, 125, 231
and politics, 134–35, 220, 225–26, 229
and public health, 219–20, 224
and slavery, 2, 270n14
*See also* New South
South Africa, 237
South America, 12, 64, 73, 144
South Asia, 18, 132, 161, 252
South Carolina (US state), 22, 27, 179, 221, 238
South Korea, 164
South Sudan, 201
Southeast Asia, 82, 104–5, 151, 161, 201, 206, 232
Southern Africa, 155, 172
sovereignty, 47, 250, 254
*See also* food sovereignty

Soviet Union, 147
Soybean, 55, 58
Spain, 16, 66, 140–44, 215, 273n35
Spanish-American War, 70, 192, 287n8
Spartanburg, South Carolina, 221, 224, 226, 229, 236
spatial fixes, 45, 143, 209, 252–53, 255
  for capital, 79, 101, 130, 211
  carceral, 45–46
  definition of, 10–11
  extraterritorial, 71
  for labor, 79, 101, 131, 211
  for overproduction, 131
  for violence, 47–48, 185
  *See also* aggregation; domestic rearrangements; segregation
Specht, Joshua, 199, 211, 293n44
speculation, 23, 44–45, 47–48, 272n16
Spies, Tom, 228
spinach, xi, 84
squash, 32, 46, 50, 58
staple crops, 153, 160–62, 169
starvation, 1, 4, 7, 117, 142, 145, 162
  and children, 108, 135
  and Indian reservations, 30, 42, 44, 187
  and plantations, 62, 168, 174
  and removal, 17, 19–20, 26, 47
  and the South, 63, 107, 245
state, 2, 20, 94, 101, 156, 164, 181
State Department, 145–47, 153, 155, 158, 170, 288n23
Steinbeck, John, 2, 8–9
sterilization, 19, 206
stockpiles, 7, 45, 107, 126, 153–55, 158, 160
stress, 112, 151, 175–76, 201, 209
strike, 53, 98–99, 127, 164, 177, 195–97, 199, 202
structure, 166–67, 183, 218, 237, 241, 257
  and class, 164, 241, 248
  and colonialism, 18, 24, 29, 142–43
  and food system, 5, 157, 238, 251
  and gender, 241, 248
  and race, 241, 248
  and revolution, 117
structural adjustment, 163, 168
students, 36, 137, 148, 258, 267
stunting, 1, 174
sub-Saharan Africa, 159, 161, 173, 252
subsidy, 101, 106, 129–30, 132–33, 139, 150
subsistence, 41, 49, 132, 143, 232, 292n17
  *See also* crops, subsistence
suburbs, 208, 234–37, 239–40, 242, 246, 250, 258
sugar, 40, 66, 140, 149, 162, 169, 192
  beet, 79, 98
  companies, 141–42, 146, 258
  in diets, 41, 48, 211, 221
  plantations, 143
  refined, 1, 3, 32
  *See also* Sugar Trust; sugarcane
Sugar Trust, 141
sugarcane, 85, 140, 221
suicide, 67, 73, 216
supermarket, 102, 107–8, 234–44, 246, 250, 252, 258

Supplemental Nutrition Assistance Program (SNAP), 247
supply, 40, 97, 122, 177, 190, 211, 271n30
  of customers, 120
  and demand, 100–1, 111, 128, 209, 249, 256
  of food, 97, 155, 162, 186
  of labor, 80, 98, 188, 201
  limits on, 21–22
  and price, 111–13
  *See also* demand
supply chain, ix, x, xi, 155, 159, 232, 255, 262
  and aggregation, 180
  bottlenecks, 213
  for fruits and vegetables, 85, 87
  middle of, 185–86, 195, 234
  and throughput, 185
Supreme Court, 54, 70, 101
surplus, 95, 97, 128, 139, 150–51, 185
  capital, 117, 120, 124, 130
  consumption of, 38, 44, 111, 155–56
  domestic, 13, 169–70
  dumping of, 32, 48, 142, 158, 170
  of food, 13, 40–42, 124, 127, 129, 152–54, 170
  labor, 122, 129–32, 196, 210
  land, 34–35
  production of, 32, 47, 142
  sale of, 10, 21
  storage of, 107–8, 126, 130, 134
  *See also* capital, surplus; commodity, surplus; surplus value
surplus value, 181–83, 192, 207–8, 213
survivance, 49–50
sustainability, x, xii, 254
swine. *See* pigs
Sydenstricker, Edgar, 226–27, 229–30

Taft, William Howard, 142, 287n15
Taíno, 51
tax, 35, 79, 93, 145, 147, 185
technology, 24, 36, 105, 251, 260, 285n43, 286n74
  adoption of, 113, 122
  development, 110, 118, 291n5
tenancy, 2, 81, 125, 131, 144
Tennessee (US state), 23, 25–27, 35, 120
Termination Act (1953), 42
territory, 72, 78, 96, 101, 109, 274n39, 278n51
  and European colonialism, 143
  and overseas expansion, 52, 54
  and settler colonialism, 17, 20, 22–26, 28, 33–34, 271n3
Texas (US state), 39, 63, 174, 189, 191, 199–200, 216
thiamine, 231–32
Third World, 110, 139, 160, 286n5
Thompson-McFadden Commission, 221, 241
Thompson, Robert. *See* Thompson-McFadden Commission
throughput, 203, 209, 213, 218
  and meat industry, 176, 179, 262, 181–86, 188, 190, 193
  and milling, 231–32
Timber Culture Act (1873), 31
Tippecanoe, 24
titles, 35, 152
tobacco, 22, 62, 64, 71, 143
Topaz, 94–98, 102

torture, 8, 50, 137–38, 141, 193, 217, 225
toxin, 216, 219, 228
tractors, 92, 118–19, 121–22, 162, 165
  development of, 107, 119
  growth in, 114, 119, 123, 125
  and sharecropping, 106, 129
  and yields, 7, 107, 159
trade, 14, 74–75, 85, 97, 110, 123, 136
  balance of, 163
  barriers to, 168
  global, 66, 140
  illicit, 61
  policies, 154–55, 254–55
  regulations, 47
  restrictions, 27, 161
  *See also* coolies; free trade agreements; guano, trade
trading houses, 67, 73
Trail of Tears, 26, 263
transporters, 2, 78, 181
treadmill, 112, 121, 126, 130–31, 134–35, 139
  agricultural, 112–14, 164, 167, 171, 199, 246, 252
  development, 163, 166, 172, 246
  of influence, 251
  sharecropping, 114
treasury, 24, 44, 87
treaties, 17, 24–25, 27, 33, 68, 274n39
  and concessions, 26, 30
  and obligations, 19, 26, 33–34, 42
tribal governments, 96, 40, 43
tribe, 17, 27, 33–34, 40–42, 45
tropical commodities. *See* commodities, and tropical
Truman, Harry S., 151, 279n6, 288n34
Trump, Donald, 247
Tseshaht, 49–50
tuberculosis, 97, 189, 216, 221
Tule Lake, 95, 97–99
Tyson Foods, 159, 177, 209

Ubico, Jorge, 145–46
undocumented migrants, 138, 200, 247
unemployment, 2, 10, 38, 43, 110, 117, 125, 196
unions, 199–202, 209, 212, 214, 258, 262
  and aggregation, 198
  and contracts, 202
  exclusion from, 82, 102, 147–48
  government support for, 197
  and inclusion, 198, 249
  leadership, 198
  and membership, 196, 199
Union Army, 30, 32
United Fruit Company (UFC), 138, 144, 147, 159, 164, 174
  and labor, 144
  and land, 144–45, 147, 149
  and power, 144, 147–50, 191
  and taxes, 173
  and violence, 13, 145
United Nations (UN), 255
United States Department of Agriculture (USDA), 87, 111, 132, 248, 281n49
  agents of, 91, 93–94, 109, 193
  census data, 133, 284n27, 299n75
  and food deserts, 218, 234, 239, 241
  and gender, 166
  and overproduction, 33, 107–8, 125–26, 150, 153, 158–60

policies of, 7, 88–89, 97, 100–1, 128, 170–2
regulations, 175
United States Department of Labor, 178, 285n64
United States Public Health Service (PHS), 224, 227
University of Chicago, 128
unrest, 11, 48, 143
Upside Foods, 207–9
US Army, 29, 30, 32, 87–89, 96, 187, 192–93, 232
USAID. *See* Agency for International Development (USAID)
USDA. *See* United States Department of Agriculture (USDA)
Utah (US state), 98, 200, 274n39

Valetti, Uma, 207–8
varieties, 49, 97, 113–14, 121, 158, 161
vegan, 14, 256
vegetables, 40, 99, 138, 169, 226, 230, 232
and California, 84, 87, 101
and fruit, 85, 88, 174, 189–90
ventilation, 204, 206
ventilation shut down (VSD), 206
vertical integration, ix, 159, 186, 209, 232
Vietnam, 105–7, 135, 199, 283n7
Vietnam War, 105–6, 199
villages
Indigenous, 2, 4, 20, 24–25, 58
in Guatemala, 136–38, 141, 157–58, 174, 263
violence, xi, 9, 12, 47, 135, 256–57, 270n15
and abundance, 7–8, 128
acts of, ix, 246
bodily, 3, 98, 200–1, 242
definition of, 4–5
direct, 29, 36
and expropriation, 255
in the food system, 3–4, 170, 253, 258–59
and geography, xii, 9–11, 13, 185, 196, 246, 255
and metaphor, 241
paradox of, 130, 214
production of, 9, 117, 122, 177, 182, 193, 246–47
proliferation of, 248–49, 254
racial, 31, 79, 102
redirection of, 46, 48, 109, 130–31, 134, 170, 180
of scale, 13, 209
state-sponsored, 20, 138
and structure, xi, 4, 11, 46–48, 104, 218
Virginia (US state), 22, 62, 68, 108
vitamin, 203, 231–32
B1. *See* thiamine
B3. *See* niacin
voting, 5, 21, 26, 28, 96, 103, 250

W.R. Grace, 67
Wadiwel, Dinesh, 207, 291n5
wages, 43, 57, 69, 80, 182, 198, 202, 246
falling, 211
higher, 162, 197, 229
low, 1, 79, 97–99
starvation, 145, 168, 174, 256
suppression of, 39, 80, 129,163–64, 181, 247
Wagner Act (1935), 197

Wakasa, James, 98
Walker County, Alabama, 229–30, 236
Wall Street, 78, 145, 243
Wallach, Jennifer, 118
Walmart, 237, 239–40, 248
war, xi, 13, 29, 30, 34–36, 77–79, 197, 246
  *See also* civil war; World War I; World War II
War Department, 33, 91, 120
War on Poverty, 106, 130
War Relocation Authority (WRA), 91, 94–101
Washington (US state), 76, 91
Washington County, Mississippi, 106–7, 127, 283n7
Washington, DC, 76–77, 85, 146, 286n1
Washington, George, 22, 26, 257
Washington, Karen, 237
water, 31, 41, 55–56, 118, 178–79, 197
  for animals, 208
  for drinking, 42, 51, 66–67, 94, 107, 127, 177, 181
  for growing crops, 121, 162, 282n65
weeding, 113
welfare, 40, 101, 127, 129, 152, 185
  and animals. *See* animals, welfare of
  *See also* welfare boards
welfare boards, 106, 108, 127, 133
West (US region), 28–30, 32, 41, 78, 98, 191
West Coast (US region), 76, 78, 83–88, 99, 101, 103
Western Defense Command, 86, 92
wheat, 58, 61, 65, 123–24, 129, 132
  as aid, 153–54, 160
  in diet, 3, 32, 154, 170, 215, 228
  exports of, 62, 72, 74, 150, 170
  markets for, 71, 128
  monocultures of, 44
  prices for, 7, 38, 112, 153
  production of, 72, 113–14
  surplus of, 40, 107–8, 126, 140
  *See also* yields
whistleblower, 73, 177, 185
white, 48, 53, 70, 223, 234, 282n62, 297n25
  businesses, 42, 76–78, 83, 100–2, 104, 242
  consumers, 234–35, 237–38
  farms, 40, 46, 78, 81–83, 88–90
  flight, 234–35, 243, 258
  landowners, 106, 127
  leaders, 106, 109
  and poor, 23, 103, 199, 218
  and rural, 199
  settlers, 17–18, 22, 26, 28–31, 95
  society, 24, 43
  supremacy, 4, 9, 13, 79, 93, 107–9, 125, 133, 142
  women, 197
  workers, 39, 80–82, 197–98, 229
  *See also* yeoman
wholesale, 78, 83, 85, 185–86, 190, 235, 240
Wickard, Claude, 88
Wilson, Woodrow, 120
Wisconsin (US state), 49
Wolfe, Patrick, 35
women, 99, 254, 300n82
  and labor, 82–83, 165–66, 188, 197

and Maya. *See* Maya, women
and pellagra, 216, 223, 228, 236
and representation, 255
and unions, 199–200
violence against, 17, 25–26, 29, 94, 136–37, 166, 189
wool, 40, 61
workers, 3, 6, 10–11, 49, 52
in agriculture, 88, 101–3, 114, 123, 131–2, 138
in Europe, 61, 188
in factories, 13, 56–57, 88, 119, 122, 131
in guano, 53–54, 68–71, 75, 262, 263
in Guatemala, 145, 168, 172, 200
incarcerated, 54, 98
issei. *See* issei, workers
in meat industry, 177–86, 188–89, 191, 193–94, 196–202, 209–14
Mexican. *See* Mexicans, labor of
nisei. *See* nisei, workers
and organization, 127, 191, 196–98, 209, 245, 247
in the South, 117, 129, 135, 226, 229
surplus of, 39, 122, 124, 129
white. *See* white, workers
*See also* laborers; unions
working class, 43, 196, 212
working conditions, 1, 3, 31, 44
in guano, 54, 66
in Guatemala, 145, 168
under incarceration, 98
in meatpacking, 196–97, 211, 220
World Bank, 161
World War I, 38, 74, 80, 96, 120
World War II, 77, 90, 103, 109, 119–24, 150, 234
Wounded Knee, 24, 29, 189
Wright, Marian, 127, 133
Wyoming (US state), 35, 98, 274n39

xenophobia, 77, 214

Yarbrough, J.F., 225
yeast, 227–28
Yellow Peril, 80, 102–3
yeoman, 21–27, 31, 46, 77, 83, 134, 261
yields, 160–61
decrease in, 62, 169, 171
increase in, 32, 113, 122, 150, 164
Yoruba, 23
Yurok, 41, 49

Zemurray, Sam, 147, 149
zero-sum game, 214, 253
Zimmern, Andrew, 239

Stian Rice is a geographer whose work examines the origins of violence in the food system. He has written about famine, genocide, and structural violence for academic and general audiences since 2012 and is the author of *Famine in the Remaking: Food System Change and Mass Starvation in Hawaii, Madagascar, and Cambodia*. He is assistant research scientist with the Center for Urban Environmental Research at the University of Maryland Baltimore County, and teaches geography at Anne Arundel Community College in Arnold, Maryland. Rice received his PhD in geography from Kent State University.

www.ingramcontent.com/pod-product-compliance
Ingram Content Group UK Ltd.
Pitfield, Milton Keynes, MK11 3LW, UK
UKHW041019250926
475962UK00004B/151

9 781959 000914